Radiochemical Syntheses

RADIOCHEMICAL SYNTHESES

Radiochemical Syntheses

VOLUME 1
RADIOPHARMACEUTICALS FOR POSITRON EMISSION TOMOGRAPHY

Edited by

PETER J. H. SCOTT
BRIAN G. HOCKLEY

WILEY
A JOHN WILEY & SONS, INC., PUBLICATION

Published by John Wiley & Sons, Inc., Hoboken, New Jersey.
Published simultaneously in Canada.

Library of Congress Cataloging-in-Publication Data:

Radiopharmaceuticals for positron emission tomography / edited by Peter J.H. Scott, Brian G. Hockley.
p. ; cm.—(Radiochemical syntheses ; v. 1)
Includes bibliographical references and index.
ISBN 978-0-470-58895-6 (cloth)
1. Radiopharmaceuticals–Synthesis. 2. Tomography, Emission. I. Scott, Peter J. H. II. Hockley, Brian G. III. Series: Radiochemical syntheses ; v.
[DNLM: 1. Radiopharmaceuticals—chemical synthesis. 2. Positron-Emission Tomography. 3. Radiopharmaceuticals—diagnostic use. WN 415]
RS431.R34R325 2011
616.07'575–dc23

2011019940

Printed in the United States of America

10 9 8 7 6 5 4 3 2 1

CONTENTS

CONTRIBUTORS

Franklin I. Aigbirhio
Wolfson Brain Imaging Centre, Department of Clinical Neurosciences, University of Cambridge, Addenbrooke's Hospital, Cambridge, UK

Amar Alfteimi
Clinic of Nuclear Medicine, Molecular Imaging and Therapy, University Hospital Schleswig-Holstein Campus Kiel, UKSH, Kiel, Germany

Marion Alvarez
CERMEP-Imagerie du Vivant, Hospices Civils de Lyon, ICBMS UMR, Université Lyon 1, Bron (Lyon), France

Jan D. Andersson
Department of Clinical Neuroscience, Psychiatry Section, Karolinska Institutet, Stockholm, Sweden

Robert Bielik
Wolfson Brain Imaging Centre, Department of Clinical Neurosciences, University of Cambridge, Addenbrooke's Hospital, Cambridge, UK

Guy Bormans
Laboratory for Radiopharmacy, Faculty of Pharmaceutical Sciences, K.U. Leuven, Leuven, Belgium

István Boros
Wolfson Brain Imaging Centre, Department of Clinical Neurosciences, University of Cambridge, Addenbrooke's Hospital, Cambridge, UK

Stefano Boschi
PET Radiopharmacy, Nuclear Medicine Unit, Policlinico S. Orsola-Malpighi, Bologna, Italy

Laurent Brichard
Wolfson Brain Imaging Centre, Department of Clinical Neurosciences, University of Cambridge, Addenbrooke's Hospital, Cambridge, UK

Roberto Canales-Candela
Wolfson Brain Imaging Centre, Department of Clinical Neurosciences, University of Cambridge, Addenbrooke's Hospital, Cambridge, UK

Assunta Carpinelli
IBFM-CNR, Nuclear Medicine Department, University of Milan Bicocca, San Raffaele Scientific Institute, Milano, Italy

Sofie Celen
Laboratory for Radiopharmacy, Faculty of Pharmaceutical Sciences, K.U. Leuven, Leuven, Belgium

Michael A. Channing
National Institutes of Health, PET Department, Clinical Center, Bethesda, Maryland

Xiaoyuan Chen
Molecular Imaging Program, Department of Radiology, Stanford University School of Medicine, Stanford, California; Laboratory of Molecular Imaging and Nanomedicine, National Institute of Biomedical Imaging and Bioengineering, National Institutes of Health, Bethesda, Maryland

Zhen Cheng
Molecular Imaging Program, Department of Radiology, Stanford University School of Medicine, Stanford, California

Frederick T. Chin
Molecular Imaging Program, Department of Radiology, Stanford University School of Medicine, Stanford, California

Robert F. Dannals
The Johns Hopkins University School of Medicine, Baltimore, Maryland

Harald Eidherr
Department of Nuclear Medicine, Medical University of Vienna, Vienna, Austria

Valentina Ferrari
Wolfson Brain Imaging Centre, Department of Clinical Neurosciences, University of Cambridge, Addenbrooke's Hospital, Cambridge, UK

Umesh B. Gangadharmath
Siemens Molecular Imaging Biomarker Research, Culver City, California

Friedrich Girschele
Department of Nuclear Medicine, Medical University of Vienna, Vienna, Austria

Tjibbe de Groot
Department of Nuclear Medicine, K.U. Leuven, Leuven, Belgium

Daniela Haeusler
Department of Nuclear Medicine, Medical University of Vienna, Vienna, Austria; Department of Pharmaceutical Technology and Biopharmaceutics, University of Vienna, Vienna, Austria

Christer Halldin
Department of Clinical Neuroscience, Psychiatry Section, Karolinska Institutet, Stockholm, Sweden

Terence G. Hamill
Discovery Imaging, Merck Research Laboratories, West Point, Pennsylvania

Bradford Henderson
University of Michigan School of Medicine, Ann Arbor, Michigan

Brian G. Hockley
Department of Radiology, University of Michigan School of Medicine, Ann Arbor, Michigan

Daniel P. Holt
The Johns Hopkins University School of Medicine, Baltimore, Maryland

Michael R. Kilbourn
Department of Radiology, University of Michigan School of Medicine, Ann Arbor, Michigan

Hartmuth C. Kolb
Siemens Molecular Imaging Biomarker Research, Culver City, California

Alexey Kostikov
McConnell Brain Imaging Centre, Montreal Neurological Institute, McGill University, Montreal, Quebec, Canada

Miriam Kovacevic
McConnell Brain Imaging Centre, Montreal Neurological Institute, McGill University, Montreal, Quebec, Canada

Raisa N. Krasikova
Department of Clinical Neuroscience, Psychiatry Section, Karolinska Institutet, Stockholm, Sweden; Institute of Human Brain, Russian Academy of Sciences, St. Petersburg, Russia

David Kryza
ISPB, Faculté de Pharmacie, Université Claude Bernard Lyon 1; Hospices Civils de Lyon, Hôpital E Herriot, Radiopharmacie; LPCML, UMR5620 CNRS, Villeurbanne, Lyon, France

Piyush Kumar
Department of Oncology, University of Alberta, Edmonton, Alberta, Canada

Jozef J. Kusmierz
National Institutes of Health, PET Department, Clinical Center, Bethesda, Maryland

Didier Le Bars
CERMEP-Imagerie du Vivant, Hospices Civils de Lyon, ICBMS UMR, Université Lyon 1, Bron (Lyon), France

Shuanglong Liu
Molecular Imaging Program, Department of Radiology, Stanford University School of Medicine, Stanford, California

Filippo Lodi
PET Radiopharmacy, Nuclear Medicine Unit, Policlinico S. Orsola-Malpighi, Bologna, Italy

Claudio Malizia
PET Radiopharmacy, Nuclear Medicine Unit, Policlinico S. Orsola-Malpighi, Bologna, Italy

Gassan Massaweh
McConnell Brain Imaging Centre, Montreal Neurological Institute, McGill University, Montreal, Quebec, Canada; Canada Imaging & Research Centre, Department of Radiology, Queen Elizabeth II Health Sciences Centre, Capital Health/Dalhousie University, Halifax, Nova Scotia, Canada

John R. Mercer
Department of Oncology, University of Alberta, Edmonton, Alberta, Canada

Markus Mitterhauser
Department of Nuclear Medicine, Medical University of Vienna, Vienna, Austria; Department of Pharmaceutical Technology and Biopharmaceutics, University of Vienna, Vienna, Austria

Vani P. Mocharla
Siemens Molecular Imaging Biomarker Research, Culver City, California

Bruce H. Mock
Department of Radiology and Imaging Sciences, Indiana University School of Medicine, Indianapolis, Indiana

Fanrong Mu
Siemens Molecular Imaging Biomarker Research, Culver City, California

John L. Musachio
National Institutes of Health, PET Department, Clinical Center, Bethesda, Maryland

Sangram Nag
Department of Clinical Neuroscience, Psychiatry Section, Karolinska Institutet, Stockholm, Sweden

Henry C. Padgett
Siemens Molecular Imaging Biomarker Research, Culver City, California

Jim Patanella
Siemens Molecular Imaging Biomarker Research, Culver City, California

Cécile Philippe
Department of Nuclear Medicine, Medical University of Vienna, Vienna, Austria; Department of Pharmaceutical Technology and Biopharmaceutics, University of Vienna, Vienna, Austria

Victor W. Pike
Molecular Imaging Branch, National Institute of Mental Health, National Institutes of Health, Bethesda, Maryland

Hayden T. Ravert
The Johns Hopkins University School of Medicine, Baltimore, Maryland

Michelle L. Richards
Department of Radiology, University of Michigan School of Medicine, Ann Arbor, Michigan

Patrick J. Riss
Wolfson Brain Imaging Centre, Department of Clinical Neurosciences, University of Cambridge, Addenbrooke's Hospital, Cambridge, UK

Adam C. Runkle
Department of Radiology, University of Michigan School of Medicine, Ann Arbor, Michigan

Peter J. H. Scott
Department of Radiology, University of Michigan School of Medicine, Ann Arbor, Michigan

Ralf Schirrmacher
McConnell Brain Imaging Centre, Montreal Neurological Institute, McGill University, Montreal, Quebec, Canada; Lady Davis Institute for Medical Research, Montreal, Quebec, Canada

Kim Serdons
Department of Nuclear Medicine, K.U. Leuven, Leuven, Belgium

Xia Shao
Division of Nuclear Medicine, Department of Radiology, University of Michigan, Ann Arbor, Michigan

Rob Smith
Wolfson Brain Imaging Centre, Department of Clinical Neurosciences, University of Cambridge, Addenbrooke's Hospital, Cambridge, UK

Scott E. Snyder
Division of Nuclear Medicine, Department of Radiological Sciences, St. Jude Children's Research Hospital, Memphis, Tennessee

Alexander Thiel
Lady Davis Institute for Medical Research, Montreal, Quebec, Canada; Department of Neurology and Neurosurgery, McGill University, Montreal, Quebec, Canada

Louis Tluczek
Division of Nuclear Medicine, Department of Radiology, University of Michigan, School of Medicine, Ann Arbor, Michigan

Amy L. Vāvere
Division of Nuclear Medicine, Department of Radiological Sciences, St. Jude Children's Research Hospital, Memphis, Tennessee

Wolfgang Wadsak
Department of Nuclear Medicine, Medical University of Vienna, Vienna, Austria; Department of Inorganic Chemistry, University of Vienna, Vienna, Austria

Joseph C. Walsh
Siemens Molecular Imaging Biomarker Research, Culver City, California

Min Wang
Department of Radiology and Imaging Sciences, Indiana University School of Medicine, Indianapolis, Indiana

Carmen Wängler
McConnell Brain Imaging Centre, Montreal Neurological Institute, McGill University, Montreal, Quebec, Canada

Yi Zhao
Clinic of Nuclear Medicine, Molecular Imaging and Therapy, University Hospital Schleswig-Holstein Campus Kiel, UKSH, Kiel, Germany

Qi-Huang Zheng
Department of Radiology and Imaging Sciences, Indiana University School of Medicine, Indianapolis, Indiana

Steve Zigler
Siemens Molecular Imaging Biomarker Research, Culver City, California

Maaz Zuhayra
Clinic of Nuclear Medicine, Molecular Imaging and Therapy, University Hospital Schleswig-Holstein Campus Kiel, UKSH, Kiel, Germany

FOREWORD

Thirty years ago, the synthesis of radiolabeled compounds for positron emission tomography (PET) was practiced by a relatively small band of dedicated chemists, alternatively challenged by the unusual requirements of synthetic speed, stochiometry, and specific activity, yet excited by the ground-breaking visualization of in vivo biochemistry in the living human subject. Twenty years ago, the number of chemists was growing rapidly, biochemical targets were increasing in complexity and scope, and the chemical and physiochemical approaches to radiopharmaceutical synthesis were becoming more numerous. Ten years ago, the technology associated with PET radiochemistry was well advanced, with an ever-widening group of chemists producing new techniques and new radiotracers at a quickening pace.

Today, PET radiochemistry has greatly matured: synthetic techniques are readily available for many structural classes of compounds, synthetic equipment can be purchased and quickly brought into use, and analytical methods have become standardized. PET is part of the clinical medicine arena, and numerous pharmaceutical companies are actively pursuing development of new radiopharmaceuticals. Research using PET imaging is now undertaken at just about every major medical center in countries around the world.

Radiochemical Syntheses: Volume 1, Radiopharmaceuticals for Positron Emission Tomography is a welcome and much-needed addition to the PET radiopharmaceutical literature, offering for the first time a description of established, reliable synthetic methods for a large number of PET radiopharmaceuticals that are routinely prepared in multiple institutions. New users of PET can rapidly find details of peer-reviewed methods suitable for syntheses of valuable PET radiopharmaceuticals, and established users may find that the methods described herein provide better or easier synthetic methods. Whether an experienced PET radiochemist, or a chemist new to this unusual field, this first volume of Radiochemical Syntheses will be a valuable guide to the exciting and challenging area of PET radiopharmaceutical chemistry.

MICHAEL R. KILBOURN, PH.D

Professor of Radiology
University of Michigan Medical School
Ann Arbor, Michigan

PREFACE

With feelings of excitement and anticipation, we present the first volume of Radiochemical Syntheses to the field, offering reliable and robust procedures for the preparation of many common radiopharmaceuticals utilized in positron emission tomographic (PET) imaging.

It is extremely likely that all of us are familiar with organic syntheses, either through the complimentary volume we receive every year from the American Chemical Society Organic Chemistry Division or through the equally powerful online edition. Certainly we are thankful when a given procedure we are trying to replicate has been published in Organic Syntheses, as it provides us with a high degree of confidence, from the outset, that the protocol will actually work in our hands! What we observed, through talking to many of our colleagues in the radiochemistry field, was that an equivalent product was in demand for the increasing numbers of radiochemical reactions and radiopharmaceutical syntheses reported in the literature. To address this need, we have established Radiochemical Syntheses (Rad. Syn.). The goal of this series is to present monographs dealing specifically with radiochemical reactions that report many of the details that are required for successful replication, but that are frequently omitted from the mainstream radiochemistry literature. Owing to the frequently very busy radiopharmaceutical production schedules for PET radiochemistry facilities, and the associated costs of laboratory operations, it is not feasible to put the monographs through the same rigorous checking that Organic Syntheses is noted for. However, to ensure reliability of the included procedures, we have adopted the common regulatory practice of requiring inclusion of complete quality control data for three successive process verification runs.

We are deeply indebted to the authors and editorial board that have made this first volume a reality. These experts from the global field of radiochemistry have responded to this series with energy and enthusiasm, whether by preparing the excellent series of monographs found herein, or through their diligent reviewing and checking of the reported synthetic procedures. Moreover, the support of the editorial team at Wiley has been invaluable. In particular, thanks are due to Jonathan Rose who has enthusiastically backed this project from the very earliest idea and patiently seen it through to publication.

Finally, this is the first volume of Radiochemical Syntheses. It is anticipated that Volume 2 will focus on new methods for preparing radiopharmaceuticals for PET imaging and Volume 3 will focus on radiopharmaceuticals for SPECT imaging. Potential authors and guest volume editors are encouraged to obtain information about preparing monographs or future volumes from the Radiochemical Syntheses Editorial Office (pjhscott@umich.edu).

PETER J. H. SCOTT, PH.D
MICHAEL R. KILBOURN, PH.D

The University of Michigan
Ann Arbor, Michigan
October 2010

ABBREVIATIONS

Ac	Acetate
AcOH	Acetic acid
ASU	Automated synthesis unit
BET	Bacterial endotoxin test
Bn	Benzyl
Boc	*tert*-Butoxycarbonyl
Bq	Becquerel
BTA	Benzothiazole
Bz	Benzoyl
CFN	Carfentanil
CHL	Choline
Ci	Curie
CMS	U.S. Centers for Medicare and Medicaid Services
CRO	Contract research organization
DAA	Diacetone alcohol
DASB	3-amino-4-[2-(*N*-methyl-*N*-methyl-amino-methyl)phenylsulfanyl]-benzonitrile
DAT	Dopamine transporter
DMAE	Dimethylamino ethanol
DMF	*N,N*-dimethyl formamide
DMSO	Dimethyl sulfoxide
DMT	Dimethoxy trityl
DOTATOC	DOTA-(Tyr^3)-octreotide acetate
DTBZ	(+)-α-Dihydrotetrabenazine
EC	Endothelial cells
EDTA	Ethylenediaminetetraacetic acid
EtOH	Ethanol
EU	Endotoxin units
Eu. Ph.	European Pharmacopeia
FA	2-(Fluoro)-3-[(2*S*)-2-azetidinylmethoxy]pyridine
FAZA	Fluoroazomycin arabinoside
FCH	Fluorocholine (or fluoromethylcholine)
FDA	6-Fluorodopamine
FDA	U.S. Food and Drug Administration

FDG	Fluorodeoxyglucose
FES	16-α-Fluoroestradiol
FET	Fluoroethyltyrosine
FID	Flame ionization detector
FLT	Fluorothymidine
FMISO	Fluoromisonidazole (1-(2-hydroxy-3-fluoropropyl)-2-nitroimidazole)
FMZ	Flumazenil
FTM	Fluid thioglycolate media
FZ	Flumazenil
GABA	γ-Aminobutyric acid
GBq	Gigabecquerel
GC	Gas chromatography
GEMS	General electric medical systems
GluT	Glucose transporters
HPLC	High-performance liquid chromatography
5-HT	5-Hydroxytryptamine (5-HT))
ICH	The International Conference on Harmonisation of Technical Requirements for Registration of Pharmaceuticals for Human Use
ICP-MS	Inductively coupled plasma mass spectrometry
IU	International units
K-2.2.2	Kryptofix-2.2.2 (4,7,13,16,21,24-hexaoxa-1,10-diazabicyclo[8.8.8]-hexacosane)
keV	Kilo electron volt
LC-MS	Liquid chromatography-mass spectrometry
LAL	Limulus amebocyte lysate
LOD	Limits of detection
MBq	Megabecquerel
mCi	Millicurie
MET	Methionine
MeCN	Acetonitrile
MeI	Iodomethane (methyl iodide)
MeOH	Methanol
MeOTf	Methyl triflate
MeV	Mega electron volt
MHED	*meta*-Hydroxyephedrine
Min	Minute
MPPF	4-(2′-Methoxyphenyl)-1-[2′-(*N*-2″-pyridinyl)-*p*-fluorobenzamido]ethylpiperazine
MPPI	4-(2′-Methoxyphenyl)-1-[2′-[*N*-(2″-pyridinyl)-*p*-iodobenzamido]ethyl]piperazine
MTO	[*O*-Methyl]metomidate
NaF	Sodium fluoride

NET	Norepinephrine transporter
NFP	4-Nitrophenyl 2-[^{18}F]-fluoropropionate
NK	Tachykinin receptors
Nos	Nosyl
n	Neutron
NITTP	1-(2-nitro-1-imidazolyl)-2-*O*-tetrahydropyranyl-3-*O*-toluenesulfonylpropanediol
OTf	Triflate
p	Proton
PET	Positron emission tomography
PIB	Pittsburgh compound B
PBR28	*N*-(2-Methoxybenzyl)-*N*-(4-phenoxypyridin-3-yl)acetamide
ppm	Parts per million
Psi	Pounds per square inch
QC	Quality control
QMA	Quaternary methyl ammonium
RCP	Radiochemical purity
RCY	Radiochemical yield
RGD	Arg-Gly-Asp
RRT	Relative retention time
Rf	Retention factor
RSA	Residual solvent analysis
Rxn	Reaction
SERT	Serotonin transporter
SA	Specific activity
SFB	*N*-Succinimidyl 4-[^{18}F]fluorobenzoate
SCDM	Soybean casein digest media
SPA-RQ	Substance-P antagonist-receptor quantifier
SPE	Solid phase extraction
SPECT	Single photon emission computed tomography
SSRI	Selective serotonin reuptake inhibitor
TBZ	Tetrabenazine
TFA	Trifluoroacetic acid
THF	Tetrahydrofuran
TLC	Thin layer chromatography
TPAH	Tetrapropylammonium hydroxide
TSTU	*O*-(*N*-Succinimidyl)-*N,N,N′,N′*-tetramethyluronium tetrafluoroborate
USP	United States Pharmacopeia
UV	Ultraviolet
VMAT2	Vesicular monoamine transporter type 2

PART I

FLUORINE-18 LABELED RADIOPHARMACEUTICALS

CHAPTER 1

SYNTHESIS OF [^{18}F]-FLUORODEOXYGLUCOSE ([^{18}F]FDG)

MICHELLE L. RICHARDS AND PETER J. H. SCOTT

Department of Radiology, University of Michigan School of Medicine, Ann Arbor, Michigan

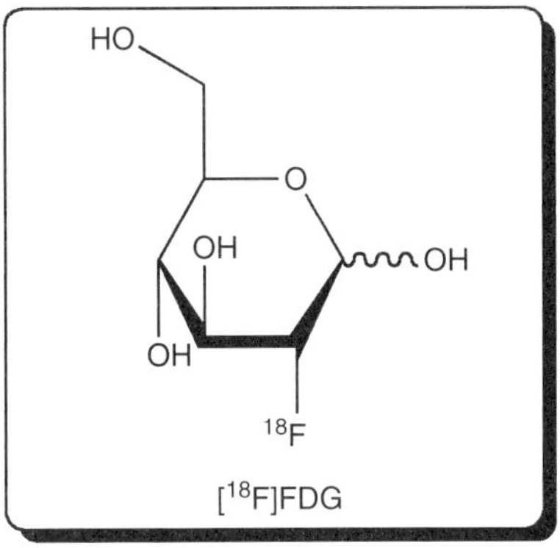

1 INTRODUCTION

Positron emission tomographic (PET) imaging provides a noninvasive, accurate diagnostic method of imaging and detecting possible diseases at a cellular, molecular, and tissue level. PET has a proportional relationship between the tissue intensity on the tomographic image and the actual radiopharmaceutical concentration in tissue [1]. [^{18}F]Fluorodeoxyglucose ([^{18}F]FDG) is a glucose analog and is known as the "work horse" of PET simply because of the multiple modalities for application. The implications of abnormal glucose metabolism are vital in evaluating a variety of diseases, and the applications of this positron emitting radiotracer are great. [^{18}F]FDG is preferentially taken up into cells with high metabolic activity by specific glucose transporters (GluT) and phosphorylated by hexokinase. It becomes metabolically trapped in the cell because [^{18}F]FDG-6-phosphate is not a substrate for glucose-6-phosphate isomerase, the enzyme that metabolizes glucose, and therefore cannot be broken down. During radioactive

Radiochemical Syntheses, Volume 1: Radiopharmaceuticals for Positron Emission Tomography, First Edition. Edited by Peter J. H. Scott, Brian G. Hockley.

decay, ^{18}F decays to ^{18}O, allowing the decayed product, 2-^{18}O-deoxyglucose -6-phosphate, to enter the normal glucose metabolic pathway.

The first synthesis of [^{18}F]FDG was based on a direct electrophilic substitution reaction by Wolf et al. in 1976 [2,3]. This fluorination reaction by electrophilic substitution refers to the addition of fluorine atoms across a double bond, producing a difluoro derivative of the parent compound [4]. However, the synthesis of [^{18}F]FDG by nucleophilic substitution fluorination (Fig. 1) proved to be a more efficient method of producing [^{18}F]FDG in higher, more reproducible yields. In 1986, Hamacher et al. utilized mannose triflate, as the precursor, and Kryptofix-[2.2.2] dissolved in acetonitrile, as a catalyst, substituting the trifluoromethane sulfonate leaving group with [^{18}F]fluoride to form [^{18}F]FDG [5]. This method of generating [^{18}F]FDG has allowed for widespread use in clinical imaging [6] in neurology [7–9], oncology [10–15], and cardiology [16]. It has been significantly studied in numerous oncology studies, including lung cancer, lymphoma, and colorectal cancer [1]. Because [^{18}F]FDG is a marker for hexokinase activity and glucose metabolism, this PET radiotracer has become a valuable tool in tumor detection, staging, and treatment evaluation because most tumors are marked by increased glucose uptake [17]. New innovative applications of [^{18}F]FDG have been applied recently as well. In 2002, Chacko et al. investigated the possible application of [^{18}F]FDG in evaluating infections, such as osteomyelitis, infected orthopedic hardware, and deep-seated soft tissue infections, and concluded that [^{18}F]FDG was an effective imaging modality in the assessment of patients with suspected infections. However numerous the possible uses for [^{18}F]FDG are, it is important to note that [^{18}F]FDG is a nonspecific radiotracer for imaging disease. This chapter provides a brief overview of the radiochemical synthesis and clinical use of [^{18}F]FDG, but because of the enormous body of literature available on [^{18}F]FDG, a detailed discussion is beyond the scope of this chapter.

2 SYNTHESIS PROCEDURES

CAUTION: All radiochemical syntheses must be carried out using the appropriate equipment in a facility authorized for the use of radioactive materials. Personal protective equipment must be worn, and all local radiation safety laws followed.

2.1 Preliminary Steps

A General Electric Medical Systems (GEMS) TRACERlab MX_{FDG} module [19] was used for the synthesis of [^{18}F]FDG at the University of Michigan Cyclotron and Radiochemistry Facility. The preliminary steps for the [^{18}F]FDG synthesis were performed before the transfer of [^{18}O]H_2O into the synthesizer. Initial tests run by the MX_{FDG} module ensured that the heater, syringe actuators, compressed air, vacuum pump, and nitrogen flow were all properly working.

The single-use TRACERlab MX_{FDG} cassette (Fig. 2) [20], manufactured by GE Healthcare, was assembled in a Horizontal LAF Hood. The cassette came prepackaged with two tC18 cartridges and one alumina-N cartridge. The cassette

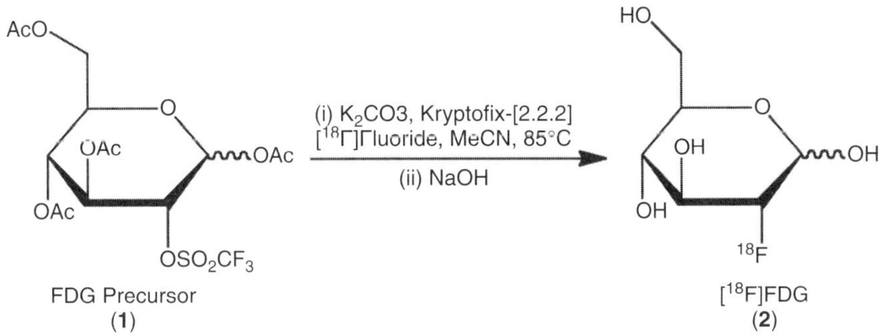

FIGURE 1 Synthesis of [^{18}F]FDG by nucleophilic aliphatic substitution on tetraacetyl mannosyl-2-triflate.

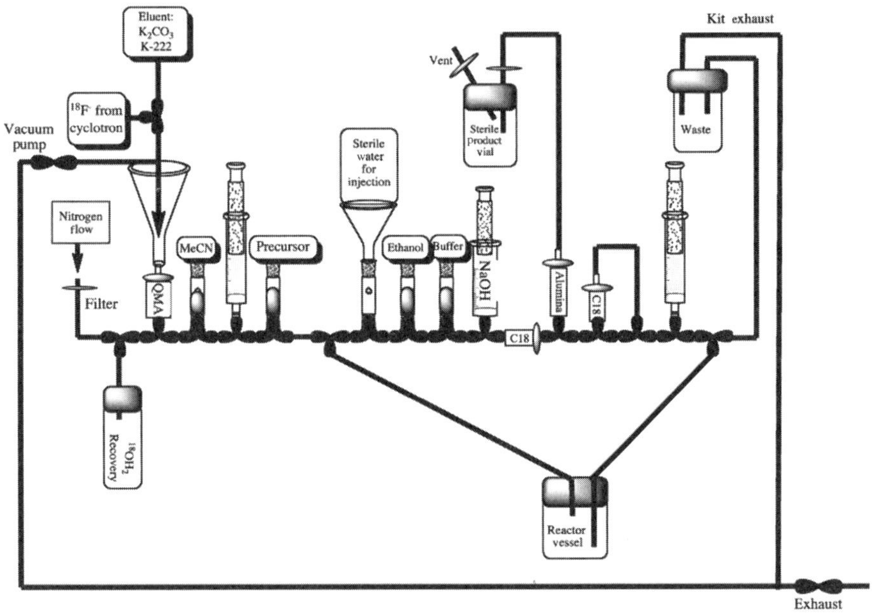

FIGURE 2 Single-use cassette for manufacture of [^{18}F]FDG.

was assembled per manufacturer recommendations, which included attaching one Millipore Millex-FG sterilizing filter [21] to the nitrogen inlet connector tubing, a conical column reservoir [22] to collect the [^{18}O]H_2O transferred from the cyclotron to the cassette, a QMA Sep-Pak Light Accell Plus cartridge [23], two single-use 30-mL syringes [24], ABX manufactured mannose triflate [25] precursor, and the ABX synthesizer reagent kit [26] containing predispensed volumes of acetonitrile, ethanol, neutralization solution, sodium hydroxide, Kryptofix-[2.2.2]/potassium carbonate solution, and sterile water for injection. The neutralization solution [26], with an approximate pH of 6.4, was a mixture of citrate buffer with hydrochloric acid, resulting in a composition of 23.5 mg of disodium hydrogen citrate-0,5-hydrate (extra pure), 144.4 mg of trisodium citrate-2-hydrate (extra pure), 5 mL of pure water, and 1 mL of HCl 2 N for one synthesis, and was manufactured by ABX.

Once the cassette was mounted to the MX_{FDG} module, additional steps performed included hooking the syringes onto the syringe actuators, dissolving the precursor in 3.5 mL of acetonitrile, pressurizing the reagent vials used during synthesis, and rinsing the single-use cassette with acetonitrile.

2.2 Production of [^{18}F]Flouride

A GEMS PETtrace cyclotron [19] was used to generate [^{18}F]fluoride by the $^{18}O(p,n)^{18}F$ nuclear reaction using a [^{18}F]fluoride target [27]. Approximately 2 mL of [^{18}O]H_2O [28] was loaded into the [^{18}F]fluoride target and bombarded at 40 μA for 25 min to generate approximately 1 Ci of activity (37 GBq).

2.3 Azeotropic Drying of [^{18}F]Fluoride

After bombardment, the [^{18}F]fluoride was transferred to a GEMS TRACERlab MX_{FDG} synthesizer [19] in approximately 2 mL of [^{18}O]H_2O [28], where it was passed through an anion exchange cartridge [23]. The fluoride ions were trapped on the cartridge and the [^{18}O]H_2O was recovered.

Once the [^{18}O]H_2O was recovered, the [^{18}F]fluoride ions were eluted from the anion exchange cartridge into the reaction vessel using a 0.6 mL solution [26] containing 7 mg of potassium carbonate, 22 mg of Kryptofix-[2.2.2], 300 μL of water, and 300 μL of acetonitrile. In the reaction vessel, the solvents were evaporated at 95°C for 8.5 min under negative pressure and nitrogen flow. Throughout the evaporation process, 80 μL of acetonitrile was added three times to the reaction vessel during the evaporation stage to maintain the H_2O/MeCN azeotrope, totaling 240 μL of acetonitrile.

2.4 Synthesis of [^{18}F]FDG

During the preliminary steps of preparing the GEMS TRACERlab MX_{FDG} module for synthesis, 40.0 mg of mannose triflate precursor [25] was dissolved in 3.5 mL of acetonitrile [26]. Following drying of the fluoride, the precursor was added to the reactor vessel. A nucleophilic aliphatic substitution reaction occurred, in which the [^{18}F]fluoride ions replaced the trifluoromethane sulfonate leaving group, creating the 2-[^{18}F]fluoro-1,3,4,6-tetra-*O*-acetyl-D-glucose molecule. This reaction occurred at 85°C for 4 min in a closed reactor vessel.

2.5 Purification of [^{18}F]FDG

The tC18 cartridges used for hydrolysis and purification were rinsed with 10 mL of sterile water for injection [26]. Afterwards, both tC18 cartridges [20] were conditioned with 3 mL of ethanol and 22 mL of water during synthesis. The labeled solution was then diluted with 26 mL of water and passed through the first tC18 cartridge, which was used for hydrolysis. The 2-[^{18}F]fluoro-1,3,4,6-tetra-*O*-acetyl-D-glucose molecule was trapped on the cartridge, while all other solvents, unreacted [^{18}F]fluoride ions, and polar byproducts were washed away into the external waste bottle. Kryptofix was trapped on this cartridge as well. The hydrolysis cartridge was washed three times with water; the first rinse was with 10 mL of water, the second with 10 mL of water, and the final with 23 mL of water.

Alkaline hydrolysis of the labeled compound occurred at room temperature, using 800 μL of 2 N sodium hydroxide. The sodium hydroxide passed through the tC18 cartridge, removing the four acetyl-protecting groups. [^{18}F]FDG was collected from the tC18 cartridge in water and pulled into a 30-mL syringe [24]. The neutralization solution [26] was added to the alkaline [^{18}F]FDG solution (10.8 mL) in order to provide a solution of physiological pH suitable for injection. The final formulation (16.8 mL) was further purified using a second tC18 cartridge [20] and an alumina-N cartridge [20], and passed through a sterile Millex-GS 0.22-μm filter [29] into a sterile dose vial [30], typically generating 500–600 mCi (18.5–22.2 GBq) of [^{18}F]FDG. The solution was then released for quality control (QC) testing. After the synthesis was complete, the cassette was rinsed with the remaining sterile water to reduce residual radioactivity in the single-use cassette before removal and disposal.

3 QUALITY CONTROL PROCEDURES

CAUTION: All radiopharmaceuticals produced for clinical use must have local regulatory approval before use in humans. Trained personnel must carry out quality control procedures, and each dose must meet all established quality control criteria before release to the clinic.

QC tests at the University of Michigan Cyclotron and Radiochemistry Facility for [^{18}F]FDG were carried out in accordance with the current US Pharmacopeia [31,32] and are summarized below. Data from QC testing for three repeat batches of [^{18}F]FDG are summarized in Table 1. Each of the three doses of [^{18}F]FDG met all established QC criteria.

3.1 Visual Inspection

The [^{18}F]FDG dose must be clear, colorless, and free of particulate matter. This must be inspected behind proper shielding. A PET L-block barrier system was used.

3.2 Radiochemical Identity and Purity

Radiochemical purity was determined by thin layer chromatography (TLC) using silica gel chromatography plates [33] and a Bioscan AR-2000 TLC scanner [34]. The plate was spotted with approximately 10 μg of FDG reference standard solution [35] and then spotted with a sample of [^{18}F]FDG dose. The FDG reference standard solution was prepared by dissolving 10 mg of 2-fluoro-2-deoxy-D-glucose [35] in 50 μL of acetonitrile and 50 μL of water. The developing solution [36], or mobile phase, contained 95% acetonitrile and 5% water. The plate was developed in this solvent system, dried using a warm laboratory hot

TABLE 1 QC Data for Three Repeat Runs of [^{18}F]FDG

QC Test	Release Criteria	Run 1	Run 2	Run 3
Yield mCi (GBq)	N/A*	528 (19.5)	530 (19.6)	507 (18.8)
Percentage yield (corrected for decay)	N/A*	68.3	66.8	69.3
Visual inspection	Clear, colorless	Clear, colorless	Clear, colorless	Clear, colorless
Radiochemical identity	Compare to standard	$rf_{stnd} = 0.40$ $rf_{FDG} = 0.40$	$rf_{stnd} = 0.384$ $rf_{FDG} = 0.390$	$rf_{stnd} = 0.32$ $rf_{FDG} = 0.32$
Radiochemical purity (%)	≥90	96.69	96.19	97.02
Residual solvent analysis	<0.5% Ethanol <0.04% Acetonitrile	0.004% Ethanol, 0.005% Acetonitrile	0.014% Ethanol, 0.008% Acetonitrile	0.023% Ethanol, 0.008% Acetonitrile
Dose pH	4.5–7.5	6.0	5.5	6.0
Residual Kryptofix-[2.2.2] (μg/mL)	≤50	<10	<10	<10
Sterile filter integrity test (psi)	>50	54	56	54
Radionuclidic identity ($t_{1/2}$) (min)	105–115	109	111	109
Endotoxin analysis (EU/mL)	≤10.9	<2.00	<2.00	<2.00
Sterility testing	No microbial growth in the product culture tube up to 14 d	Pass	Pass	Pass

*Release criteria not available.

plate [37] (temperature setting 5), and placed on the TLC scanner for analysis. The Bioscan TLC Scanner measured the radiochemical purity, which must be greater than or equal to 90%.

Radiochemical identity was determined by TLC and analysis of the retention factor (Rf). The retention factor of the standard was compared to the [^{18}F]FDG dose sample placed on the TLC plate. One plate was developed and dried, as described above, for both radiochemical purity and identity analysis. Once the standard became visibly brown after heating, the plate was removed and placed on the TLC scanner for analysis. The Rf was calculated for the hot [^{18}F]FDG sample. The "cold" FDG standard retention factor was calculated manually and compared with the [^{18}F]FDG sample, the radioactivity distribution of which is determined by use of the Bioscan AR-2000 TLC scanner [34]. The Rf value of the [^{18}F]FDG, free [^{18}F]fluoride, and acetylated [^{18}F]FDG are about 0.45, 0.0, and 0.85, respectively [4].

3.3 Radionuclidic Identity

A Capintec CRC®-15R Radioisotope Dose Calibrator [38] was used to measure activities. A sample of the [^{18}F]FDG product was placed in the Capintec Dose Calibrator located behind an L-block lead shield and measured, noting both the time and activity. After a period of at least 15 min, time and activity was noted again. The half-life was calculated using Equation 1, which must be within the range of 105–115 min for the dose to be released to the clinic.

$$t_{1/2} = -\ln 2(\text{Time difference}/[\ln(\text{ending activity}/\text{starting activity})]) \quad (1)$$

3.4 Residual Solvent Analysis

Residual solvent analysis of [^{18}F]FDG was performed on a Shimadzu GC-2010 Gas Chromatograph [39]. Acetonitrile and ethanol were tested for concentration levels and compared with a standard solution in accordance with the US Pharmacopeia [31]. The standard solution contained known concentrations of acetonitrile and ethanol, 0.01% and 0.1%, respectively. A Restek Stabliwax with Integra-Guard 30 m × 0.25 mm GC column [40] was used with helium as the carrier gas. Concentration percentages were calculated, and no more than 0.5% ethanol and 0.04% acetonitrile can be found in the dose, according to the US Pharmacopeia [31]. Acetonitrile, a class 2 solvent, has a limit on concentrations of <410 ppm per patient per day, imposed by the ICH Harmonised Tripartite Guidelines [41]. Ethanol, a class 3 solvent, has a limit of <5000 ppm/day.

3.5 Dose pH

ColorpHast® nonbleeding pH strips [42] were used to analyze a small amount of [^{18}F]FDG dose. The specification range of the colorpHast nonbleeding pH strips used were 2.0–9.0, and the determination of pH was done by visual comparison of the pH strip with the reference card supplied with the strips. The pH must be within the range of 4.5–7.5 to meet the release criteria.

3.6 Residual Kryptofix-[2.2.2] Analysis

Kryptofix-[2.2.2] is a phase transfer agent used to facilitate the nucleophilic fluorination reaction for [^{18}F]FDG, and because of its toxicity, verification of the absence of Kryptofix-[2.2.2] in the dose must take place before its release [43]. Residual Kryptofix-[2.2.2] levels in [^{18}F]FDG were analyzed using the established spot test. Kryptofix standards [44] were prepared by serial dilution of a 1.0 mg/mL stock solution using deionized water. Both water (negative control) and a 50-μg/mL standard (positive control) were used in this test. Stock iodoplatinate reagent was prepared to create iodoplatinate indicator strips used in spot testing. This reagent was generated using 5 mL of water mixed with 0.26 g of chloroplatinic acid [45]. This solution was then mixed with 45 mL of water and 5 g of potassium iodide [46] and diluted with an additional 100 mL of water. Precut strips of thin-layer silica gel [47] were immersed in the iodoplatinate reagent for 5–10 s. Once the strips had been completely saturated by the reagent, they were dried overnight and then stored at room temperature in a vented glass jar.

A sample of the [^{18}F]FDG dose was applied to the indicator strip using a single droplet. On the same strip, both the negative water control and the Kryptofix standard [44] (50 µg/mL) were applied and allowed to dry for 5–10 min. The three spots were analyzed, and a visual determination of the relative concentration of Kryptofix-[2.2.2] was made for the [^{18}F]FDG dose.

The presence of Kryptofix-[2.2.2] in the dose is confirmed by the appearance of a blue-black ring around a white circle of silica surrounded by a pink ring of unreacted reagent. At very low or no concentrations of Kryptofix, a faint pink core remains in the center of the white circle, with no visible blue-black ring. The dose must contain <50 µg/mL for the dose to be acceptable for release.

3.7 Sterile Filter Integrity Test

Validation of the integrity of the sterile filter going into the dose vial must also be performed before the release of the dose to the clinic. The filter integrity test is essential in testing [^{18}F]FDG product sterility, as the actual sterility testing results will not be available until 14 days later [4]. The Millex-GS sterile filter [29] from the [^{18}F]FDG synthesis was attached to the nitrogen supply through a calibrated regulator. The tip of the filter was submerged in water. The nitrogen flow was gradually turned on to increase the nitrogen pressure on the filter. The pressure was increased to greater than 50 psi, which is the acceptance criterion for the Millipore-GS filter. If the pressure reaches 50 psi or more without resulting in a stream of bubbles in the water, the filter is considered intact. If a steady stream of bubbles is seen before the minimum bubble point pressure is reached, the test fails and the dose is rejected.

3.8 Endotoxin Analysis

The Charles River Portable Testing System [48,49] was used to determine the endotoxin content in the [^{18}F]FDG doses. Limulus Amebocyte Lysate (LAL) test cartridges [50], purchased from Charles River, were used in this test. A volume of 3.9 mL of Endosafe LAL Reagent water [51] was pipetted into a test tube. A volume of 0.1 mL of the [^{18}F]FDG dose was then added to this tube, mixed for 10 s using a vortex [52], and 25 µL of the diluted [^{18}F]FDG sample was added to each well on the Charles River test cartridge. This was performed in accordance with the US Pharmacopeia [53]. The dose must contain $\leq$175 Endotoxin Units (EU) per dose.

3.9 Sterility Testing

Sterility testing must also be performed on all doses of [^{18}F]FDG produced to verify that the process by which each PET radiopharmaceutical is produced is consistently sterile and suitable for human use. Sterility testing was performed by inoculating samples of [^{18}F]FDG into a fluid thioglycolate media (FTM) [54] and a soybean casein digest media (SCDM) [55]. SCDM is used for the detection of aerobic bacteria and fungi, while the general purpose of FTM is to detect common aerobic, facultative, and anaerobic microorganisms.

The [^{18}F]FDG medium was incubated along with positive and negative controls for 14 days. Incubation temperatures were 32°C and 22°C for FTM and SCDM, respectively, according to current USP guidelines [56] for sterility testing. The media was visually inspected on the third, eigth, and fourteenth days of the incubation test period and compared to the positive and negative controls. The positive control must show growth (cloudiness or turbidity) in the medium, and the [^{18}F]FDG samples and negative control must show no growth after 14 days of incubation in order to be considered sterile.

CHEMICAL ABSTRACTS NOMENCLATURE (REGISTRY NUMBER)

Mannose triflate (92051-23-5)
[^{18}F]Fluorodeoxyglucose ([^{18}F]FDG) (128441-61-2)
Carabonic acid, potassium salt (1:2) (584-08-7)
Sodium hydroxide (1310-73-2)
Acetonitrile (75-05-8)
Ethanol (64-17-5)
2-Fluoro-2-deoxy-D-glucose (29702-43-0)
Kryptofix-[2.2.2] (23978-09-8)
Chloroplatinic acid (16941-12-1)
Potassium iodide (7681-11-0)

REFERENCES AND NOTES

1. Hoh CK. Nucl Med Biol 2007;34:737–742.
2. Miller PW, Long NJ, Vilar R, Gee AD. Angew Chem Int Ed 2008;47:8998–9033.
3. Fowler JS, Ido T. Semin Nucl Med 2002;32:6–12.
4. Yu S. Biomed Imaging Interv J 2006;2(4):e57.
5. Hamacher K, Coenen HH, Stocklin G. J Nucl Med 1986;27:235–238.
6. Beuthien-Baumann B, Hamacher K, Oberdorfer F, Steinbach J. Carbohydr Res 2000; 327:107–118.
7. Masdeu JC, Zubieta JL, Arbizu J. J Neurol Sci 2005;236:55–64.
8. Silverman DHS, Small GW, Chang CY, Lu CS, Kung de Aburto MA, Chen W, et al. Neurology 2001;286:2120–2127.
9. Herholz K, Heiss WD. Mol Imaging Biol 2004;6:239–269.
10. Vallabhajosula S. Semin Nucl Med 2007;37:400–416.
11. Vansteenkiste JF, Stroobants SS. J Thorac Oncol 2006;1:71–73.
12. Pelosi E, Deandreis D. Eur J Surg Oncol 2007;33:1–6.
13. Macapinlac HA. Cancer J 2004;10:262–270.
14. Hutchings M, Eigtved AI, Specht L. Crit Rev Oncol/Hematol 2004;52:19–32.
15. Larson SM, Schwartz LH. J Nucl Med 2006;47:901–903.
16. Machac J. Semin Nucl Med 2005;35:17–36.

17. Black NF, McJames S, Kadrmas DJ. IEEE Trans Nucl Sci 2009;56:2750–2758.
18. Chacko TK, Zhuang H, Stevenson K, Moussavian B, Alavi A. Nucl Med Commun 2002;23: 851–855.
19. GE Healthcare, USA.
20. TRACERlab MX_{FDG} Cassettes were purchased from General Electric Healthcare (part no. P5150ME), and used as received. The cassettes came prepackaged and assembled with two tC18 cartridges and one alumina-N cartridge.
21. Millex-FG sterile 0.20 μm filters were purchased from Millipore (part no. SLFG025LS) and used as received.
22. Conical Column 10mL Reservoir was purchased from ABX (part no. 9180) and used as received.
23. QMA-Light Sep-Pak® preconditioned cartridges (part no. K-920) were purchased from Advanced Biochemical Compounds (ABX) and used as received.
24. 30mL syringes were purchased from Fischer Scientific (part no. 309650) and used as received.
25. Mannose triflate ultrapure precursor was purchased from ABX (part no. 101.0040) and used as received.
26. Coincidence FDG18 Synthesizer Reagent Kit was purchased from ABX (part no. K-153TM) and used as received.
27. GEMS silver high-yield [^{18}F]fluoride target.
28. Virgin [^{18}O]H_2O was purchased from ABX, Amic, Rotem, or Medical Isotopes, and used as received.
29. Millex-GS sterile 0.22-μm filters were purchased from Millipore (part no. SLGSV255F) and used as received.
30. Sterile 30-mL dose vial was purchased from Hospira (part no. 5829-30), and used as received.
31. U.S. Pharmacopeia <823> Radiopharmaceuticals for positron emission tomography compounding. USP 32-NF 27; 2009.
32. U.S. Pharmacopeia Official Monographs/Fludeoxyglucose. USP 29-NF 24; 2006.
33. Whatman Thin Layer Chromatography Plates (part no. 4861-110) MK6F Silica Gel 60 Å, Dimensions 2.5 × 7.5 cm, layer thickness 250 μm.
34. Bioscan, Inc., USA.
35. The FDG reference standard solution contained both 2-fluoro-2-deoxy-D-glucose reference standard, purchased from Sigma Aldrich (part no. F5006) and used as received, and anhydrous acetonitrile, purchased from Acros (part no. 61096-1000) and used as received.
36. The TLC developing solution contained anhydrous acetonitrile, purchased from Acros (part no. 61096-1000) and used as received and MilliQ water, in concentrations of 95% and 5%, respectively.
37. Laboratory Hot Plate, manufactured by Corning, Model no. PC-200, and used at temperature setting 5.
38. Capintec, Inc., USA.
39. Shimadzu Corporation, USA.
40. GC Column: Restek Stabliwax with Integra-Guard; 30 m × 0.25 mm, purchased from Restek (part no. 10623-124).
41. International Conference on Harmonisation of Technical Requirements for Registration of Pharmaceuticals for Human Use (ICH).
42. EMD Chemicals Inc., USA (part no. 9578-3).
43. Mock BH, Winkle W, Vavrek MT. Nucl Med Biol 1997;24:193–195.
44. Kryptofix–[2.2.2] was purchased from Acros (part no. 29195-0010) and used as received for preparation of reference standard.
45. Chloroplatinic acid was purchased from Acros (part no. 363590010) and used as received.
46. Potassium iodide was purchased from Acros (part no. 19379-0500) and used as received.
47. Thin-layer silica gel sheets were purchased from EMD Chemicals (part no. 57487), 20 × 20 cm sheets, silica gel 60 Å, and cut into squares 4 × 5 cm.
48. Charles River Laboratories, USA.
49. Endosafe Portable Testing System (model no. PTS 100), purchased and calibrated according to manufacture guidelines by Charles River Laboratories, USA.

50. Limulus Amebocyte Lysate (LAL) Test Cartridges (part no. PTS20F) purchased from Charles River Laboratories and used as received.
51. Endosafe Limulus Amebocyte Lysate (LAL) Reagent water (part no. W110) purchased from Charles River Laboratories and used as received.
52. VWR Vortexer 2 (model no. G-560), vortex speed setting 8.
53. US Pharmacopeia <85>. Bacterial endotoxins test. USP 32-NF 27, 2009.
54. Fluid Thioglycolate Media (part no. 220889) purchased from Beckon, Dickson and Company, USA, and used as received.
55. Soybean Casein Digest Media (part no. 221823) purchased from Beckon, Dickson and Company, USA, and used as received.
56. US Pharmacopeia <71> Sterility tests. USP 32-NF 27, 2009.

CHAPTER 2

SYNTHESIS OF SODIUM [^{18}F]FLUORIDE (Na[^{18}F]F)

BRIAN G. HOCKLEY AND PETER J. H. SCOTT

Department of Radiology, University of Michigan School of Medicine, Ann Arbor, Michigan

Na^{18}F

[^{18}F]Sodium Fluoride

1 INTRODUCTION

Sodium [^{18}F]fluoride (Na[^{18}F]F) [1] is a positron emission tomographic (PET) imaging agent that can be used clinically to detect and monitor metastatic malignancies in bone. The use of sodium [^{18}F]fluoride (Na[^{18}F]F) for bone imaging was approved by the FDA in 1972 (NDA 17–042). However, with the introduction and widespread distribution of technetium-99m bone imaging agents in 1975 the demand for Na[^{18}F]F dropped steadily, and the original NDA (new drug application) holder stopped marketing the drug [2]. The technetium bone imaging agents offered the advantages of a longer half-life, wider distribution options, and lower cost because of centralized production operations with little compromise in image and data quality. The US Centers for Medicare and Medicaid Services (CMS) and the nuclear medicine community at large embraced the new imaging agents causing Na[^{18}F]F to become obsolete, although no less effective. Decline in marketing, manufacturing, and clinical use of Na[^{18}F]F naturally led to the end of reimbursements from CMS and private insurers.

By 2009 and 2010, there were only three reactors in the world still producing molybdenum-99, the parent isotope of technetium-99m for medical imaging. However, because of a combination of routine maintenance and unscheduled safety-related shutdowns, a worldwide shortage of technetium became a reality with only one reactor online. The isotope shortage coupled with technological

Radiochemical Syntheses, Volume 1: Radiopharmaceuticals for Positron Emission Tomography, First Edition. Edited by Peter J. H. Scott, Brian G. Hockley.

advances and more routine clinical use of PET imaging since the 1970s reinvigorated interest in Na[^{18}F]F within the nuclear medicine community. In February 2011, the FDA approved a National Cancer Institute NDA for a new strength of Na[^{18}F]F for use in bone scans, and at the time of publication of this book CMS was considering limited reimbursement for Na[^{18}F]F bone scans if used during a proof of efficacy clinical trial, of which there were several active in North America. In addition to technological advancement, there have been many regulatory changes since 1972, and the Academy of Molecular Imaging has initiated a multicenter clinical trial to reexamine the clinical use of Na[^{18}F]F. It is for these reasons that this brief chapter detailing the synthesis and quality control (QC) of Na[^{18}F]F has been included in this volume despite not having broad clinical application at this time.

2 SYNTHESIS PROCEDURES

CAUTION: All radiochemical syntheses must be carried out using appropriate equipment in a facility authorized for the use of radioactive materials. Personal protective equipment must be worn and all local radiation safety laws followed.

2.1 Production of [^{18}F]Fluoride

[^{18}O]H_2O (2 mL) [3] was loaded into the [^{18}F]fluoride target [4] of a General Electric Medical Systems (GEMS) PETTrace cyclotron [5]. The target was bombarded (40 μA beam for 2 min) to generate ~100 mCi (3.7 GBq) of [^{18}F]fluoride by the $^{18}O(p,n)^{18}F$ nuclear reaction.

2.2 Preparation and Formulation of Na[^{18}F]F

Fluoride-18 as a solution in [^{18}O]H_2O (2 mL) was delivered through an in-line Plus-CM cation exchange cartridge (as a precaution to remove positively charged recoil nuclei generated during irradiation) into the dilution flask of a TRACERlab FX_{FN} that was precharged with 10 mL Sterile Water for Injection, USP. The resulting solution was then passed through a QMA-light Sep-Pak [6] to extract the fluoride-18. The QMA-light Sep-Pak was washed with additional Sterile Water for Injection (10 mL) to remove any residual [^{18}O]H_2O and then dried with an argon stream. The fluoride-18 was then eluted from the Sep-Pak with 0.9% sodium chloride, USP (10 mL) into the TRACERlab FX_{FN} collection vial. This solution was passed through a sterile filter (Millipore Millex GS pn: SLGSV255F) into a 10-mL sterile, empty vial (Hollister-Stier pn: 7515ZA) vented with a needle and Millex-FG sterile filter (MIllipore pn: SLFG025LS). The yield of sodium [^{18}F]fluoride was recorded (typically 95 mCi (3.5 GBq)), and the product was released for QC.

3 QUALITY CONTROL PROCEDURES

CAUTION: All radiopharmaceuticals produced for clinical use must have local regulatory approval before use in humans. Trained personnel must carry out QC

TABLE 1 QC Data for Three Repeat Runs of $Na[^{18}F]F$

QC Test	Release Criteria	Run 1	Run 2	Run 3
Visual inspection	Clear, colorless	Clear, colorless	Clear, colorless	Clear, colorless
Radiochemical identity	RRT = 0.9 1.1	1.0	1.0	1.0
Radiochemical purity (%)	≥95	98.9	99.4	99.4
Specific activity	No limit established	NA	NA	NA
Dose pH	4.5–8	5	5	5
Sterile filter integrity test (psi)	>50	>50	>50	>50
Radionuclidic identity ($t_{1/2}$) (min)	105–115	105	114	107
Endotoxin analysis (EU/mL)	≤17.5	≤2	≤2	≤2
Sterility testing	No visible colony growth at 14 d	Pass	Pass	Pass

procedures, and each dose must meet all established QC criteria before release to the clinic.

QC procedures for $Na[^{18}F]F$ are summarized below. Complete QC data for three repeat batches of $Na[^{18}F]F$ produced using the method disclosed herein are summarized in Table 1. Each of the three doses met all of the established QC criteria.

3.1 Visual Inspection

The $Na[^{18}F]F$ dose is examined in a clear glass vial and must be clear, colorless, and free of particulate matter.

3.2 Radiochemical Identity

Radiochemical identity is determined using a Shimadzu HPLC system with the following components: SCL-10Avp system controller, DGU-14A in-line degassing unit, LC-10ADvp pump, CDD-10Avp conductivity detector with temperature controlled cell, CTO-20A oven, and Bioscan FC3400 flow count radioactivity detector. The system is not equipped with an ion suppressor. Chromatographic separation is performed on a Phenomonex Rezex RHM-monosaccharide (hydrogen form), 300 × 7.8 mm; mobile phase 0.0015 M aqueous sulfuric acid; flow rate = 0.8 mL/min. The retention time of $Na[^{18}F]F$ is compared to that of the $[^{18}F]$fluoride reference standard (sodium fluoride, Aldrich pn: 450022) and must be ±10% (relative retention time (RRT) must be 0.9–1.1).

3.3 Radiochemical Purity

The radiochemical purity of each batch is determined using the IC (Ion chromatography) system outlined in Section 3.2. Radiochemical purity must be >95%. Representative analytical HPLC traces are displayed in Fig. 1.

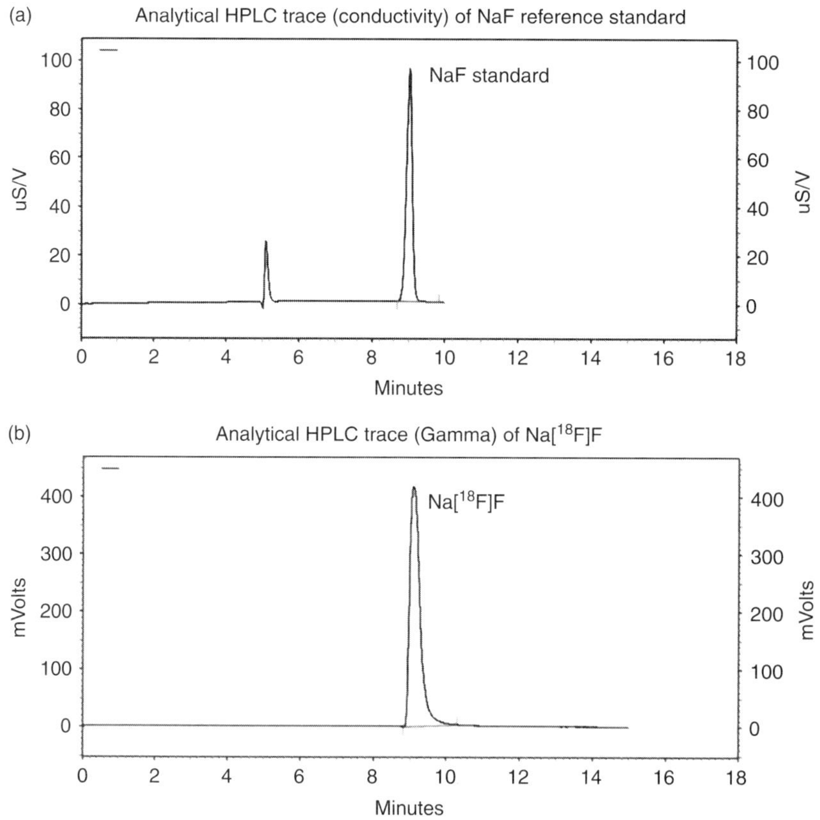

FIGURE 1 Analytical HPLC trace of NaF reference standard (a) and Na[^{18}F]F (b).

3.4 Specific Activity

The concentration of fluoride in each batch can be determined using the HPLC system described in Section 3.2. Although absolute specific activity cannot always be calculated using this method, a minimum specific activity can be determined based on the detection limit. There is no specific activity release limit for Na[^{18}F]F in place at our institution.

3.5 Dose pH

The pH of the Na[^{18}F]F dose was analyzed by applying a small amount of the dose to colorpHast® pH 2.0–9.0 nonbleeding pH-indicator strips [7] and determined by visual comparison to the scale provided. Dose pH must be 4.5–7.5.

3.6 Sterile Filter Integrity Test

The sterile filter from the Na[^{18}F]F (with needle still attached) was connected to a nitrogen supply via a regulator. The needle was submerged in water, and the nitrogen pressure was gradually increased. If the pressure was raised above the filter acceptance pressure (50 psi) without seeing a stream of bubbles, the filter was considered intact. If a stream of bubble occurs <50 psi, the test fails.

3.7 Radionuclidic Identity

Activities were measured using a Capintec CRC®-15R Radioisotope Dose Calibrator, and half-life was calculated using Equation 1. Calculated half-life must be 105–115 min.

$$t_{1/2} = -\ln 2(\text{time difference}/(\ln(\text{ending activity}/\text{starting activity}))) \quad (1)$$

3.8 Endotoxin Analysis

Endotoxin content in doses of Na[^{18}F]F was analyzed by a Charles River Laboratories EndoSafe® Portable Testing System and according to the US Pharmacopeia (USP) [8]. A dose of Na[^{18}F]F must contain ≤175 endotoxin units (EU), which is adjusted to 17.5 EU/mL, as our formulation volume is 10 mL.

3.9 Sterility Testing

Samples of Na[^{18}F]F were added to tubed fluid thioglycolate media (FTM) and tubed soybean casein digest media (SCDM). FTM is used to test for anaerobes, aerobes, and microaerophiles, while SCDM is used to test for nonfastidious and fastidious microorganisms. The media tubes were incubated along with positive and negative controls for 14 days. FTM is incubated at 30–35°C, and SCDM is incubated at 20–25°C according to current USP guidelines [9]. Tubes were visually inspected on the third, eighth, and fourteenth days of the test period and compared to the positive and negative standards. The test article (Na[^{18}F]F) and negative controls must have no culture growth after 14 days to be indicative of sterility. Positive controls can be established separately as a growth promotion test [10] according to USP guidelines.

CHEMICAL ABSTRACTS NOMENCLATURE (REGISTRY NUMBER)

Sodium fluoride (7681-49-4)

REFERENCES AND NOTES

1. U. S. Pharmacopeia Monograph: Sodium Fluoride F18 Injection.
2. US Federal Register/Vol. 65, No. 48/Friday, March 10, 2000/ Notices.
3. Virgin [^{18}O]H_2O purchased from ABX, Amic, Rotem, or Medical Isotopes, and used as received.
4. GEMS silver high yield [18F]fluoride target.
5. GE Healthcare, USA.
6. QMA-Light Sep-Pak® cartridges were purchased from Waters (part no. WAT023525) and conditioned with 10mL ethanol, 10mL water, 10mL 0.5M sodium bicarbonate, and a further 10mL water before use.
7. EMD Chemicals Inc., USA (part no. 9578-3).
8. U. S. Pharmacopeia <85>. Bacterial Endotoxins Test. USP 32–NF 27, 2009.
9. U. S. Pharmacopeia <71>. Sterility Tests. USP 32–NF 27, 2009.
10. U. S. Pharmacopeia <71>. Sterility Tests. USP 32–NF 27, 2009.

CHAPTER 3

RADIOSYNTHESIS OF 3′-DEOXY-3′-[^{18}F]FLUOROTHYMIDINE ([^{18}F]FLT)

Hartmuth C. Kolb, Henry C. Padgett, Steve Zigler, Jim Patanella, Fanrong Mu, Umesh B. Gangadharmath, Vani P. Mocharla, Peter J. H. Scott, and Joseph C. Walsh

Siemens Molecular Imaging, Biomarker Research, Culver City, California

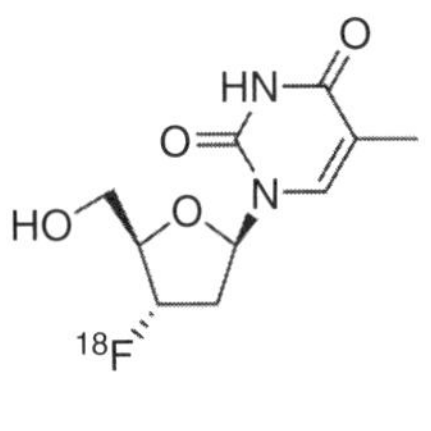

1 INTRODUCTION

Despite the initial development of the nucleoside 3′-deoxy-3′-fluorothymidine as a potential antiviral therapeutic for HIV-1-infected patients [1], its radioactive analog 3′-deoxy-3′-[^{18}F]fluorothymidine ([^{18}F]FLT) has gained increased use as a tracer for positron emission tomographic (PET) imaging. [^{18}F]FLT is widely used for a multitude of oncology-related preclinical studies and is currently undergoing several investigational studies for its clinical use in detecting proliferating cells in various tissues in vivo [2]. [^{18}F]FLT localizes in lung, breast, and brain tumors, providing a sensitive means for detecting these tumors [3–5]. [^{18}F]FLT is currently undergoing evaluation as a surrogate marker for a patient's response to both chemo- and radiotherapies [2,6]. [^{18}F]FLT does not significantly accumulate in inflamed tissues, and is, therefore, less susceptible to inflammation-based tissue uptake as compared to 2-deoxy-2-(^{18}F)fluoro-D-glucose ([^{18}F]FDG) [4,7,8].

Several precursors have been used for producing [^{18}F]FLT, including cyclic anhydrides and 3′-sulfonate esters, on both manual and automated platforms

Radiochemical Syntheses, Volume 1: Radiopharmaceuticals for Positron Emission Tomography, First Edition. Edited by Peter J. H. Scott, Brian G. Hockley.

[9–17]. The method used for this disclosed synthesis of [^{18}F]FLT is unique in that it employs a precursor ("BocBocNos") bearing a 5′-*O*-Boc protecting group in place of the traditional 5′-*O*-dimethoxy trityl (DMT) protecting group. This protecting group change led to a number of distinct advantages over traditional FLT precursors. BocBocNos is highly crystalline, possesses a long shelf life, and reliably produces FLT in an average yield of 34.5% (decay corrected to end of synthesis) in 45 minutes on an automated synthesis module [18].

2 SYNTHESIS PROCEDURES

CAUTION: All radiochemical syntheses must be carried out using appropriate equipment in a facility authorized for the use of radioactive materials. Personal protective equipment must be worn and all local radiation safety laws followed.

2.1 Production of [^{18}F]Fluoride

[^{18}O]H_2O (2 mL) [19] is loaded into the [^{18}F]fluoride tantalum target of a Siemens RDS-111 cyclotron [20]. The target is bombarded (60-μA beam for 45–60 min) to generate approximately 1.5–2.0 Ci (56–74 GBq) of [^{18}F]fluoride by the ^{18}O(p,n)^{18}F nuclear reaction. The [^{18}F]fluoride is delivered through polyethylene, polypropylene, or polyetheretherketone (PEEK) tubing into a v-vial within a shielded area before delivery into the automated synthesis module. If high specific activity [^{18}F]FLT is required, the target and delivery lines are pre-rinsed with [^{18}O]H_2O (2 mL) through the target and delivery lines into a separate collection vial for recycling.

2.2 Azeotropic Drying of [^{18}F]Fluoride

Aqueous [^{18}F]fluoride ion is trapped on an anion exchange resin cartridge (Machery Nagel PS-HCO_3 Chromafix) [21] located on a Siemens Explora RN chemistry module, and the [^{18}O]H_2O is collected for recycling. The fluoride ion is then eluted into a glassy carbon reaction vessel using a solution of aqueous potassium carbonate (3 mg in 0.4 mL of water) [22]. A solution of Kryptofix® [2.2.2] (20 mg in 1 mL of acetonitrile) [23] is added directly into the glassy carbon reaction vessel after the elution of the [^{18}F]fluoride ion. The solution is concentrated to dryness to remove both the acetonitrile and water. Typically, the drying is carried out at 68°C for 5 min followed by 95°C for 30 s under reduced pressure [24] (250 mbar) and a stream of argon [25]. After removal of the solvents, a hard vacuum is pulled for 30 s to 2 min at 70°C until a low, steady-state pressure is obtained [26].

2.3 Synthesis of [^{18}F]FLT

A solution of the precursor, 1(2H)-Pyrimidinecarboxylic acid, 3-[2-deoxy-5-*O*-[(1,1-dimethylethoxy)carbonyl]-3-*O*-[(4-nitrophenyl)sulfonyl]-β-D-threo-pentofuranosyl]-3,6-dihydro-5-methyl-2,6-dioxo-, 1,1-dimethylethyl ester [18] (15.0

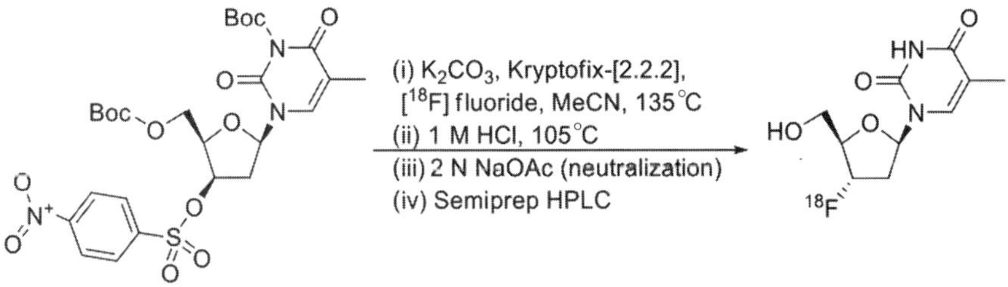

"BocBocNos"
(**1**)

[^{18}F]FLT
(**2**)

FIGURE 1 Radiosynthesis of [^{18}F]FLT using BocBocNos.

–18.0 mg, 24–29 μmol), in anhydrous acetonitrile (0.8–1.0 mL) is added to the activated [^{18}F]fluoride, while the reaction vessel remains at 70°C. The mixture is then heated from 70 to 135°C, which takes approximately 1–2 min. When the reaction reaches 135°C, the allotted time for fluorination is 3 min [14]. After cooling and evaporating most of the acetonitrile [27] at 55°C under simultaneous reduced pressure and a stream of argon (3 min), the pressure is equalized and aqueous hydrochloric acid [28] (1.0 N, 0.8–0.9 mL) is added. The mixture is heated to 105°C for 4 min. The reaction mixture is then cooled to 40°C, and the pH is adjusted with aqueous sodium acetate [29] (2.0 M, 0.4–0.6 mL) (Fig. 1).

2.4 Purification and Formulation of [^{18}F]FLT

The crude [^{18}F]FLT solution is manually transferred to the high performance liquid chromatography (HPLC) load loop (1.5 mL) and purified by semipreparative HPLC using either a Macherey-Nagel Nucleodur C18 Pyramid or a Phenomenex Synergi Hydro-RP 250 × 10 mm HPLC column [30] and 8% EtOH:92% 21 mM sodium phosphate [31] (v/v) mobile phase with a flow rate of 5.5 mL/min. The product is collected from the column as monitored by flow-through radioactivity and UV (254 nm) detectors (Fig. 2). The HPLC fraction containing the [^{18}F]FLT (eluting between 14 and 17 min) is passed through an activated Alumina N Sep-Pak® Light cartridge [32] (to capture any residual [^{18}F]fluoride) and then through a Pall® Supor AEF sterile filter (0.22 μm) [33] and collected directly into a sterile empty vial for quality control (QC) testing.

2.5 Reformulation of [^{18}F]FLT

When it is necessary to prepare [^{18}F]FLT in concentrations greater than the concentration obtained in the reverse-phase high performance liquid chromatography (RP-HPLC) fraction (typically for preclinical studies), the [^{18}F]FLT solution may be reformulated. The collected [^{18}F]FLT fraction is diluted with water (10 times the volume of the HPLC fraction), passed through an Oasis® HLB Plus cartridge [34], and then eluted with 50% EtOH:water solution (1 mL) into a collection vial. The resultant solution is adjusted to 10% EtOH:water by the addition of water

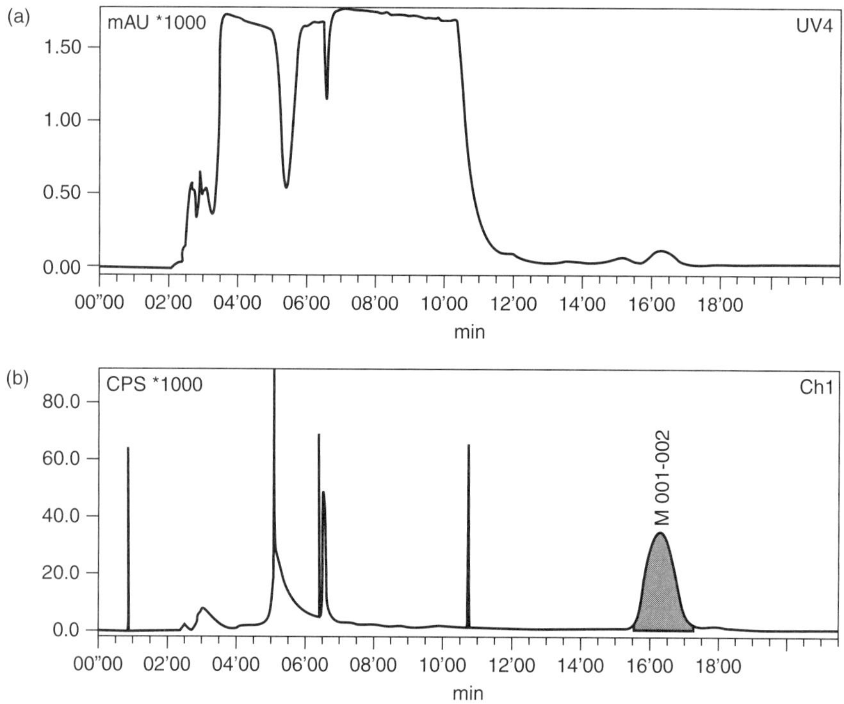

FIGURE 2 Semipreparative HPLC UV (a) and radioactive γ (b) traces for [^{18}F]FLT. The shaded area in (b) denotes the collection of [^{18}F]FLT.

(4 mL). The average recovery is 78.5%, with losses related to loading (14.4%) and eluting inefficiencies (7.1%).

3 QUALITY CONTROL PROCEDURES

CAUTION: All radiopharmaceuticals produced for clinical use must have local regulatory approval before use in humans. Trained personnel must carry out QC procedures, and each dose must meet all established QC criteria before release to the clinic.

Three batches of [^{18}F]FLT were produced according to the above conditions and analyzed according to the QC tests summarized in Table 1. Each of the three batches met all of the established QC criteria.

3.1 Radioactivity Concentration

The radioactivity concentration is assessed by measuring the radioactivity of a sample of the [^{18}F]FLT final product contained in a low dead volume syringe. Our internal specification for radioactivity concentration is within the range of 1–75 mCi/mL (0.037–2.8 GBq/mL) [35]. The radioactivity concentration for all three batches was within the range of 1–75 mCi/mL (0.037–2.8 GBq/mL).

TABLE 1 QC Specifications and Data for Three Batches of [^{18}F]FLT

QC Test	Specification	Batch 1	Batch 2	Batch 3
Radioactivity concentration mCi/mL (GBq/mL)	1–75 (0.037–2.8)	27.3 (1.0)	35.6 (1.3)	42.3 (1.6)
Visual inspection	Clear, colorless solution, free of particulate matter	Pass	Pass	Pass
Radiochemical identity	Retention time must correspond to standard peak	Pass	Pass	Pass
Radiochemical purity (%)	≥95	99.9	99.9	99.9
Specific activity (Ci/μmol)	≥0.4	9.0	11.9	18.3
Acetonitrile concentration (% wt/wt)	≤0.04	Pass	Pass	Pass
Ethanol concentration (% wt/wt)	4.0–8.8	Pass	Pass	Pass
pH	5.5–8.0	6	6	6
Residual Kryptofix[2.2.2] (μg/mL)	<50	Pass	Pass	Pass
Membrane filter integrity test	Meets or exceeds manufacturer's specifications	Pass	Pass	Pass
Radionuclidic identity ($t_{1/2}$) (min)	105–115	Pass	Pass	Pass
Bacterial endotoxins	<175 EU/injected volume	Pass	Pass	Pass
Sterility testing	Absence of microbials after 14-d incubation in two media	Pass	Pass	Pass

3.2 Visual Inspection

The [^{18}F]FLT final product is visually examined through leaded glass. The solution must be clear, colorless, and free of particulate matter. All three batches met this specification.

3.3 Radiochemical Identity and Purity

The radiochemical identity and purity of the [^{18}F]FLT final product is assessed by analytical RP-HPLC [36]. The system is equipped with a γ-detector (NaI), a UV detector (267 nm), and a Phenomenex Gemini column [37] operated in isocratic mode (10% EtOH:water at 1 mL/min). The radiochemical identity of the tracer is confirmed by use of an [^{19}F]FLT standard and by comparing the retention times between the UV peak for FLT from the FLT standard and the radioactive peak for [^{18}F]FLT from the [^{18}F]FLT final product solution (Fig. 3). The average offset between the UV and γ detectors is 0.15 min. Our internal specification for radiochemical purity is ≥95%. All three batches met this specification.

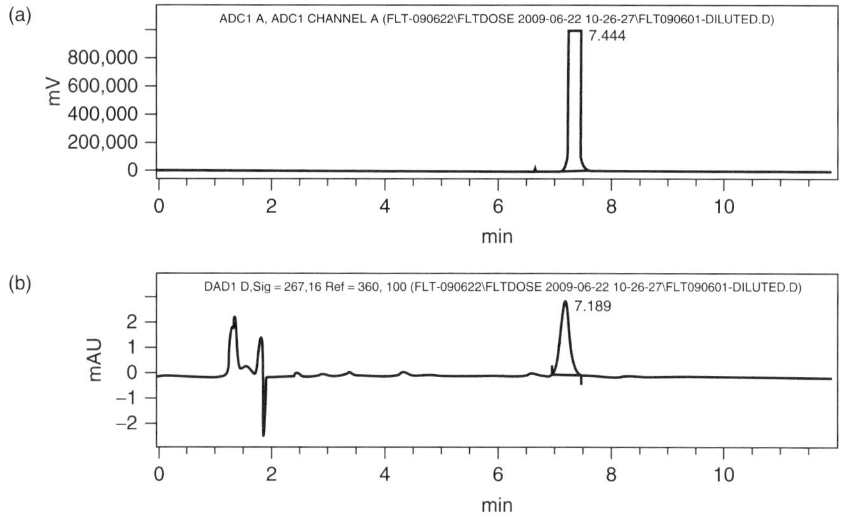

FIGURE 3 Analytical RP-HPLC traces for both the γ (a) and UV (b) traces of undiluted [^{18}F]FLT.

3.4 Specific Activity

The mass of [^{19}F]FLT in the [^{18}F]FLT final product solution is determined by RP-HPLC (see above). Our internal specification for specific activity is $\geq$0.4 Ci/μmol at the end of synthesis. The specific activity of the three [^{18}F]FLT batches ranged from 9.0 to 18.3 Ci/μmol, with [^{19}F]FLT mass concentrations ranging from 0.56 to 0.74 μg/mL.

3.5 Residual Solvents and Ethanol Concentration

The presence of residual solvents (acetonitrile) and the concentration of ethanol in the [^{18}F]FLT final product is assessed by gas chromatography (GC). The system is equipped with a flame ionization detector (FID) and a polyethylene glycol column (helium carrier gas at 10 mL/min). Two injections are made into the GC [38] for analysis: first, an undiluted sample is injected to determine the acetonitrile concentration and second, a diluted (60:1) sample to determine the ethanol concentration. Our internal specification for acetonitrile is $\leq$0.04% (w/w), which is also the United States Pharmacopoeia (USP) limit concentration for acetonitrile, and the allowed concentration range of ethanol is 4.0–8.8% (w/w). The acetonitrile concentration in all three [^{18}F]FLT batches was $<$0.04% (w/w). The ethanol concentration in the three batches ranges from 4.7% to 8.7% (w/w). These ethanol concentrations are considered within the tolerances for IV injection [39].

3.6 pH

The pH of the [^{18}F]FLT final product is assessed with pH strips. Our internal specification for pH is within the range of 5.5–8.0. The pH values for all three [^{18}F]FLT batches fell within this range.

3.7 Residual Kryptofix[2.2.2]

The concentration of Kryptofix[2.2.2] in the [^{18}F]FLT final product is assessed with an iodine spot test. Aluminum oxide plates are spotted with a Kryptofix[2.2.2] standard (50 μg/mL) and the final product solution, followed by drying and visualization of the plate using iodine vapor. The product contains <50 μg/mL of Kryptofix[2.2.2] if the spot for the standard is darker than that for the product. Our internal specification for Kryptofix[2.2.2] is that the concentration must be <50 μg/mL. All three [^{18}F]FLT batches met this specification.

3.8 Membrane Filter Integrity Test

The integrity of the membrane filter used for sterilization of the [^{18}F]FLT is assessed by the bubble point test [40]. Before performing this test, the membrane filter is carefully rinsed with 10 mL of water to eliminate ethanol present in the membrane. A gas supply is then attached to the inlet of the membrane filter and the pressure slowly increased until a continuous stream of bubbles appears at the outlet. Our internal specification is that the bubble point must meet or exceed the manufacturer's specification. The filters for all three [^{18}F]FLT batches met this specification.

3.9 Sterility Testing

The sterility of the [^{18}F]FLT final product is assessed by inoculation of the product into fluid thioglycollate media (FTM) and tryptic soy broth (TSB). In each case, 0.25 mL of the inoculum is added into a media volume of ~10 mL [40]. The inoculated samples are incubated for 14 days at either 30–35°C (FTM) or 20–25°C (TSB) and visually inspected daily during the test period. Owing to the length of the sterility test, the [^{18}F]FLT final product is released for administration before completing the test. All three [^{18}F]FLT batches demonstrated the absence of microbial growth at the completion of the 14-day incubation period.

3.10 Radionuclidic Identity

The radionuclidic identity of the [^{18}F]FLT final product is assessed by half-life determination over a period of 10 min. The calculated half-life must be within a range of 105–115 min. All three [^{18}F]FLT batches met this specification.

3.11 Bacterial Endotoxin Test

The presence of bacterial endotoxins in the [^{18}F]FLT final product is assessed with the limulus amebocyte lysate (LAL) reagent. The test is performed by the gel-clot method or the kinetic chromogenic method according to USP requirements and specifications [41]. All three [^{18}F]FLT batches met USP specifications of <175 EU/dose.

3.12 Radionuclidic Purity

The radionuclidic purity of the [^{18}F]FLT final product is assessed on a decayed sample for the presence of long-lived radionuclides by γ-spectrometry using a multichannel analyzer (MCA) [42]. Our specification for radionuclidic purity is that 99.5% of the γ-emissions are due to F-18. All three [^{18}F]FLT batches met this specification.

CHEMICAL ABSTRACTS NOMENCLATURE (REGISTRY NUMBER)

Acetonitrile (75-05-8)

1(2H)-Pyrimidinecarboxylic acid, 3-[2-deoxy-5-O-[(1,1-dimethylethoxy)carbonyl]-3-O-[(4-nitrophenyl)sulfonyl]-β-D-threo-pentofuranosyl]-3,6-dihydro-5-methyl-2,6-dioxo-, 1,1-dimethylethyl ester "BocBocNos" (853782-87-3)

Carbonic acid, potassium salt (1:2) (584-08-7)

Ethanol (64-17-5)

4,7,13,16,21,24-Hexaoxa-1,10-diazabicyclo[8.8.8]hexacosane (23978-09-8)

Hydrochloric acid (7647-01-0)

Sodium acetate (127-09-3)

Sodium phosphate, dibasic (7558-79-4)

Sodium phosphate, monobasic (7558-80-7)

REFERENCES AND NOTES

1. Kong XB, Zhu QY, Vidal PM, Watanabe KA, Polsky B, Armstrong D, Ostrander M, Lang SA Jr, Muchmore E, Chou TC. Comparisons of anti-human immunodeficiency virus activities, cellular transport, and plasma and intracellular pharmacokinetics of 3′-fluoro-3′-deoxythymidine and 3′-azido-3′-deoxythymidine. Antimicrob Agents Chemother 1992;36:808–818.
2. Buck AK, Herrmann K, Shen C, Dechow T, Schwaiger M, Wester H-J. Molecular imaging of proliferation in vivo: positron emission tomography with [18F]fluorothymidine. Methods 2009;48:205–215.
3. Buck AK, Halter G, Schirrmeister H, Kotzerke J, Wurziger I, Glatting G, Mattfeldt T, Neumaier B, Reske SN, Hetzel M. Imaging proliferation in lung tumors with PET: 18F-FLT versus 18F-FDG. J Nucl Med 2003;44:1426–1431.
4. Chen W, Cloughesy T, Kamdar N, Satyamurthy N, Bergsneider M, Liau L, Mischel P, Czernin J, Phelps ME, Silverman DH. Imaging proliferation in brain tumors with 18F-FLT PET: comparison with 18F-FDG. J Nucl Med 2005;46:945–952.
5. Kenny LM, Vigushin DM, Al-Nahhas A, Osman S, Luthra SK, Shousha S, Coombes RC, Aboagye EO. Quantification of cellular proliferation in tumor and normal tissues of patients with breast cancer by [18F]fluorothymidine-positron emission tomography imaging: evaluation of analytical methods. Cancer Res 2005;65:10104–10112.
6. Buck AK, Schirrmeister H, Hetzel M, Von Der Heide M, Halter G, Glatting G, Mattfeldt T, Liewald F, Reske SN, Neumaier B. 3-deoxy-3-[(18)F]fluorothymidine-positron emission tomography for noninvasive assessment of proliferation in pulmonary nodules. Cancer Res 2002;62:3331–3334.
7. van Waarde A, Cobben DC, Suurmeijer AJ, Maas B, Vaalburg W, de Vries EF, Jager PL, Hoekstra HJ, Elsinga PH. Selectivity of 18F-FLT and 18F-FDG for differentiating tumor from inflammation in a rodent model. J Nucl Med 2004;45:695–700.
8. Reske SN, Deisenhofer S. Is 3′-deoxy-3′-(18)F-fluorothymidine a better marker for tumour response than (18)F-fluorodeoxyglucose? Eur J Nucl Med Mol Imaging 2006;33 (Suppl. 1):38–43.

9. Teng B, Wang S, Fu Z, Dang Y, Wu Z, Liu L. Semiautomatic synthesis of 3′-deoxy-3′-[18F]fluorothymidine using three precursors. Appl Radiat Isot 2006;64:187–193.
10. Machulla H-J, Blocher A, Kuntzsch M, Piert M, Wei R, Grierson JR. Simplified labeling approach for synthesizing 3′-deoxy-3′-[18F]fluorothymidine ([18F]FLT). J Radioanal Nucl Chem 2000;243:843–846.
11. Wodarski C, Eisenbarth J, Weber K, Henze M, Haberkorn U, Eisenhut M. Synthesis of 3′-deoxy-3′-[18F]fluoro-thymidine with 2,3′-anhydro-5′-O-(4,4′-dimethoxytrityl)-thymidine. J Label Compd Radiopharm 2000;43:1211–1218.
12. Reischl G, Blocher A, Wei R, Ehrlichmann W, Kuntzsch M, Solbach C, Dohmen BM, Machulla H-J. Simplified, automated synthesis of 3′[18F]fluoro-3′-deoxy-thymidine ([18F]FLT) and simple method for metabolite analysis in plasma. Radiochim Acta 2006;94:447–451.
13. Grierson JR, Shields AF. Radiosynthesis of 3′-deoxy-3′-[(18)F]fluorothymidine: [(18)F]FLT for imaging of cellular proliferation in vivo. Nucl Med Biol 2000;27:143–156.
14. Martin SJ, Eisenbarth JA, Wagner-Utermann U, Mier W, Henze M, Pritzkow H, Haberkorn U, Eisenhut M. A new precursor for the radiosynthesis of [18F]FLT. Nucl Med Biol 2002;29: 263–273.
15. Eisenbarth J, Martin SJ, Wagner-Utermann U, Eisenhut M. Synthesis of 1-(2,5′-anhydro-lyxo-furanosyl)deoxynucleosides as intermediates for the synthesis of 3′-18-F-deoxynucleotides as PET agents. PCT International Application WO 2003018599. 2003.
16. Yun M, Oh SJ, Ha HJ, Ryu JS, Moon DH. High radiochemical yield synthesis of 3′-deoxy-3′-[18F]fluorothymidine using (5′-O-dimethoxytrityl-2′-deoxy-3′-O-nosyl-beta-D-threo pentofuranosyl)thymine and its 3-N-BOC-protected analogue as a labeling precursor. Nucl Med Biol 2003;30:151–157.
17. Oh SJ, Mosdzianowski C, Chi DY, Kim JY, Kang SH, Ryu JS, Yeo JS, Moon DH. Fully automated synthesis system of 3′-deoxy-3′-[18F]fluorothymidine. Nucl Med Biol 2004;31:803–809.
18. Walsh JC, Padgett HC, Ysaguirre T. Method for preparing radiolabeled thymidine having low chromophoric byproducts. US Utility patent 7,419,653. Siemens Medical Solutions USA, Inc. 2008.
19. Virgin ^{18}O-water with an enrichment of greater than 98% was used for these runs.
20. Siemens Medical Solutions USA, Inc. Molecular imaging, 2501 N. Barrington Road, Hoffman Estates, IL 60192-5203, USA.
21. Chromafix cartridges were purchased from ABX. The cartridges were washed with deionized water and dried under a flow of Argon before use.
22. Potassium carbonate was purchased from Sigma (ACS reagent grade, 99+%) and used directly. Sterile water was purchased from Hospira, Inc.
23. Kryptofix-[2.2.2] was purchased from Sigma (98%) and used directly. Acetonitrile was purchased from Sigma (anhydrous, ≥99%).
24. A vacuum pump from ILMVAC GmbH (type: 400861) was used for these runs.
25. Owing to the delay in temperature ramps, the drying times for this step may be longer than 5.5 min.
26. This low and constant pressure reading provides an indication that the reaction vial is sufficiently free of moisture to begin the fluorination process. Labeling failures were most often attributed to reagent valve failures, which can introduce unwanted moisture into the reaction vessel.
27. The [^{18}F]FLT peak shape on the semi-prep column, and hence quality of the final product, is very sensitive to the residual amount of acetonitrile remaining in the crude reaction mixture before purification.
28. Hydrochloric acid (1 N) from Sigma (volumetric standard solution) was used.
29. Sodium acetate was purchased from Sigma (ACS grade, 99+%).
30. Macherey-Nagel Nucleodur C18 Pyramid, 7 μm, 250 × 10 mm, part number 762290; Phenomenex Synergy 4 μm Hydro-RP 18, 250 × 10 mm, part number 00G-4375-E0.
31. Sodium hydrogenphosphate was purchased from Sigma (ACS grade). Sodium dihydrogenphosphate was purchased from Sigma (ACS grade).
32. The cartridge was washed with sterile water (10 mL) before use.

33. Care must be taken to ensure that the sterile filters contain low amounts of leachable materials, which may give rise to UV impurities in the final dose.
34. The cartridge was washed with EtOH (5 mL) followed by water (10 mL) before use.
35. This specification is based on stability studies and the ease of handling doses. The radioactivity concentration could be increased to values beyond this specification, but this would require additional stability studies to demonstrate the shelf life of the product.
36. An Agilent 1200 series HPLC was used for the analysis.
37. Phenomenex Synergy 4 μm Hydro-RP 18, 150 × 4.6 mm, part number 00F-4375-E0.
38. Varian model 3800 gas chromatograph using either a DB-Wax or CP-Wax 52CB column (0.53 mm × 30 m, 1.0–1.2 μm film thickness).
39. Strickley RG. Solubilizing excipients in oral and injectable formulations. Pharm Res 2004;21:201–230.
40. Yu S. Review of ^{18}F-FDG synthesis and quality control. Biomed Imaging Intervent J 2006;2:e57–e67.
41. The kits were purchased from Charles River Endosafe, Inc. Before performing the pyrogen test, the EtOH:sodium phosphate formulation was found to neither enhance nor inhibit the test for pyrogens.
42. Canberra Multi-Channel Analyzer with Germanium detector and PROcount-2000 software.

CHAPTER 4

SYNTHESIS OF [^{18}F]FLUOROAZOMYCIN ARABINOSIDE ([^{18}F]FAZA)

BRIAN G. HOCKLEY AND PETER J. H. SCOTT

Department of Radiology, University of Michigan School of Medicine, Ann Arbor, Michigan

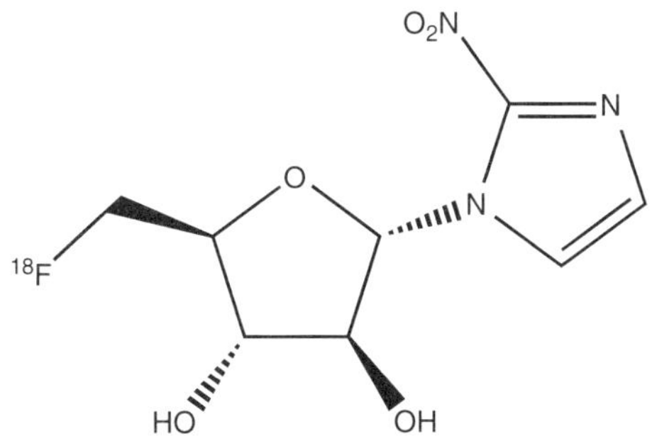

[^{18}F]FAZA

1 INTRODUCTION

Tumor hypoxia is a condition that results when tumor cells have been deprived of oxygen and frequently occurs when a tumor has outgrown its blood supply. It is important to determine as early as possible if a tumor is hypoxic, because such tumors are typically resistant to radiotherapy and chemotherapy and require alternative treatment strategies. 2-Nitroimidazoles are an important class of compounds because they are retained in higher levels in hypoxic tumor cells than the corresponding normoxic cells [1]. The reason for the increased uptake is that in a hypoxic environment 2-nitroimidazoles undergo intracellular reduction, which leads to fragmentation into species that are unable to pass back out of the cell. An analogous process is also known to occur in myocardial ischemia.

Radiochemical Syntheses, Volume 1: Radiopharmaceuticals for Positron Emission Tomography, First Edition. Edited by Peter J. H. Scott, Brian G. Hockley.

2-Nitroimidazoles have therefore been adapted for use in molecular imaging of hypoxia. For example, radioactive 2-nitroimidazole derivatives have been prepared, such as [^{18}F]FAZA (**2**) [2–6] and [^{18}F]FMISO (Chapter 5) which allow for positron emission tomographic (PET) imaging of tumor hypoxia and myocardial ischemia.

2 SYNTHESIS PROCEDURES

CAUTION: All radiochemical syntheses must be carried out using appropriate equipment in a facility authorized for the use of radioactive materials. Personal protective equipment must be worn, and all local radiation safety laws followed.

2.1 Production of [^{18}F]Fluoride

[^{18}O]H_2O (2 mL) [7] was loaded into the [^{18}F]fluoride target [8] of a General Electric Medical Systems (GEMS) PETTrace cyclotron [9]. The target was bombarded (60 μA beam for 30 min) to generate ~1.5 Ci (55.5 GBq) of [^{18}F]fluoride by the $^{18}O(p,n)^{18}F$ nuclear reaction.

2.2 Azeotropic Drying of [^{18}F]Fluoride

The [^{18}F]fluoride was delivered to a GEMS TRACERlab FX_{FN} synthesis module [9] as a solution in [^{18}O]H_2O (2 mL). This solution was passed through a QMA-Light Sep-Pak® cartridge [10] to trap the [^{18}F]fluoride and recycle the [^{18}O]H_2O. The [^{18}F]fluoride was then eluted into the TRACERlab FX_{FN} glassy-carbon reaction vessel using a solution of aqueous potassium carbonate (3.5 mg in 0.5 mL H_2O) [11]. A solution of Kryptofix-[2.2.2] (15 mg in 1 mL MeCN) [12] was added, and the reaction mixture was azeotropically dried, initially at 80°C under vacuum for 4 min and subsequently at 60°C with both vacuum and argon flow for an additional 4 min.

2.3 Synthesis of [^{18}F]FAZA

A solution of [^{18}F]FAZA precursor [13] (**1**; 5 mg) in anhydrous DMSO [14] (0.7 mL) was added to the dried [^{18}F]fluoride, and the reaction was heated to 100°C with stirring for 10 min. After this time, the reaction was cooled to 40°C and 0.1 M aqueous sodium hydroxide [15] (1 mL) was added. The reaction was stirred for 5 min at 40°C to hydrolyze the acetate protecting groups (Fig. 1).

2.4 Purification and Formulation of [^{18}F]FAZA

After hydrolysis, the crude reaction mixture was diluted with 0.25 M aqueous sodium dihydrogen phosphate solution [16] (0.5 mL) and 3% ethanol:97% 10 mM sodium dihydrogenphosphate semipreparative HPLC mobile phase [17] (1 mL). The diluted reaction mixture was passed through an alumina Sep-Pak cartridge [18] and then purified by semipreparative HPLC (Econosphere column [19], flow rate = 4 mL/min) as shown in Fig. 2.

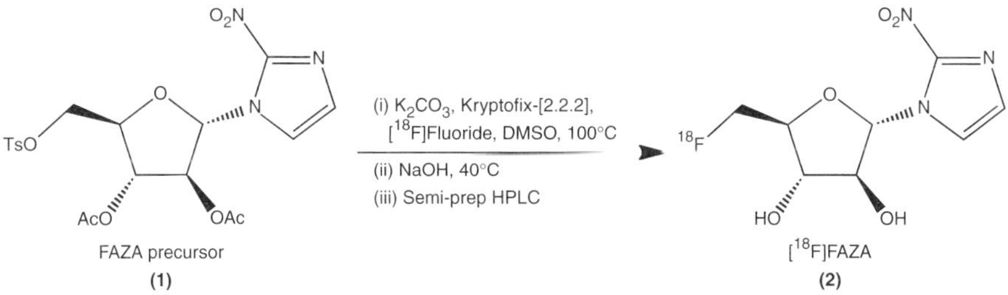

FIGURE 1 Synthesis of [^{18}F]FAZA.

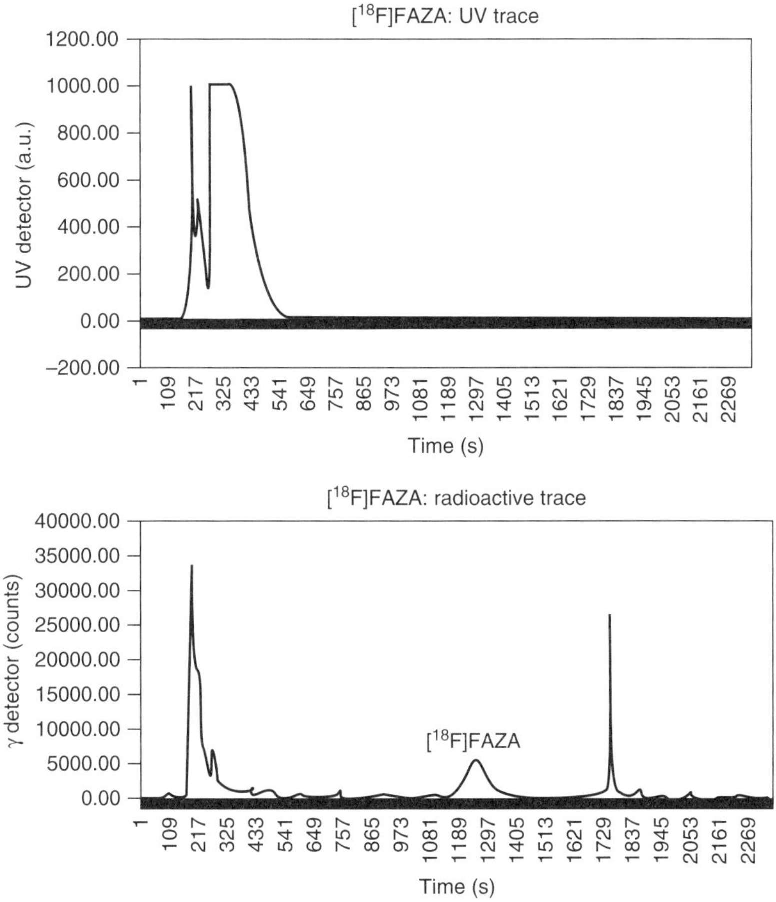

FIGURE 2 Semipreparative UV and radioactive HPLC traces for [^{18}F]FAZA.

The fraction corresponding to [^{18}F]FAZA (typically eluting between 20 and 23 min) was collected for 1 min into a vial precharged with 0.9% sodium chloride, USP [20] (3.8 mL) and sodium phosphates, USP [21] (0.2 mL). The final formulation (8 mL) was then passed through a sterile filter [22] into a sterile vial [23] to provide [^{18}F]FAZA (typically 75 mCi (2.78 GBq), n = 25) in an isotonic solution released for quality control (QC). After synthesis was complete, the semipreparative HPLC column was flushed with 70% ethanol.

3 QUALITY CONTROL PROCEDURES

CAUTION: All radiopharmaceuticals produced for clinical use must have local regulatory approval before use in humans. Trained personnel must carry out QC procedures, and each dose must meet all established QC criteria before release to the clinic.

QC procedures for [^{18}F]FAZA, based on the current requirements for radiopharmaceuticals laid out in the US Pharmacopeia [24], are summarized below. Complete QC data for three repeat batches of [^{18}F]FAZA produced using the method disclosed herein are summarized in Table 1. Each of the three doses met all of the established QC criteria.

TABLE 1 QC Data for Three Repeat Runs of [^{18}F]FAZA

QC Test	Release Criteria	Run 1	Run 2	Run 3
Yield/mCi (GBq)	N/A	107 (4.0)	96 (3.6)	39 (1.4)
Visual inspection	Clear, colorless	Clear, colorless	Clear, colorless	Clear, colorless
Radiochemical identity	RRT = 0.9–1.1	1.0	1.0	1.0
Radiochemical purity (%)	≥95	99	100	99
Specific activity (Ci/μmol)	No limit established	4627	4583	5066
Residual solvent analysis (in ppm)	DMSO <5000	224.8	198.4	61.6
	Acetone <5000	1.0	1.9	1.0
	MeCN <410	151.3	0.2	0.6
Dose pH	4.5–7.5	6.5	5.5	6.0
Residual Kryptofix-[2.2.2] (μg/mL)	≤50	≤10	≤10	≤10
Sterile filter integrity test (psi)	>50	>50	>50	>50
Radionuclidic identity ($t_{1/2}$) (min)	105–115	108	106	107
Endotoxin analysis (EU/mL)	≤17.5	≤2	≤2	≤2
Sterility testing	No colony growth out to 14 d	Pass	Pass	Pass

3.1 Visual Inspection

The [^{18}F]FAZA dose is examined behind a PET L-block and must be clear, colorless, and free of particulate matter.

3.2 Radiochemical Identity

HPLC analysis of radiochemical identity was conducted using a Shimadzu LC-2010A$_{HT}$ Liquid Chromatograph [25] fitted with UV detectors and Bioscan γ-detectors [26] (column: Phenomenex Luna 5μ C8, 100A, 100 × 2 mm [27]; mobile phase: 5% acetonitrile: 95% 20 mM aqueous ammonium acetate, pH 4.5 [28]; flow rate: 0.5 mL/min, $\lambda = 254$ nm). The retention time of [^{18}F]FAZA is compared to that of the [^{19}F]FAZA reference standard [13] and must be ±10% (relative retention time (RRT) must be 0.9–1.1).

3.3 Radiochemical Purity

HPLC analysis of radiochemical purity was conducted using a Shimadzu LC-2010A$_{HT}$ Liquid Chromatograph [25] fitted with a UV detector and a Bioscan γ-detector [26] (column: Phenomenex Luna 5μ C8, 100A, 100 × 2 mm [27]; mobile phase: 45% acetonitrile: 55% 20 mM aqueous ammonium acetate, pH 4.5 [28]; flow rate: 0.5 mL/min). Radiochemical purity must be >95%. Representative analytical HPLC traces are displayed in Fig. 3.

3.4 Specific Activity

There is no specific activity release limit for [^{18}F]FAZA in place at our institution. The injection volume is based on the injectable mass limit for FAZA and is calculated on a dose-by-dose basis by the administering study team.

3.5 Residual Solvent Analysis

Gas chromatography (GC) is used to determine the residual solvents levels in the dose. The GC system used was a Shimadzu GC-2010 with an AOC-20i autoinjector, split/splitless inlet, a flame ionization detector (FID), and a Restek column (Stabilwax 30 m × 0.25 mm, 0.25 μm G16 stationary phase) [30]. Limits imposed by the ICH Harmonised Tripartite Guidelines are <410 ppm/day for acetonitrile and <5000 ppm per day for each class 3 solvent (DMSO, acetone) [29]. The method conditions are as follows:

Oven temperature Profile

1. hold at 40°C for 3 min
2. increase 40–180°C at 60°C/min
3. hold at 180°C for 5 min

Inlet temperature: 180°C

Column Linear velocity: 42.7 cm/sec

FID temperature: 250°C

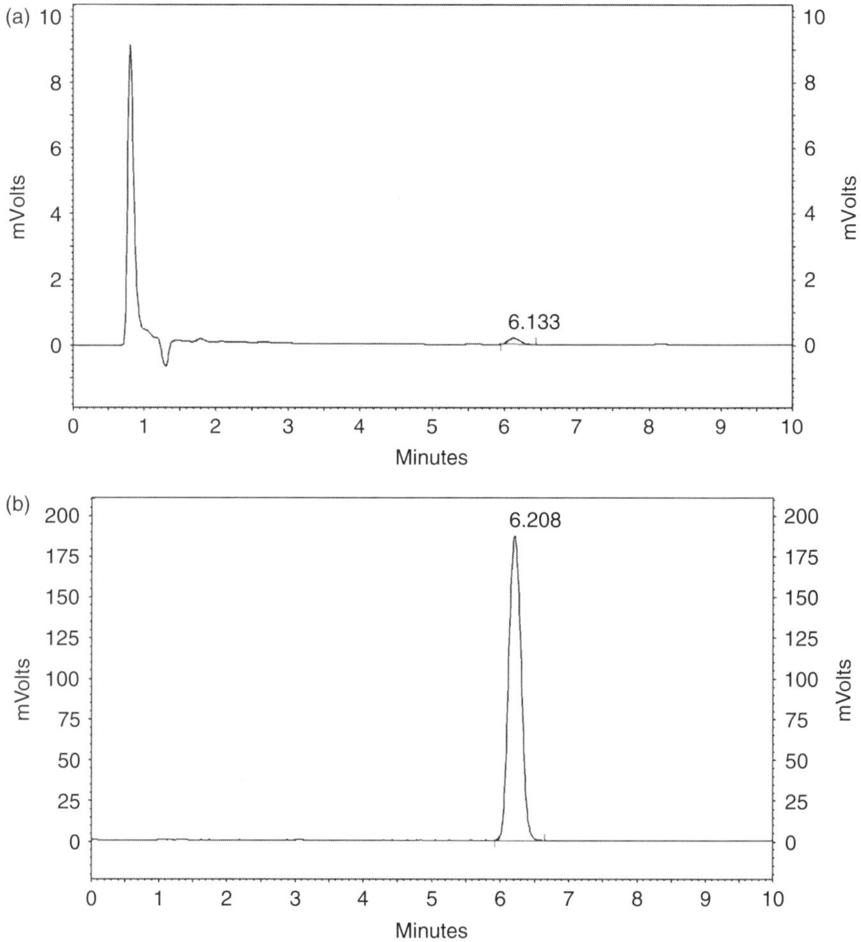

FIGURE 3 Analytical UV (a) and radioactive (b) HPLC trace for [^{18}F]FAZA.

3.6 Dose pH

The pH of the [^{18}F]FAZA dose was analyzed by applying a small amount of the dose to colorpHast® pH 2.0–9.0 nonbleeding pH-indicator strips [31] and determined by visual comparison to the scale provided. Dose pH must be 4.5–7.5.

3.7 Residual Kryptofix-[2.2.2] Analysis

Residual Kryptofix-[2.2.2] levels in [^{18}F]FAZA doses were analyzed using the established spot test [32]. Strips of plastic-backed silica gel thin layer chromatography (TLC) plates saturated with iodoplatinate reagent [32] were spotted with water (negative control), 50 μg/mL Kryptofix-[2.2.2] standard (positive control), and [^{18}F]FAZA dose. If Kryptofix-[2.2.2] is present in a sample, a blue-black ring appears around the pink center of the spot. Spots for the three samples were compared, and a visual determination of residual Kryptofix-[2.2.2] in the [^{18}F]FAZA dose was made. A value of <50 μg/mL is acceptable.

3.8 Sterile Filter Integrity Test

The sterile filter from the [^{18}F]FAZA (with needle still attached) was connected to a nitrogen supply via a regulator. The needle was submerged in water, and the nitrogen pressure was gradually increased. If the pressure was raised above the filter acceptance pressure (50 psi) without seeing a stream of bubbles, the filter was considered intact. If a stream of bubble occurs at <50 psi, the test fails.

3.9 Radionuclidic Identity

Activities were measured using a Capintec CRC®-15R Radioisotope Dose Calibrator [33], and half-life was calculated using Equation 1. Calculated half-life must be 105–115 min.

$$t_{1/2} = -\ln 2(\text{time difference}/(\ln(\text{ending activity}/\text{starting activity}))) \quad (1)$$

3.10 Endotoxin Analysis

Endotoxin content in doses of [^{18}F]FAZA was analyzed by a Charles River Laboratories EndoSafe® Portable Testing System [34] and according to the US Pharmacopeia [35]. Doses must contain ≤17.5 Endotoxin Units (EU)/mL.

3.11 Sterility Testing

Samples of [^{18}F]FAZA were placed onto fluid thioglycolate media (FTM) plates and soybean casein digest media (SCDM) plates. FTM plates are used to test for anaerobes, aerobes, and microaerophiles while SCDM plates are used to test for nonfastidious and fastidious microorganisms [36]. [^{18}F]FAZA plates were incubated along with positive and negative controls for 14 days. FTM plates were incubated at 32°C and SCDM plates were incubated at 22°C according to current USP guidelines [37]. Plates were visually inspected on the third, eighth, and fourteenth days of the test period and compared to the positive and negative standards. Positive standards must show growth (turbidity) on the plates, and [^{18}F]FAZA/negative controls must have no culture growth after 14 days to be indicative of sterility.

CHEMICAL ABSTRACTS NOMENCLATURE (REGISTRY NUMBER)

Acetonitrile (75-05-8)

Carbonic acid, potassium salt (1:2)(584-08-7)

Ethanol (64-17-5)

4,7,13,16,21,24-Hexaoxa-1,10-diazabicyclo[8.8.8]hexacosane (23978-09-8)

1H-Imidazole,1-[2,3-di-*O*-acetyl-5-*O*-[(4-methylphenyl)sulfonyl]-α-D-arabinofuranosyl]-2-nitro- (494775-35-8)

1H-Imidazole, 1-[5-deoxy-5-(fluoro-^{18}F)-α-D-arabinofuranosyl]-2-nitro- (864084-85-5)

Methane, 1,1′-sulfinylbis- (67-68-5)

Phosphoric acid, sodium salt (1:1) (7558-80-7)

Sodium hydroxide (1310-73-2)

For detailed supplier information, see Appendix 1 (page 335).

REFERENCES AND NOTES

1. Hodgkiss R. Anticancer Drug Des 1998;13:687–702.
2. Kumar P, Wiebe LI, Asikoglu M, Tandon M, McEwan AJB. Appl Radiat Isot 2002;57:697–703.
3. Reischl G, Ehrlichmann W, Bieg C, Solbach C, Kumar P, Wiebe LI, Machulla H-J. Appl Radiat Isot 2005;62:897–901.
4. Piert M, Machulla HJ, Picchio M, Reischl G, Ziegler S, Kumar P, Wester HJ, Beck R, McEwan AJB, Wiebe LI, Schwaiger M. J Nucl Med 2005;46:106–113.
5. Reischl G, Dorow DS, Cullinane C, Katsifis A, Roselt P, Binns D, Hicks RJ. J Pharm Pharm Sci 2007;10:203–211.
6. Beck R, Roeper B, Carlsen JM, Huisman MC, Lebschi JA, Andratschke N, Picchio M, Souvatzoglou M, Machulla HJ, Piert M. J Nucl Med 2007;48:973–980.
7. Virgin [^{18}O]H_2O purchased from ABX, Amic, Rotem, or Medical Isotopes, and used as received.
8. GEMS silver high yield [^{18}F]fluoride target.
9. GE Healthcare, USA.
10. QMA-Light Sep-Pak cartridges were purchased from Waters (part no. WAT023525) and conditioned with 10 mL ethanol, 10 mL water, 10 mL 0.5 M sodium bicarbonate, and a further 10 mL water before use.
11. Potassium carbonate purchased from Aldrich (part no. 209619) and used as received. Milli-Q water used for diluent from Millipore Milli-Q water filtration system.
12. Kryptofix-[2.2.2] was purchased from Acros (part no. 29195-0010) and used as received. Anhydrous acetonitrile was purchased from Acros (part no. 61096-1000) and used as received.
13. [^{18}F]-FAZA precursor and reference standard were purchased from Prof. Friedrich Hammeschmidt (Universität Wien, Austria) and Prof. Hans-Jürgen Machulla (Steinbeis Transfer Center Radiopharmacy, Germany) and used as received.
14. Anhydrous DMSO was purchased from Acros (part no. 61042-10000) and used as received. We notice a gradual decline in yield throughout the lifetime of a bottle of DMSO and stress the importance of using anhydrous solvents in this synthesis.
15. Sodium hydroxide, 97%, was purchased from Sigma-Aldrich (part no. 221465) and used as received.
16. Sodium dihydrogenphosphate (sodium phosphate, monobasic, 98% anhydrous) was purchased from Sigma-Aldrich (part no. S3139) and used as received.
17. Semipreparative HPLC mobile phase: 3% ethanol:97% 10 mM sodium dihydrogenphosphate. Preparation: sodium dihydrogenphosphate (X g) was dissolved in water (970 mL). Ethanol (30 mL) was added, and the mobile phase was stirred to mix and then passed through a sterile filter before use.
18. Alumina Sep-Pak® cartridges were purchased from Waters (part no. WAT051810) and conditioned with 10 mL ethanol and 10 mL water before use.
19. Semipreparative HPLC column: Alltech Econosphere C18, 10 μ, 250 × 10 mm, purchased from Grace (part no. 28085); flow rate = 4 mL/min; λ = 254 nm; semipreparative column was equilibrated with 200–300 mL of mobile phase before synthesis.
20. Sodium Chloride, USP, 0.9%, was purchased from Hospira (part no. 0409-4888-50) and used as received.

21. Sodium phosphates, USP was purchased from Hospira (part no. 0409-7391-72) and used as received.
22. Millex-GS sterile 0.22-μm filters were purchased from Millipore (part no. SLGSV255F) and used as received.
23. Ten-milliliter sterile vials were purchased from Hollister-Stier (part no 7515ZA) and vented with a sterile Millex-FG vent filter (part no. SLFG025LS).
24. US Pharmacopeia <823> Radiopharmaceuticals for Positron Emission Tomography-Compounding. USP 32–NF 27. 2009.
25. Shimadzu Corporation, USA.
26. Bioscan, Inc., USA.
27. Analytical HPLC column: Luna C18, 150 × 2 mm, purchased from Phenomenex (part no. 00F-4041-B0); flow rate = 0.5 mL/min; λ = 254 nm; analytical column was equilibrated with mobile phase for 30–60 min before use.
28. Analytical mobile phase. 5% MeCN in 20 mM NH_4OAc, pH adjusted to 4.5 with acetic acid.
29. International Conference on Harmonisation of Technical Requirements for Registration of Pharmaceuticals for Human Use (ICH).
30. GC column: Restek Stabilwax with Integra-Guard; 30 m × 0.25 mm, purchased from Restek (part no. 10623-124).
31. EMD Chemicals Inc., USA (part no. 9578-3).
32. Mock BH, Winkle W, Vavrek MT. Nucl Med Biol 1997;24:193–195.
33. Capintec, Inc., USA.
34. Charles River Laboratories, USA.
35. U. S. Pharmacopeia <85> Bacterial Endotoxins Test. USP 32–NF 27, 2009.
36. Becton, Dickinson and Company, USA.
37. U. S. Pharmacopeia <71> Sterility tests. USP 32–NF 27, 2009.

CHAPTER 5

SYNTHESIS OF [^{18}F]FLUOROMISONIDAZOLE (1-(2-HYDROXY-3-[^{18}F]FLUOROPROPYL)-2-NITROIMIDAZOLE, [^{18}F]FMISO)

PATRICK J. RISS, VALENTINA FERRARI, ROBERT BIELIK, ROBERTO CANALES-CANDELA, ROB SMITH, AND FRANKLIN I. AIGBIRHIO

Wolfson Brain Imaging Centre, Department of Clinical Neurosciences, University of Cambridge Addenbrooke's Hospital, Cambridge, UK

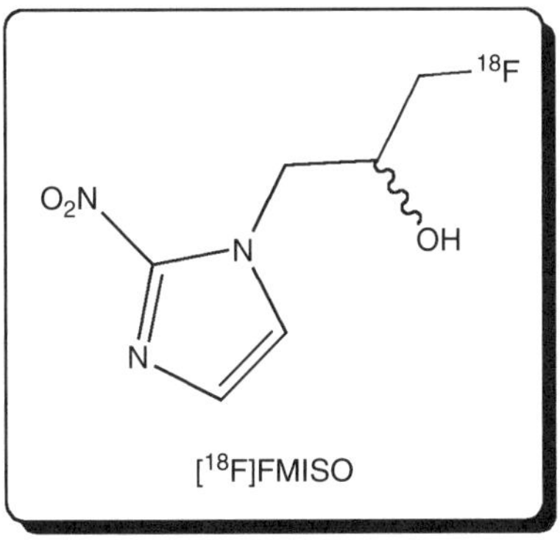

1 INTRODUCTION

[^{18}F]Fluoromisonidazole, 1-(2-hydroxy-3-[^{18}F]fluoropropyl)-2-nitroimidazole ([^{18}F]FMISO), is the most widely applied nitroimidazole for positron emission tomographic (PET) imaging of hypoxia [1]. Applications include the determination of hypoxia in various cancers [2], in myocardial infarction [3], and in ischemic tissue in stroke [4]. The radiotracer is synthesized in two subsequent radioactive steps. The [^{18}F]fluoride label is introduced via aliphatic nucleophilic substitution of a leaving group. Cleavage of a hydroxyl protective group then affords the title compound.

Radiochemical Syntheses, Volume 1: Radiopharmaceuticals for Positron Emission Tomography, First Edition. Edited by Peter J. H. Scott, Brian G. Hockley.

2 SYNTHESIS PROCEDURE

CAUTION: All radiochemical syntheses must be carried out using appropriate equipment in a facility authorized for the use of radioactive materials. Personal protective equipment must be worn and local radiation safety rules followed.

The production of [^{18}F]-FMISO is based on the synthesis procedure developed by Oh et al. [5] and Lim et al. [6]. [^{18}F]FMISO is synthesized in a two-step process via aliphatic nucleophilic substitution of a 4-methylbenzenesulfonate-leaving group by [^{18}F]fluoride ion on the tetrahydropyranyl-protected precursor 1-(2-nitro-1-imidazolyl)-2-*O*-tetrahydropyranyl-3-*O*-toluenesulfonylpropanediol (NITTP) followed by hydrolysis of the protecting group (Fig. 1).

The labeling step contains nucleophilic ^{18}F-fluorination of 10 mg of precursor in acetonitrile heated at 100°C for 10 min, followed by prepurification on Sep-Pak® tC18 cartridge. The trapped intermediate is eluted with ethanol, and the THP ether is hydrolyzed using 1 M HCl at 100°C for 3 min. Neutralization is achieved using a mixture of 2 M NaOH and citrate buffer. Subsequently, the reaction mixture is passed through a Sep-Pak C18 plus SPE cartridge and a Sep-Pak Alumina N SPE cartridge and purified by preparative high performance liquid chromatography (HPLC). The total duration of synthesis is about 75 min. Approximately 10 mL of 5% ethanol solution containing 81 mCi (3 GBq) of [^{18}F]FMISO is produced per batch.

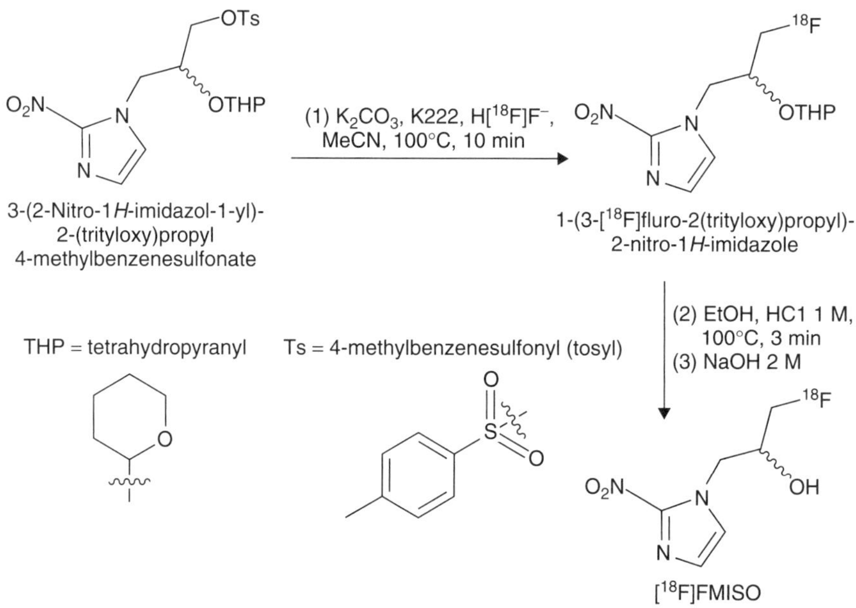

FIGURE 1 Scheme for the radiosynthesis of [^{18}F]FMISO.

2.1 Radiosynthesis Preparation

2.1.1 MX_{FDG} Disposable Cassette Setup Scheme

The preparation uses the GE MX_{FDG} synthesizer cassette designed for automated FDG production [7]. The scheme in Fig. 2 represents the screen of the Synthesiser control software that reflects the hardware setup. The front panel of the TRACERLab MX_{FDG} matches the cassette and the reagent containers mounted onto the synthesizer.

2.1.2 Production of [^{18}F]Fluoride ion

[^{18}F]Fluoride ion was produced via proton bombardment of a 98% enriched $H_2[^{18}O]O$ liquid target using a General Electrics (GE) PETtrace Cyclotron. A beam current of 40 μA is applied to the liquid target to facilitate the $^{18}O(p,n)^{18}F$ nuclear reaction at a particle energy of 16→3 MeV. For [^{18}F]FMISO, the target is irradiated at 40 μA for 20 min. The [^{18}F]fluoride ion is transferred to the module and trapped on a Sep-Pak light QMA strong anion exchanger cartridge [8], preconditioned to the carbonate form using potassium carbonate solution (1 M, 10 mL) and water (20 mL). The [^{18}F]fluoride is eluted as a $[K \subset 222]^+[^{18}F]^-$ cryptate complex in a mixture of acetonitrile (300 μL) and water (300 μL) containing Kryptofix® K222 (22 mg, 58 μmol) and K_2CO_3 (7 mg, 50 μmol) directly into the reactor.

2.1.3 Azeotropic Drying of the [^{18}F]Fluoride

The eluate from the QMA cartridge (2.1.2.) is concentrated to dryness at 95°C in a gentle stream of nitrogen at reduced pressure. MeCN (1 mL) is added to the reactor after 1 min to remove excess free water from the $[K \subset 222]^+[^{18}F]^-$ cryptate complex via azeotropic distillation [9]. After 4 min, the reactor is cooled to 70°C.

2.1.4 Synthesis of [^{18}F]FMISO

1-(2-Nitro-1-imidazolyl)-2-*O*-tetrahydropyranyl-3-*O*-toluenesulfonylpropanediol (NITTP) (10 mg, 23.5 μmol) in MeCN (1 mL) is added to the reactor and heated to 100°C [9–10]. After keeping the reaction mixture at 100°C for 10 min, the reactor is cooled to 70°C. Water (5 mL) is added and the mixture is passed through a Sep-Pak tC18 cartridge [8–11]. The trapped intermediate is eluted into the reactor with ethanol, and HCl solution (1 M, 1 mL) is added [12]. The reactor is heated to 100°C, and the temperature is kept constant for 3 min. Neutralization is achieved via the addition of NaOH solution (2 M, 0.5 mL) and sodium citrate (M, 0.5 mL) [8]. Subsequently, the reaction mixture is passed through a Sep-Pak C18 plus SPE cartridge and a Sep-Pak Alumina N SPE cartridge and purified by preparative HPLC [8]. The purified product is collected in a receiving vial, sterile filtered through a Millex GS filter, and transferred into a multiinjection vial [13–14].

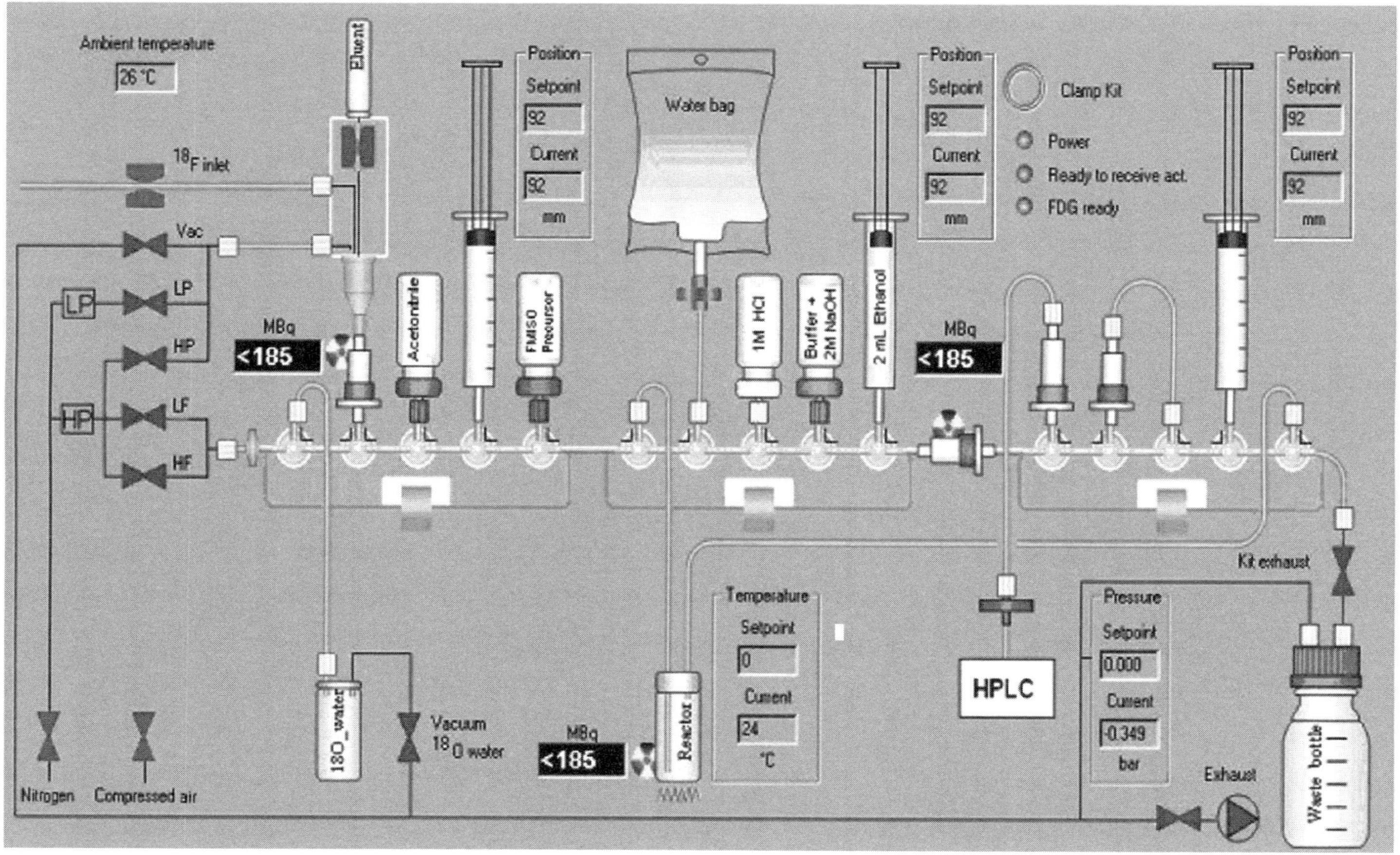

FIGURE 2 Scheme of [^{18}F]FMISO production setup using MX_{FDG} synthesizer.

2.1.5 Preparative HPLC

A Phenomenex® Luna RP18 5 μm 250 × 10 mm 100 Å column equipped with a SecurityGuard precolumn of the same type is used as the stationary phase [15]. The mobile phase consists of a mixture of 5% ethanol in water. The flow rate is 4 mL/min.

The TracerLab MX_{FDG} module has a fitted preparative HPLC unit for purification of [^{18}F]FMISO with a NaI(Tl) radiation detector and a UV detector. The UV detection wavelength is set to 254 nm.

During preparative HPLC, the product peak is obtained manually by operating the open/close selection valve (Fig. 4).

Preparation of Mobile Phase, 5% Ethanol/Water (v/v) Dilute 50 mL of absolute EtOH, BP, with Water, BP, in a 1-L volumetric flask and mix well. Transfer the solution into a 1 L glass solvent bottle and degas under vacuum in an ultrasonic bath for 15 min. The shelf life of the prepared solvent is one week.

Filter the solution using Corning Filter System, a 0.5-L bottle, under vacuum before use. The shelf life of the filtered solvent is one day [16].

Preparation of Cleaning Solution, 70% Ethanol/Water (v/v) Dilute 700 mL of Absolute Ethanol, BP, with Water, BP, in a 1-L volumetric flask and mix well. Transfer the solution into a 1-L glass solvent bottle and degas the solvent under

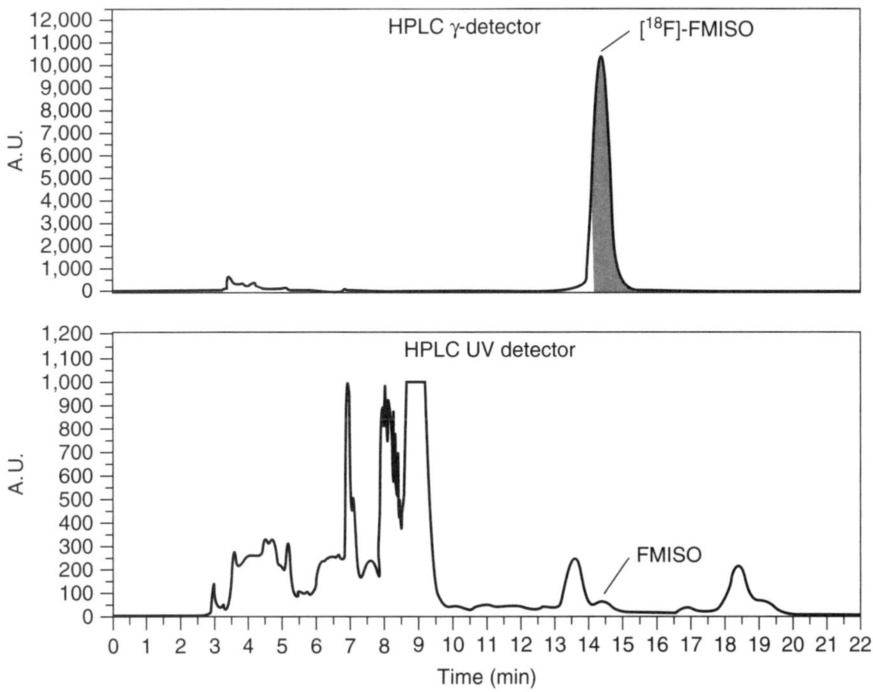

FIGURE 3 HPLC purification of [^{18}F]FMISO.

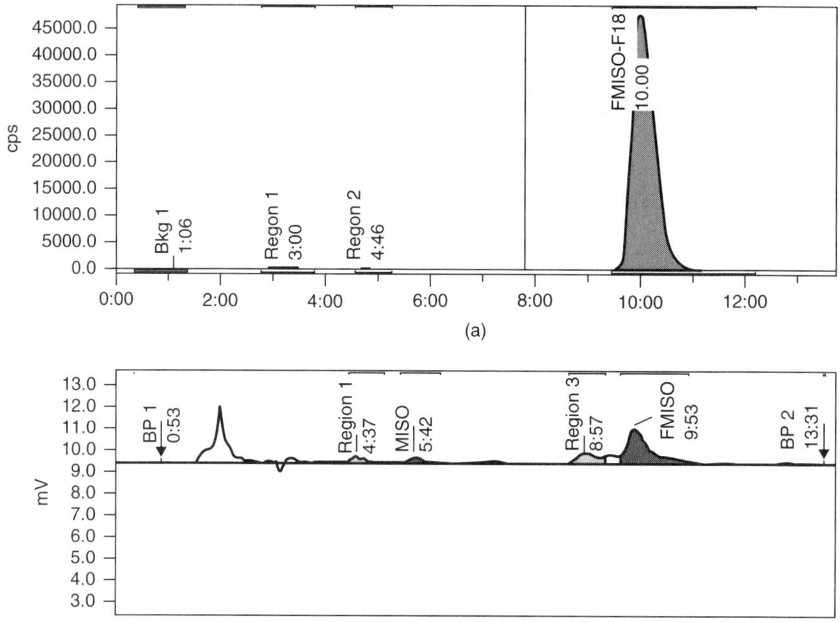

FIGURE 4 [^{18}F]FMISO quality control: radioactivity (a) and UV (b) traces.

vacuum in an ultrasonic bath for 15 min. The shelf life of the prepared solvent is one week.

Filter the solution using Corning Filter System, a 0.5-L bottle, under vacuum before production. The shelf life of the filtered solvent is one day [16].

3 QUALITY CONTROL (QC) PROCEDURES

CAUTION: All radiopharmaceuticals produced for clinical use must have local regulatory approval before use in humans. Trained personnel must carry out QC procedures, and each dose must meet all established QC criteria before release to the clinic.

Owing to the short half-life of ^{18}F (109.7 min) and high levels of radioactivity, the QC method is divided in prerelease tests and postrelease tests. Prerelease tests have to be performed before release and all tests have to be passed for release of the production batch to the clinic.

3.1 Prerelease Tests

3.1.1 Visual Inspection

A visual check of the product solution is performed behind a lead window; the product must be clear, colorless, and free of particulate matter (Table 1).

TABLE 1 Quality control of [18F]FMISO: release criteria and experimental results from three subsequent productions.

QC Test	Release Criteria	Run 1	Run 2	Run 3
Yield mCi (GBq)	≥13.5 in 10 mL	79 (2.93)	80 (2.97)	73 (2.71)
Visual inspection	Clear colorless	Pass	Pass	Pass
Radiochemical identity	t_R[^{18}F]FMISO = ($t_{R\ STD}$ + 15 s) ± 10 s	8:00	7:48	7:55
Radiochemical purity (%)	≥95	99.96	100	99.75
Chemical amount in μg/dose	FMISO ≤7 mg/dose	2.9	1.2	0.9
	MISO ≤14 mg/dose	0.9	0.0	1.0
Specific activity (MBq/nmol)	NA	447.5	646.3	613.9
Residual solvent analysis	Ethanol ≤10%	6.1	5.4	6.3
	MeCN <410 ppm	Pass	Pass	Pass
Dose pH	6–8	7	6.5	6.5
Sterile filter integrity test (bar)	≥3.4	4.8	4.9	4.8
Radionuclidic identity ($t_{1/2}$) (min)	109.7 ± 0.5	nd	nd	nd
Endotoxin analysis (EU/mL)	LAL <17.5	<1	<1	<1
Sterility testing	Sterile	Sterile	Sterile	Sterile
Residual Kryptofix K222 (μg/mL)	<220	<10	<10	<10

nd - not determined, i.e. only for validation.

3.1.2 pH Measurement

The pH of the product is determined using pH indicator sticks and should be within the range 6–8 pH units (Table 1).

3.1.3 Radiochemical Identity and Chemical Purity

Radiochemical purity should be >95% and is determined using the following HPLC method (Fig. 4, Table 1):

Column: Phenomenex Primesphere C_{18} 5μ 250 × 4.6 mm.

Injection loop: 20 μL

Injection volume: 25 μL

Mobile phase: ethanol–water (5:95 v/v)

Flow rate: 1.0 mL/min

Radioactivity detector: Bioscan flow-count with NaI detector

UV detector: $\lambda = 320$ nm; AUFS = 0.05

The suitability of the system is checked before each production by injecting a standard mix solution of 5 μg/mL of FMISO and 2 μg/mL 1-(2-nitro-1-imidazolyl)-2-*O*-tetrahydropyranyl-3-*O*-toluenesulfonylpropanediol. The mass concentrations of FMISO and 1-(2-nitro-1-imidazolyl)-2-*O*-tetrahydropyranyl-3-*O*-toluenesulfonylpropanediol are determined by quantification of the UV absorbance peaks of an injection of the QC sample. A spiked sample is prepared with FMISO standard solution (20 μg/mL) and the QC sample

(50:50) for radiochemical identity determination. The total retention times of the radioactivity signal of [^{18}F]FMISO and the UV signal of the reference compound peaks should differ by the dead time between the UV and γ-detectors.

3.2 Postrelease Tests

3.2.1 Filter Integrity Test

The integrity of the filter is tested using the bubble point test; to pass the test the bubble point has to be $\geq$3.86 bar (56 psi). (Ph Eur method 5.1.1. Methods of preparation of sterile products 5.1.1) (Table 1).

3.2.2 Sterility Test

The test for sterility is performed by an external contractor on decay of the radioactivity in the product. The sterility of the final product is tested by the membrane filtration method (Ph Eur method 2.6.1) (Table 1).

3.2.3 Bacterial Endotoxins

Endotoxin testing is performed using a chromogenic technique with Endosafe Limulus Amebocyte Lysate (LAL) system in a Portable Test System (PTS) Kinetic Reader (Charles River Laboratories, Charleston, SC). The LAL test is performed immediately after release of the product, and the acceptance criteria for release is that the endotoxin concentration is below 17.5 EU/mL (Ph Eur method 2.6.14) Table 1.

3.2.4 Analysis of Formulation and Residual Solvents

Residual solvents in the final product are analyzed using a gas chromatography instrument using the following conditions:

Column: Alltech Gas Chrom 254 80/100, 6′ × 1/8″ × 0.85″

Detector: flame ionization detector (FID)

Carrier gas: helium at ~25 mL/min

FID: hydrogen at ~45 mL/min

FID: air at ~350 mL/min

Specification limits: residual solvent MeCN $\leq$0.88 mg/mL and formulation solvent ethanol $\leq$10%.

CHEMICAL ABSTRACTS NOMENCLATURE (REGISTRY NUMBER)

Acetonitrile [75-05-8]

Ethanol [64-17-5]

[^{18}F]Fluoromisonidazole: (1-(2-Hhydroxy-3-[^{18}F]fluoropropyl)-2-nitro-imidazole [13551-89-8]

Hydrochloric acid [7647-01-0]

Kryptofix K222 4,7,13,16,21,24-Hhexaoxa-1,10-diazabicyclo[8.8.8]hexacosane [23978-09-8]

NITTP: 1-(2-nitro-1-imidazolyl)-2-*O*-tetrahydropyranyl-3-*O*-toluenesulfonyl-propanediol [150196-34-2]
Potassium carbonate [584-08-7]
Sodium hydroxide [1310-73-2])

REFERENCES AND NOTES

1. Lee ST, et al. Semin Nucl Med 2007;37:451–461.
2. Krohn KA, et al. J Nucl Med 2008;49Suppl 2:129S–148S.
3. Martin GV, et al. J Nucl Med 1992;33:2202–2208.
4. Takasawa M, et al. J Cereb Blood Flow Metab 2007;27(4):679–689.
5. Oh SJ, et al. Nucl Med Biol 2005;32:899–905.
6. Lim J, et al. Appl Radiat Isot 1993;44:1085–1091.
7. GE TRACERlab MX_{FDG} Synthesiser Kit Ph.Eur., GE Medical Systems/P5150ME; stored at RT.
8. GE TRACERlab MX_{FDG} Reagent Kit, ABX/K-100TM; store at RT.
9. MeCN anhydrous >99.5%, Sigma-Aldrich 00695-250 ML; store at RT.
10. 1-(2-Nitro-1-imidazolyl)-2-*O*-tetrahydropyranyl-3-*O*-toluenesulfonylpropanediol >97%, 10 mg per vial, ABX/1401.0010; storage: desiccated at 2–8°C, protected from light.
11. Water for Injections Ph. Eur. for i.v. injection, Hameln Pharmaceuticals/PL 01502/0003R; store at RT.
12. Hydrogen chloride 1 M standard solution Ph. Eur., Fluka/35328; keep tightly closed at room temperature.
13. TechneVial (11 mL) or 10 mL nitrogen filled, sterile, Mallinckrodt Amersham; store at RT.
14. Millex GV sterile, 0.22-μm MF-durapore membrane, Millipore/SLGV013SL; store at RT.
15. Phenomenex® Luna RP18 5 μm 250 × 10 mm, Phenomenex/00G-4041N0; store at RT under 70% EtOH.
16. Mobile phase filter, sterile, 0.22-μm PES membrane, Corning/431097; store at RT.

CHAPTER 6

SYNTHESIS OF [^{18}F]FPPRGD2

SHUANGLONG LIU, FREDERICK T. CHIN, AND ZHEN CHENG

Molecular Imaging Program, Department of Radiology, Stanford University School of Medicine, Stanford, California

XIAOYUAN CHEN

Molecular Imaging Program, Department of Radiology, Stanford University School of Medicine, Stanford, California
Laboratory of Molecular Imaging and Nanomedicine, National Institute of Biomedical Imaging and Bioengineering, National Institutes of Health, Bethesda, Maryland

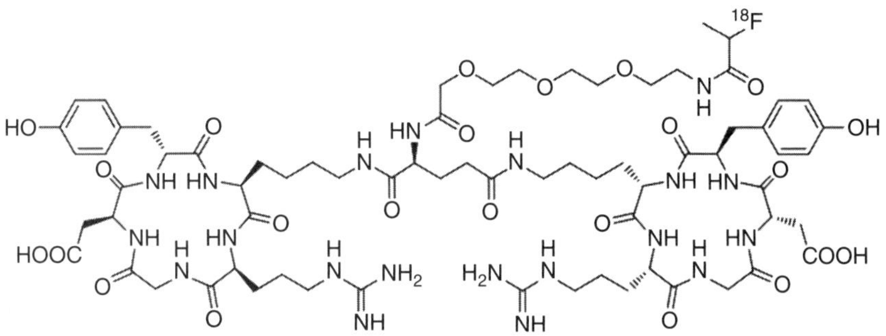

[^{18}F]FPPRGD2

1 INTRODUCTION

Angiogenesis is a physiological process involving the growth of new blood vessels from preexisting vessels. This is not only required for tumor spread and metastasis but is also necessary for natural healing after ischemic injury [1]. Integrins are cell surface receptors responsible for the regulation of cellular activation, migration, proliferation, survival, and differentiation. One of the most

Radiochemical Syntheses, Volume 1: Radiopharmaceuticals for Positron Emission Tomography, First Edition. Edited by Peter J. H. Scott, Brian G. Hockley.

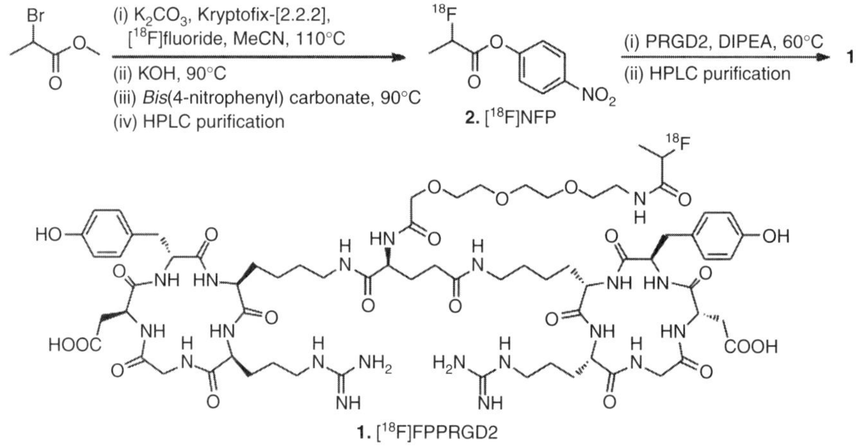

FIGURE 1 Synthesis of [^{18}F]FPPRGD2.

important members of this receptor class is integrin $\alpha_v\beta_3$, which is involved in many pathological processes. The $\alpha_v\beta_3$ integrin is a protein target chosen for designing diagnostic and therapeutic agents aimed at tumor angiogenesis because it is highly expressed on activated endothelial cells (ECs) of tumor neovasculature but not on normal cells or quiescent ECs [2]. Radiolabeled RGD (Arg-Gly-Asp) peptides and their analogs have been pursued for visualizing and quantifying $\alpha_v\beta_3$ integrin expression noninvasively in tumors using positron emission tomography (PET) [3–5]. Dimeric RGD has been [^{18}F]-labeled by 4-nitrophenyl 2-[^{18}F]-fluoropropionate ([^{18}F]NFP) to make the peptide tracer [^{18}F]FPPRGD2, which is now in clinical trial. Compared to the other monomer RGD peptides that have been in clinical trial [6–8], the dimeric has higher tumor uptake because of the polyvalency effect. The detailed synthesis protocol for [^{18}F]FPPRGD2 is given herein (Fig. 1).

2 SYNTHESIS PROCEDURES

CAUTION: All radiochemical syntheses must be carried out using appropriate equipment in a facility authorized for the use of radioactive materials. Personal protective equipment must be worn and all local radiation safety laws followed.

2.1 Production of [^{18}F]Fluoride

[^{18}O]H_2O (1.6 mL) [9] was loaded into the [^{18}F]fluoride target [10] of a General Electric Medical Systems (GEMS) PETTrace cyclotron [11]. The target was bombarded (36-μA beam for 30 min) to generate ~1.0 Ci (37.0 GBq) of [^{18}F]fluoride by the ^{18}O(p,n) ^{18}F nuclear reaction.

2.2 Azeotropic Drying of [^{18}F]Fluoride

The [^{18}F]fluoride was delivered to a GEMS TRACERlab FX_{FN} synthesis module [11] as a solution in [^{18}O]H_2O (1.6 mL). This solution was passed through a Chromafix® 30-PS-HCO_3 cartridge [12] to trap the [^{18}F]fluoride. The [^{18}F]fluoride was then eluted into the TRACERlab FX_{FN} glassy-carbon reaction vessel using a solution of potassium carbonate (3.0 mg) and Kryptofix-[2.2.2] (15.0 mg) in 1.0 mL acetonitrile (MeCN) and sterile water (9:1, v/v) [13]. The mixture was dried at 65°C under helium flow and vacuum for 5 min, followed by heating at 88°C under vacuum for 5 min.

2.3 Synthesis of [^{18}F]FPPRGD2

A solution of methyl 2-bromopropionate (5 mg) in anhydrous MeCN (1 mL) was added to the dried K[^{18}F]F-Kryptofix-[2.2.2] complex, and the reaction was heated to 110°C with stirring for 10 min. After cooling the reactor down to 40°C, 0.1 N KOH (0.5 mL) was added and the reaction mixture was heated at 90°C for 10 min. A second azeotropic drying step was carried out with MeCN (2 mL) to remove the residual water. Once the dried [^{18}F]fluoride complex was obtained, a solution of *bis*(4-nitrophenyl) carbonate in MeCN (2 mL, 20 mg/mL) was added to the residue and the mixture was heated at 90°C for 20 min. The reaction mixture was then diluted with sterile water (1.5 mL) and purified by semipreparative HPLC [14]. The radioactive fraction corresponding to [^{18}F]NFP (4-nitrophenyl 2-[^{18}F]-fluoropropionate, Rt = 22–23 min) was collected in a round bottom flask containing sterile water (20 mL) and then transferred to a customized module for solid phase extraction (SPE) using a C-18 Sep-Pak (Fig. 2) [15].

The trapped [^{18}F]NFP on C-18 cartridge was eluted through a Na_2SO_4 cartridge [16] with diethyl ether (2 mL) into a 5-mL V-vial in the customized module, diethyl ether was removed under helium stream at ambient temperature, and the dried labeling agent was reconstituted with H-AEEEA-Glu[cyclo(Arg-Gly-Asp-D-Tyr-Lys)]$_2$ (denoted as PRGD2) [17] in anhydrous dimethyl sulfoxide (DMSO). The conjugation between [^{18}F]NFP and peptide was executed using 2 mg PRGD2 in anhydrous DMSO (200 μL) with diisopropylethylamine (DIPEA, 40 μL) at 60°C for 20 min (Fig. 1).

2.4 Purification and Formulation of [^{18}F]FPPRGD2

After cooling down to room temperature, the crude reaction mixture was quenched by adding 5% acetic acid (AcOH, 0.7 mL). The crude [^{18}F]FPPRGD2 product solution was purified on a second semipreparative HPLC [18]. The [^{18}F]FPPRGD2 fraction was collected (Rt = 28 min) as shown in Fig. 3, diluted with sterile water, and sent back to TRACERlab FX_{FN} module for formulation. The diluted [^{18}F]FPPRGD2 fraction was passed through a second C-18 Sep-Pak cartridge [15]. The cartridge was then washed with sterile water (10 mL). Finally, [^{18}F]FPPRGD2 was eluted with ethanol (United States Pharmacopeia, USP, 1 mL) and saline (9 mL) and pushed through a sterile Millex Gs filter (0.22 μm, 4 mm) into a sterile dose vial (30 mL).

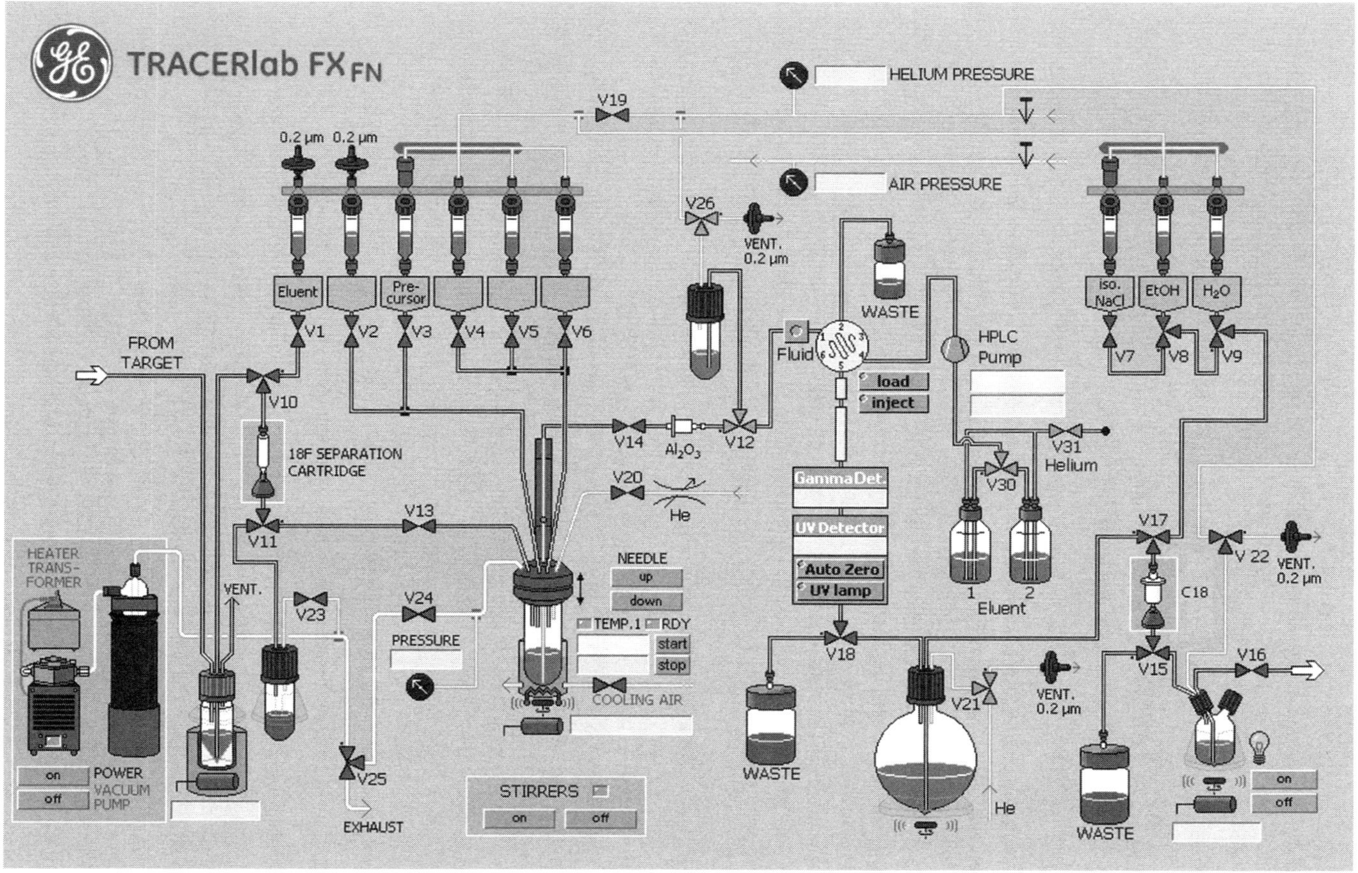

FIGURE 2 Schematic illustration of modified GE TRACERlab FX_{FN} for the fully automated synthesis of [^{18}F]NFP. The HPLC-purified [^{18}F]NFP was transferred to another GE TRACERlab FX_{FN} through a tube via valve 18 (V18) for coupling with PRGD2 by a remote control.

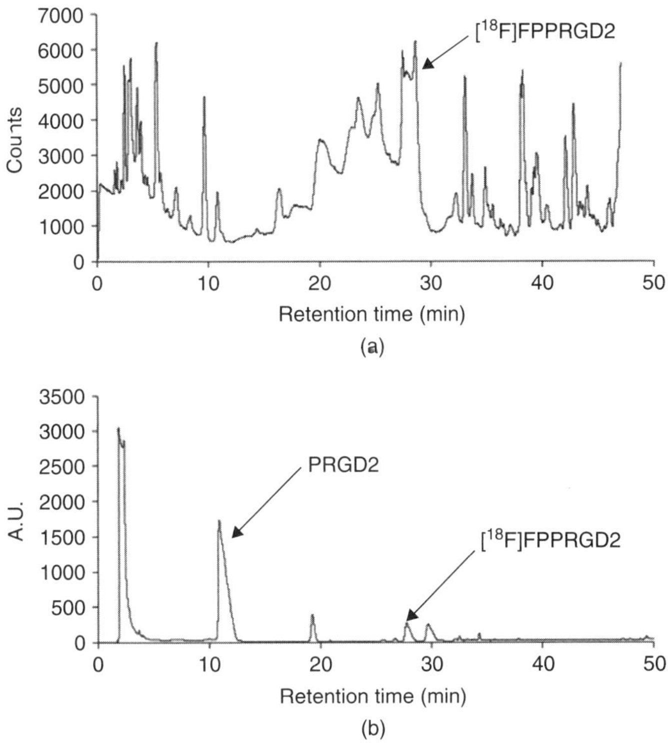

FIGURE 3 Semipreparative (a) radioactive and (b) UV HPLC traces for [^{18}F]FPPRGD2.

3 QUALITY CONTROL PROCEDURES

CAUTION: All radiopharmaceuticals produced for clinical use must have local regulatory approval before use in humans. Trained personnel must carry out quality control (QC) procedures, and each dose must meet all established QC criteria before release to the clinic.

QC procedures for [^{18}F]FPPRGD2, based on the current requirements for radiopharmaceuticals laid out in the US Pharmacopeia [19], are summarized below. Complete QC data for three consecutive batches of [^{18}F]FPPRGD2 produced using the method described herein are summarized in Table 1. Each of the three doses met all the established QC criteria.

3.1 Visual Inspection

The [^{18}F]FPPRGD2 dose is examined behind a PET L-block and must be clear, colorless, and free of particulates.

TABLE 1 QC Data for Three Repeat Runs of [^{18}F]FPPRGD2

QC Test	Release Criteria	Run 1	Run 2	Run 3
Product mCi (GBq)	N/A*	18.9 (0.7)	23.3 (0.9)	22.4 (0.8)
Visual inspection	Clear, colorless	Clear, colorless	Clear, colorless	Clear, colorless
Radiochemical identity	RRT = 0.9–1.1	1.0	1.0	1.0
Radiochemical purity (%)	>90	>95	>95	>95
Specific activity (mCi/μmol)	>100	269	400	220
Residual solvent analysis (% wt/V)	Acetonitrile ≤ 0.04	Pass	Pass	Pass
	Ethyl alcohol ≤ 10	Pass	Pass	Pass
Dose pH	4.5–7.5	6.0	6.0	6.0
Sterile filter integrity test (psi)	>45	>45	>45	>45
Radionuclidic identity ($t_{1/2}$) (min)	105–115	109	108	109
Endotoxin analysis (EU/mL)	≤17.5	<1.25	<1.25	<1.25
Sterility testing	No colony growth out to 14 days	Pass	Pass	Pass

*N/A, not applicable.

3.2 Radiochemical Identity

Analytical HPLC was performed using a Lab Alliance (LA; State College, PA) system equipped with an LA Series III HPLC pump (1 mL/min), LA Model 500 UV detector (214 nm), Carroll & Ramsey Associates Model 105S pin-diode radioactivity detector (Berkeley, CA), and an SRI Instruments Model 202 four-channel serial port chromatography data system using PeakSimple software (Torrance, CA). The final [^{18}F]FPPRGD2 was analyzed on a Vydac column (218TP54, C-18, 5 μm, 250 × 4.6 mm) eluted with a mixture of solvent (82.5% solvent A and 17.5% solvent B; solvents A and B are as described in Ref. [14] at 1 mL/min. The retention time of [^{18}F]FPPRGD2 is compared to that of the [^{19}F]FPPRGD2 reference standard [20] and must be ±10% (relative retention time (RRT) must be 0.9–1.1).

3.3 Radiochemical Purity

HPLC analysis of radiochemical purity was conducted as above. Radiochemical purity must be >90%. Representative analytical HPLC chromatograms are shown in Fig. 4.

3.4 Specific Activity

The specific activity (SA) release limit for [^{18}F]FPPRGD2 is 100 mCi/μmol [21]. The SA was calculated at the end of synthesis (EOS) using the HPLC

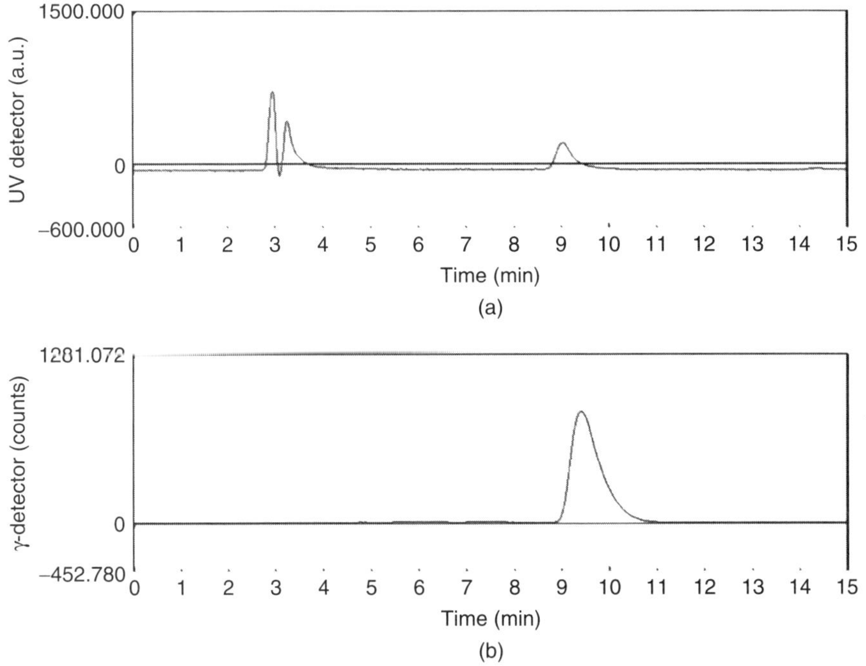

FIGURE 4 Analytical UV (a) and radioactive (b) HPLC trace for [^{18}F]FPPRGD2.

calibration curve obtained by the multiinjection of [^{19}F]FPPRGD2 with different concentrations.

3.5 Residual Solvent Analysis

Analysis of residual solvent levels (MeCN and ethyl alcohol) in [^{18}F]FPPRGD2 doses was conducted using an Agilent 6850 GC (Agilent Technologies Inc., USA. The initial temperature is 40°C at 0–4 min. With 1°C/min ramp, the temperature increases to 43°C at 7 min and to 130°C at 12 min.) with a capillary column (length: 30 m, inside tube diameter (ID): 0.320 mm, liquid phase film thickness (DF): 0.25 μm; Agilent Technologies Inc., USA). The specification for acetonitrile is ≤0.04% (wt/v), and the specification for ethyl alcohol is ≤10% (wt/v).

3.6 Dose pH

The pH of the [^{18}F]FPPRGD2 dose was analyzed by applying a small amount of the dose to colorPHast® pH 0–14 nonbleeding pH-indicator strips and determined by visual comparison to the scale provided. Dose pH must be 4.5–7.5.

3.7 Sterile Filter Integrity Test

The sterile filter from [^{18}F]FPPRGD2 (with needle still attached) was connected to nitrogen supply via a regulator. The needle was submerged in water, and the nitrogen pressure was gradually increased. If the pressure could be raised above the filter acceptance pressure (45 psi) without seeing a stream of bubbles, the filter was considered intact. If a stream of bubble occurs at <45 psi, the test fails.

3.8 Radionuclidic Identity

Radioactivity was measured using a Capintec CRC®-15PET Radioisotope Dose Calibrator [21], and the half-life was calculated using Equation 1. Calculated half-life must be 105–115 min.

$$t_{1/2} = -\ln 2(\text{time difference}/(\ln(\text{ending activity}/\text{starting activity}))) \quad (1)$$

3.9 Endotoxin Analysis

Endotoxin content in doses of [^{18}F]FPPRGD2 was analyzed by Charles River Laboratories EndoSafe® Portable Testing System and according to the US Pharmacopeia [22]. Dose must contain <17.5 endotoxin units (EU)/mL.

3.10 Sterility Test

Samples of [^{18}F]FPPRGD2 were placed onto fluid thioglycolate media (FTM) plates and trypticase soy broth (TSB) plates. FTM plates were used to test for anaerobes, aerobes, and microaerophiles, while, TSB plates were used to test for nonfastidious and fastidious microorganisms. [^{18}F]FPPRGD2 plates were incubated along with negative controls for 14 days. FTM plates were incubated at 32°C and TSB plates were incubated at 22°C according to the current USP guidelines [23]. Plates were visually inspected on the third, eighth, and fourteenth days of the test period and compared to the negative standards. [^{18}F]FPPRGD2/negative controls must have no culture growth after 14 days to be indicative of a sterile product.

CHEMICAL ABSTRACTS NOMENCLATURE (REGISTRY NUMBER)

Acetic acid (64-19-7)
Acetonitrile (75-05-8)
Bis(4-nitrophenyl) carbonate (5070-13-3)
Diethyl ether (60-29-7)
N,*N*-diisopropylethylamine (7087-68-5)
Dimethylsulfoxide (67-68-5)
Ethyl alcohol (64-17-5)
4,7,13,16,21,24-Hexaoxa-1,10-diazabicyclo[8.8.8]hexacosane (23978-09-8)
Methyl 2-bromopropionate (5445-17-0)
Potassium carbonate (584-08-7)
Potassium hydroxide (1310-58-3)

Sodium chloride (7647-14-5)
Trifluoroacetic acid (76-05-1)

REFERENCES AND NOTES

1. Tonini T, Rossi F, Claudio PP. Oncogene 2003;22:6549–6556.
2. Tucker GC. Curr Opin Investig Drugs 2003;4:722–731.
3. Beer AJ, Niemeyer M, Carlsen J, Sarbia M, Nahrig J, Watzlowik P, Wester HJ, Harbeck N, Schwaiger M. J Nucl Med 2008;49:255–259.
4. Decristoforo C, Hernandez Gonzalez I, Carlsen J, Rupprich M, Huisman M, Virgolini I, Wester HJ, Haubner R. Eur J Nucl Med Mol Imaging 2008;35:1507–1515.
5. Zhang X, Xiong Z, Wu Y, Cai W, Tseng JR, Gambhir SS, Chen X. J Nucl Med 2006;47:113–121.
6. Haubner R, Weber WA, Beer AJ, Vabuliene E, Reim D, Sarbia M, Becker KF, Goebel M, Hein R, Wester HJ, Kessler H, Schwaiger M. PLoS Med 2005;2:e70.
7. Wester HJ, Schottelius M, Scheidhauer K, Meisetschlager G, Herz M, Rau FC, Reubi JC, Schwaiger M. Eur J Nucl Med Mol Imaging 2003;30:117–122.
8. Kenny LM, Coombes RC, Oulie I, Contractor KB, Miller M, Spinks TJ, McParland B, Cohen PS, Hui AM, Palmieri C, Osman S, Glaser M, Turton D, Al-Nahhas A, Aboagye EO. J Nucl Med 2008;49:879–886.
9. Virgin $[^{18}O]H_2O$ >95% isotopic enrichment was purchased from Rotem Industries Ltd., Israel, and used as received.
10. GEMS silver high yield $[^{18}F]$fluoride target.
11. GE Healthcare, USA.
12. SPE cartridge Chromafix 30-PS-HCO_3 cartridges (catalog No. 731876) were purchased from Macharey-Nagel, Germany, and conditioned with 1 mL ethanol and 1 mL water before use and dried over air flow.
13. Potassium carbonate was purchased from Aldrich (catalog No. 367877) and used as received. Kryptofix-[2.2.2] was purchased from Aldrich and (catalog No. 291110) and used as received. Acetonitrile was purchased from Aldrich (catalog No. 271004) and used as received. Sterile water was purchased from Hospipra, Inc., USA, (catalog No. NDC 0409-4887-10) and used as received.
14. The semipreparative reversed-phase high performance liquid chromatography (RP-HPLC) using a Vydac protein and peptide column (218TP510; 5 μm, 250 × 10 mm) was performed on Dionex 680 pump with KANUR UV detector K-2001. With a flow rate of 5 mL/min, the mobile phase was changed from 95% solvent A [0.1% trifluoroacetic acid (TFA) in water] and 5% B [0.1% TFA in acetonitrile (MeCN)] (0–2 min) to 35% solvent A and 65% solvent B at 32 min.
15. C-18 Sep-Pak cartridges were purchased from Waters, USA, (part no. WAT020515) and conditioned with 5 mL ethanol and 10 mL sterile water before use.
16. Sodium sulfate cartridges were purchased from Varian, USA, (part No. WAT054265) and used without any conditioning.
17. PRGD2 was purchased from Peptides International, USA (catalog No. PCS-37458-PI) and used as received.
18. The semipreparative reversed-phase high performance liquid chromatography (RP-HPLC) using Phenomenex column (Gemini, 5 μm, C18, 250 × 10 mm) with the flow rate at 7.5 mL/min was performed on Dionex Ultimate 3000 chromatography system with a UVD 340U absorbance detector. The mobile phase was started at 90% solvent A and 10% solvent B (0–2 min) and changed to 86.5% solvent A and 13.5% solvent B at 20 min, was kept at this percentage for 5 min, and then again gradually changed to 70% solvent A and 30% solvent B in 20 min.
19. U. S. Pharmacopeia <823> radiopharmaceuticals for Positron Emission Tomography Compounding, USP 32-NF 27. 2009.

20. The preparation of standard [^{19}F]FPPRGD2: To the solution of 2-fluoropropionic acid (7.8 mg, 84.5 μmol) in 1.0 mL anhydrous MeCN was added *O*-(*N*-succinimidyl)-1,1,3,3-tetramethyluronium tetrafluoroborate (TSTU) (17.6 mg, 58.5 μmol). The pH of the solution was adjusted to 8.5–9.0 by *N*,*N*-diisopropylethylamine (DIPEA). The reaction mixture was stirred at room temperature for 0.5 h, and peptide PRGD2 (3 μmol) in DMF was added in one aliquot. After being stirred at room temperature for 2 h, the product [^{19}F]FPPRGD2 was isolated by semipreparative HPLC in 85% yield. MALDI-TOF-MS was *m/z* 1613.2 for $[MH]^+$ ($C_{70}H_{106}FN_{20}O_{23}$, calculated molecular weight, 1613.8).
21. Radioactivity measurements were carried out with a CRC-15PET dose calibrator (Capintec Inc., USA), which was calibrated daily using Cs-137 (Bench/mark, USA) and Co-57 source (Isotope products laboratories, USA).
22. U. S. Pharmacopeia <85> Bacterial Endotoxins Test. USP 32-NF 27. 2009.
23. U.S. Pharmacopeia <71> Sterility Tests. USP 32-NF 27. 2009.

CHAPTER 7

SYNTHESIS OF [^{18}F]FLUOROCHOLINE ([^{18}F]FCH)

DAVID KRYZA

ISPB, Faculté de Pharmacie, Université Claude Bernard Lyon 1
Hospices Civils de Lyon, Hôpital E Herriot, Radiopharmacie
LPCML, UMR5620 CNRS, Villeurbanne, Lyon, France

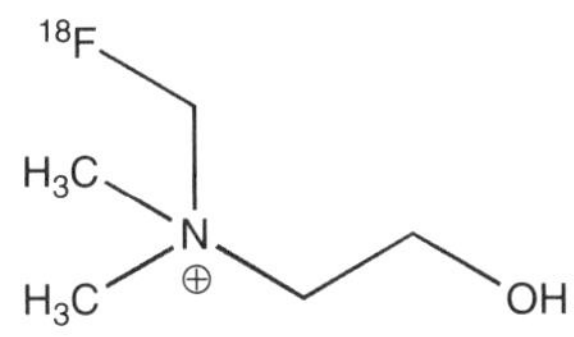

[^{18}F]FCH

1 INTRODUCTION

Many tumors are characterized by an enhanced cell proliferation. This is usually associated with an elevated choline uptake. The rationale of using positron emission tomographic (PET) imaging with radiolabeled choline analogs is the role of choline as a precursor for the biosynthesis of phosphatidylcholine and other phospholipids that are the major components of cell membranes [1]. It has been demonstrated that in most cancers, there is a high content of phosphorylcholine compared to the very low levels present in normal tissues. Several lines of research suggest the metabolic pathway to be a promising imaging target, and it has been well established that [^{11}C]choline is a very effective tracer in imaging tumors in human brain, lung, urinary bladder, or prostate [2,3].

Owing to the short half-life of carbon-11 (20 min), it is difficult to use [^{11}C]choline in routine imaging. To circumvent the cumbersome synthesis of [^{11}C]choline, [^{18}F]-labeled choline has been developed [4,5]. To meet the strong demand for [^{18}F]fluorocholine, a reliable automated synthesis has been developed [6], without any demanding purification steps (Fig. 1).

Radiochemical Syntheses, Volume 1: Radiopharmaceuticals for Positron Emission Tomography, First Edition. Edited by Peter J. H. Scott, Brian G. Hockley.

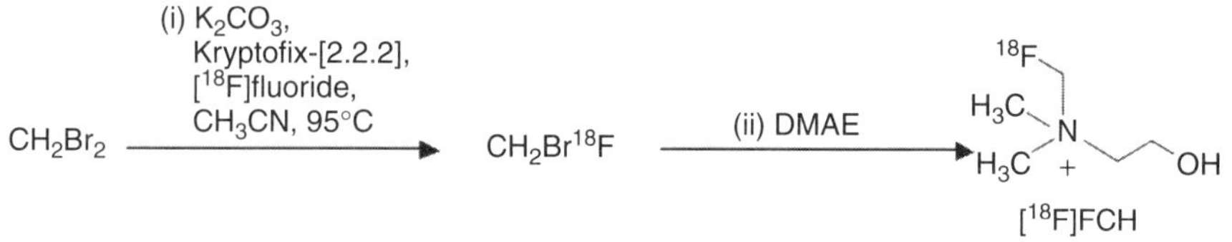

FIGURE 1 Synthesis of [^{18}F]fluorocholine.

2 SYNTHESIS PROCEDURES

CAUTION: All radiochemical syntheses must be carried out using appropriate equipment in a facility authorized for the use of radioactive materials. Personal protective equipment must be worn and all local radiation safety laws followed.

2.1 Production of [^{18}F]Fluoride

No-carrier-added [^{18}F]fluoride was produced by $^{18}O(p,n)^{18}F$ nuclear reaction with a General Electric Medical Systems PET Trace cyclotron [7] using a 16.5-MeV proton beam. The target was bombarded (75-μA beam for 120 min) to generate approximately 8.1 Ci (300 GBq) of [^{18}F]fluoride.

2.2 Azeotropic Drying of [^{18}F]Fluoride

The [^{18}F]fluoride was transported via PEEK tubing to the TRACERlab MX_{FDG} synthesis module [8], where it was trapped on a QMA-Light Sep-Pak® cartridge [9]. [^{18}F]Fluoride was then eluted to the reactor by a mixture of K_2CO_3 (22 mg in 0.3 mL H_2O) [10] and Kryptofix-[2.2.2] (7 mg in 0.3 mL acetonitrile) [11]. The [^{18}F]fluoride was dried by azeotropic distillation at 85°C using three successive additions of acetonitrile (0.7 mL).

2.3 Synthesis of [^{18}F]Fluorocholine

A solution of dibromomethane [12] (300 μL) in acetonitrile (3 ml) was added to the dried [^{18}F]fluoride. The reactor was then heated at 95°C for 300 s to produce [^{18}F]fluorobromomethane. To avoid escape of too much activity from the reactor, especially at high temperatures (boiling point of [^{18}F]fluorobromomethane, 8°C), two normally closed pinch valves [13] were added at the entrance and exit of the reactor at each run. These pinch valves were remote controlled by the TRACERlab MX_{FDG} software. The valves were open throughout synthesis, except during fluoration.

[^{18}F]Fluorobromomethane was then purified from dibromomethane and fluoride by distillation over three Sep-Pak Plus silica cartridges [14] under nitrogen (35 mL/min) and vacuum (−100 mbar).

[^{18}F]Fluorobromomethane was reacted with dimethylamino ethanol (DMAE) [15] (400 μL) previously loaded on a *t*C18 cartridge [16]. The synthesized [^{18}F]fluorocholine was removed from the *t*C18 to a cation-exchange CM light cartridge [17] by 10 mL ethanol [18] (the CM light cartridge was conditioned with 10 mL of 0.5 M HCl). The cartridge was washed with 15 mL of water, and the final product was recovered in a vial through a sterile filter by eluting with 3 mL of isotonic saline [19] released for quality control (QC). Total synthesis time was about 35 min.

The optimized procedure resulted in a non-decay-corrected yield of 20–25%.

3 QUALITY CONTROL PROCEDURES

CAUTION: All radiopharmaceuticals produced for clinical use must have local regulatory approval before use in humans. Trained personnel must carry out QC procedures, and each dose must meet all established QC criteria before release to the clinic.

QC procedures for [^{18}F]fluorocholine, based on the current requirements for radiopharmaceuticals laid out in the European Pharmacopoeia, are summarized below. Complete QC for three repeat batches of [^{18}F]fluorocholine produced using the method described herein are summarized in Table 1. Each of the 3 doses met all the established QC criteria.

TABLE 1 QC Data for Three Repeat Runs of [^{18}F]Fluorocholine

QC Test	Release Criteria	Run 1	Run 2	Run 3
Yield non decay corrected (%)	N/A	22	17	24
Visual inspection	Clear, colorless	Clear, colorless	Clear, colorless	Clear, colorless
Radiochemical identity	RTT = 0.9–1.1	1.0	1.0	1.0
Radiochemical purity (%)	≥95	99	99	99
Specific activity (GBq/μmol)	No limit established	36	38	34
Residual solvent analysis (in ppm)	MeCN <410 Ethanol <5000	1.4 42.6	3.6 22.6	0.5 54.5
Chemical purity (in ppm)	Dimethylaminoethanol <100	60.2	52.6	66.8
Dose pH	4.5–7.5	6.5	7	6.0
Residual Kryptofix-[2.2.2] (μg/mL)	≤50	≤50	≤50	≤50
Sterile filter integrity test (psi)	>50	>50	>50	>50
Radionuclidic identity ($t_{1/2}$) (min)	105–115	112	110	111
Endotoxin analysis (EU/mL)	≤17.5	≤17.5	≤17.5	≤17.5
Sterility testing	No colony growth out to 14 days	Pass	Pass	Pass

3.1 Visual Inspection

The [^{18}F]fluorocholine solution must be clear, colorless, and free from particulate matter when observed under adequate light.

3.2 Radiochemical Identity

Radiochemical identity was achieved using a Dionex [20] GP 50 gradient pump and an EP 50 conductimeter, equipped with a CSRS® autosuppressor-II in series with a β^+ flow count detector (Bioscan) [21], using an IonPac® CS12A 4 × 250 mm column [22] plus 4 × 50 mm guard column. (Mobile phase, 9 mM sulphuric acid; flow rate, 1.2 mL/min).

The retention time of [^{18}F]fluorocholine was compared to and found to be approximately the same as that of the cold reference ([^{19}F]fluorocholine [23]) (10–11 min).

Relative retention time (RTT) must be 0.9–1.1.

3.3 Radiochemical Purity

Radiochemical purity has been achieved using a Dionex [20] GP 50 gradient pump and an EP 50 conductimeter, equipped with a CSRS autosuppressor-II in series with a $\beta+$ flow count detector (Bioscan) [21] using an IonPac CS12A 4 × 250 mm column [22] plus 4 × 50 mm guard column. (Mobile phase, 9 mM sulphuric acid; flow rate, 1.2 mL/min).

Radiochemical purity must be >95%. A typical chromatogram is shown in Fig. 2.

3.4 Specific Activity

A typical run leads to a specific activity >37 GBq/μmol. Nevertheless, there is no specific activity release limit for [^{18}F]fluorocholine.

3.5 Residual Solvent Analysis

Gas chromatography analysis was conducted using a Varian 3900 GC system [24] (column: Rtx-200 silica fused capillary column, 30 × 0.3 mm [25]).

According to the ICH Harmonised Tripartite Guidelines [26], the following specifications are applied: acetonitrile, <410 ppm; ethanol, <5000 ppm.

Dibromomethane should be below 10 μg/mL.

3.6 Chemical Purity

Radiochemical purity has been achieved using a Dionex [20] GP 50 gradient pump and an EP 50 conductimeter, equipped with a CSRS autosuppressor-II using an IonPac CS12A 4 × 250 mm column [22] plus 4 × 50 mm guard column. (Mobile phase, 9 mM sulphuric acid; flow rate, 1.2 mL/min).

No more than 100 ppm dimethylaminoethanol must be found in the total volume of the batch of [^{18}F]fluorocholine injection produced.

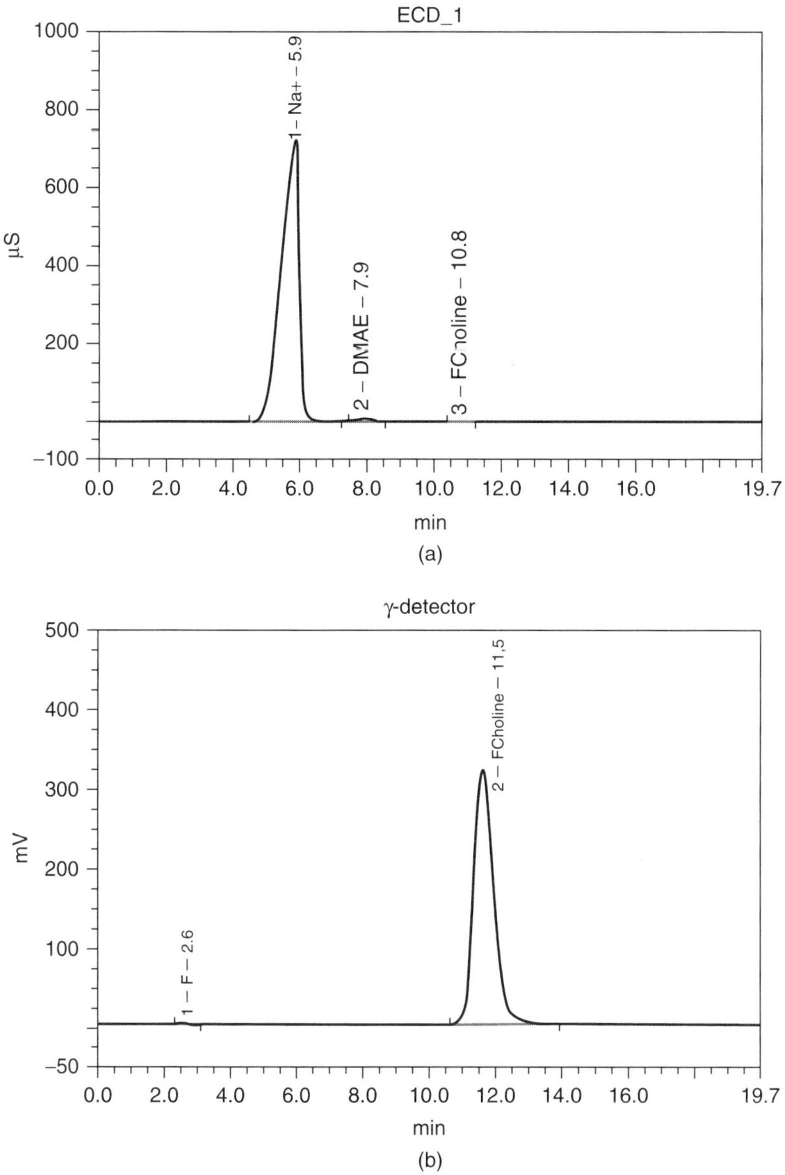

FIGURE 2 Analytical ECD (a) and radioactive (b) HPLC trace for [^{18}F]fluorocholine.

3.7 Dose pH

The pH of [^{18}F]fluorocholine was measured by depositing a small drop of [^{18}F]fluorocholine on paper pH-indicator strips [27].

Dose pH must be 4.5–8.5.

3.8 Residual Kryptofix-[2.2.2] Analysis

Residual Kryptofix-[2.2.2] levels in [^{18}F]fluorocholine solution were determined using a spot test [28]. Thin layer chromatography-silica gel (TLC-SG) strip plates were exposed to iodoplatinate reagent. Samples of the various solutions (water as the negative control, 50 μg/mL Kryptofix-[2.2.2] standard as the positive control, and [^{18}F]fluorocholine solution) were applied to the precoated TLC strips. A blue-black circular spot is clearly visible if Kryptofix-[2.2.2] is present. The size and intensity of the spot in [^{18}F]fluorocholine solution does not exceed that of the standard solution that corresponds to the 50 μg/mL Kryptofix-[2.2.2] reference standard spot; $<$50 μg/mL is acceptable.

3.9 Sterile Filter Integrity Test

A stream of nitrogen is passed through the devices to the filter from [^{18}F]fluorocholine, then to a reservoir of water. The filter membrane should be able to stand the maximum pressure indicated in the specification of the filter (50 psi). If the membrane is broken, the air stream will pass through the membrane into the water. Air bubbles will then be seen, and the test fails.

3.10 Radionuclidic Identity

Radionuclide was identified by calculating the half-life of the isotope from two radioactivity measurements of the product vial in the dose calibrator [29] using Equation 1.

Half-life must be 105–115 min.

$$t_{1/2} = \ln 2 \text{ (time difference/(ln(ending activity/starting activity)))} \qquad (1)$$

3.11 Endotoxin Analysis

The bacterial endotoxin level in doses of [^{18}F]fluorocholine is tested using the gel-clot technique (limulus amebocyte lysate (LAL) test). The formation of a gel-clot indicates the presence of endotoxin in a sample. Test for pyrogen was performed using a commercial LAL kit (Charles River Endosafe® [30]). Doses must contain $\leq$17.5 endotoxin units (EU)/mL.

3.12 Sterility Testing

Sterility is to be tested by incubating samples of [^{18}F]fluorocholine with both soybean casein digest medium (SCDM) and fluid thioglycolate medium (FTM) for 14 days at 37°C. SCDM plates are used to test for aerobic bacteria and fungi, whereas FTM plates are used for anaerobic bacteria [31]. Growth promotion tests should be performed simultaneously. Plates are regularly examined during the incubation period (on the third, seventh, and fourteenth days of the test period) for a macroscopic evidence of microbial growth. If this visual test does not indicate the presence of any growth, the product is considered sterile.

ACKNOWLEDGMENT

This work was supported by Advanced Accelerator Applications (AAA).

CHEMICAL ABSTRACTS NOMENCLATURE (REGISTRY NUMBER)

Acetonitrile (75-05-8)
Carbonic acid, potassium salt (584-08-7)
Dibromomethane (74-95-3)
Dimethylaminoethanol (108-01-0)
Ethanol (64-17-5)
4,7,13,16,21,24-Hexaoxa-1,10-diazabicyclo[8.8.8]hexacosane (23978-09-8)

REFERENCES AND NOTES

1. Hara T. 18F-Fluorocholine: a new oncologic PET tracer. J Nucl Med 2001;42(12):1815–1817.
2. Hara T, Kosaka N, Kondo T, Kishi H, Kobori O. Imaging of brain tumor, lung cancer, esophagus cancer, colon cancer, prostate cancer, and bladder cancer with [C-11]choline. J Nucl Med 1997; 38 Suppl:250P [abstract].
3. Hara T, Kosaka N, Kishi H. PET imaging of prostate cancer using carbon-11-choline. J Nucl Med 1998;39:990–995.
4. DeGrado TR, Baldwin SW, Wang S, Orr MD, Liao RP, et al. Synthesis and evaluation of (18)F-labeled choline analogs as oncologic PET 336 tracers. J Nucl Med 2001;42(12):1805–1814.
5. DeGrado TR, Coleman RE, Wang S, et al. Synthesis and evaluation of 18F-labeled choline as an oncologic tracer for positron emission tomography: initial findings in prostate cancer. Cancer Res 2001;61:110–117.
6. Kryza D, Tadino V, Filannino MA, Villeret G, Lemoucheux L. Nucl Med Biol 2008;35(2): 255–260.
7. GE Healthcare, USA.
8. From the website of GE Healthcare. Available at: http://www.gehealthcare.com/usen/fun_img/radiopharmacy/products/tracerlabmx.html
9. QMA-light Sep-Pak cartridges were purchased from Waters (part no. WAT023525) and conditioned with 10 mL ethanol, 10 ml water, 10 mL 0.5 M sodium bicarbonate, and a further 10 mL water before use.
10. Potassium carbonate was obtained from Sigma Aldrich (part no. 209619) and used as received. Sterile water (part no. 3120495) was purchased from Lavoisier and used as received.
11. Kryptofix-[2.2.2] was purchased from ABX (part no.800.1000) and used as received. Anhydrous acetonitrile (part no. 49,444-5) was purchased from Sigma-Aldrich and used as received.
12. Dibromomethane was obtained from Sigma Aldrich (part no. 239992) and used as received.
13. NReshearch, Switzerland.
14. Sep-Pak Plus silica cartridges were purchased from Waters (part no. WAT020520) and used as received.
15. Dimethylaminoethanol was obtained from Sigma Aldrich (part no. 391253) and used as received.
16. Sep-Pak Plus tC18 cartridges were purchased from Waters (part no. WAT036810) and used as received.

17. Accell™ plus CM Cartridges were purchased from Waters (part no. WAT020550) and conditioned with 10 mL of 0.5M HCl.
18. Ethanol was purchased from Sigma Aldrich (part no. 32294) and used as received.
19. Sodium chloride 0.9% was purchased from Lavoisier (part no. 305-784-5) and used as received.
20. Dionex, Germany.
21. Bioscan, Inc., USA.
22. Analytical HPLC column: IonPac CS12A 4 × 250 mm column was purchased from Dionex (part no. P/N 44001) and guard column 4 × 50 mm was purchased from Dionex (part no. P/N 44002); mobile phase, 9 mM sulphuric acid; flow rate = 1.2 mL/min; analytical column was equilibrated with mobile phase for 30–60 min before use.
23. [^{19}F]Fluorocholine was purchased from ABX (part no. 6130.0010) and used as received.
24. Varian, Inc., USA.
25. Analytical GC Column: Rtx® -200 was purchased from Restek (part no. 15024).
26. International Conference on Harmonisation of Technical Requirements for Registration of Pharmaceuticals for Human Use (ICH).
27. Merck chemicals, Germany (part no. 109584).
28. Mock BH, Winkle W, Vavrek MT. A color spot test for the detection of Kryptofix 2.2.2 in [18F] FDG preparations. Nucl Med Biol 1997;24:193–195.
29. Dose calibrator: Capintec CRC®-15R, Capintec, Inc., USA.
30. Charles River Laboratories, France.
31. Biomerieux Laboratories, France.

CHAPTER 8

CLINICAL MANUFACTURING OF [^{18}F]-16-α-FLUOROESTRADIOL ([^{18}F]FES)

PIYUSH KUMAR AND JOHN R. MERCER

Department of Oncology, University of Alberta, Edmonton, Alberta, Canada

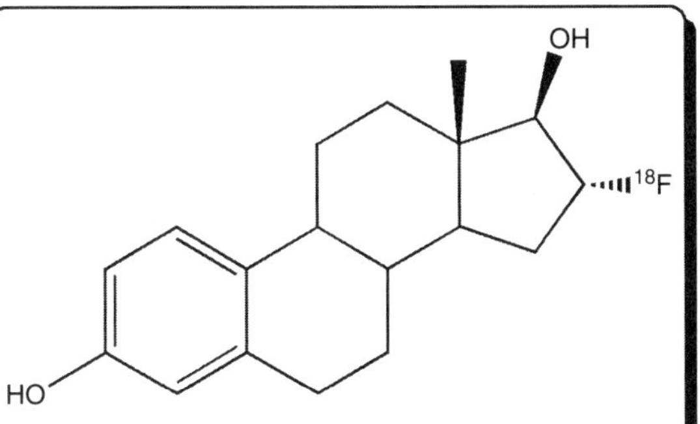

[^{18}F]FES

1 INTRODUCTION

Positron emission tomography (PET) is a valuable molecular imaging approach primarily used to diagnose and stage oncological and neurological disorders. It is often used to plan appropriate therapy for patients and determine treatment response [1]. In this context, [^{18}F]fluoroestradiol ([^{18}F]FES), a steroidal fluorinated tracer, developed in the early 1980s [2,3], has demonstrated proven clinical utility for patients identified with breast cancer by determining their tissue levels of the estrogen receptor and helping in predicting which patients with primary, recurrent, or metastatic breast cancer will respond to antiestrogen therapy [4,5]. A robust automated synthesis is a key requirement for routine production of clinical PET radiopharmaceuticals since this minimizes radiation exposure to laboratory personnel, increases the efficiency of reactions by reducing the production time, and provides a consistency to the synthesis process of a radiopharmaceutical product that will be more acceptable to regulatory agencies.

Radiochemical Syntheses, Volume 1: Radiopharmaceuticals for Positron Emission Tomography, First Edition. Edited by Peter J. H. Scott, Brian G. Hockley.

We have developed and optimized the automated radiosynthesis of [^{18}F]FES (Fig. 1), providing a reliable product for clinical imaging [6]. The synthesis procedure is based on improvements to existing literature methods [7–13], is more rapid and simplified, that provides the pure product as a sterile injectable solution ready for clinical use.

2 SYNTHESIS PROCEDURES

CAUTION: All radiochemical syntheses must be carried out using appropriate equipment in a facility authorized for the use of radioactive materials and by personnel with appropriate training. Personal protective equipment must be worn at all times during the synthesis and all local radiation safety laws followed.

Note 1: This synthesis is sensitive to moisture. The reaction components (reaction vial, reagents reservoirs, and the transfer lines) of the automated synthesis unit (ASU) must be thoroughly cleaned and dried using a defined set of procedures and appropriate solvents before performing the radiopharmaceutical manufacture.

2.1 Production and Trapping of [^{18}F]Fluoride

[^{18}F]Fluoride was produced from [^{18}O]H_2O [14] via an ^{18}O(p,n)^{18}F nuclear reaction. A 19-MeV cyclotron [15] was used to generate proton beams (65 μA) that bombarded [^{18}O]H_2O (3 mL), contained in a niobium target [16], for 35 min, to produce ~55 GBq of [^{18}F]fluoride. No-carrier-added radiofluoride was transferred, after complete decay of ^{13}N[NH_3] (generated as a side nuclear product of this reaction), to the ASU (GEMS TRACERlab FX_{FDG} module) [17] contained in a Class 100 isolated radiation shielded cell [18], and adsorbed on an ion exchange cartridge [19] (Chromafix® 30-PS-HCO_3). The cartridge was preactivated by the sequential passage of potassium carbonate [20] (0.5 M, 10 mL), sterile water for injection (SWFI, 10 mL) [21], and nitrogen gas [22] (to dryness) and stored sterile before use.

2.2 Azeotropic Drying of [^{18}F]Fluoride

The trapped [^{18}F]fluoride was transferred from the cartridge to the reaction vial by passage of an eluent (1 mL) containing 4,7,13,16,21,24-hexaoxa-1,10-diazabicyclo[8.8.8]hexacosane (Kryptofix®-2.2.2 (K-2.2.2) [23]; 15 mg) and potassium carbonate (3.5 mg) in acetonitrile [24]: SWFI (90:10, respectively, v/v). The solvents were evaporated at 85°C under reduced pressure (22 KPa) and nitrogen gas flow (100 KPa) for 3 min. Dry acetonitrile (1 mL) was added, followed by a single azeotropic drying sequence (85°C, 3 min, 22 KPa followed by 95°C, 3 min, ~5 KPa) to remove residual traces of water. The temperature of the reaction vessel was lowered to 50°C prior to the addition of the precursor solution to minimize any losses that might happen because of solvent splashes at higher temperatures.

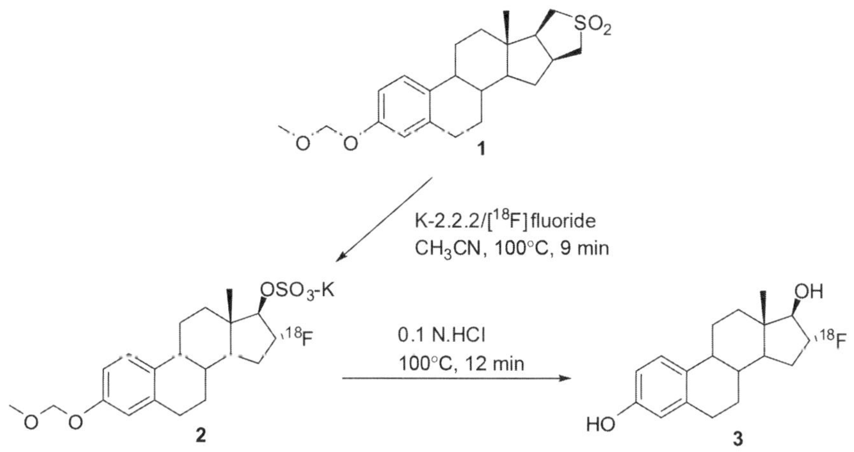

FIGURE 1 Synthesis of [^{18}F]FES, **3**.

2.3 Synthesis of [^{18}F]FES

The sulfone precursor (**1**) [25] (1–2 mg) dissolved in anhydrous acetonitrile (1 mL) was added to the reaction vial, the temperature raised to 100°C, and the reaction allowed to proceed for 10 min followed by cooling to 55°C. The reaction afforded the radiofluorinated intermediate product 3-*O*-methoxymethyl (MOM)-[^{18}F]FES sulfate (**2**) that was then subjected to two evaporative acid hydrolyses using two separate additions of 0.1 N HCl [26] in acetonitrile : water (2.0 mL; 9:1 v/v) at 100°C under reduced pressure (22 KPa) and nitrogen flow. Total hydrolysis time was approximately 12 min (Figure 1). The crude [^{18}F]FES (**3**) was dissolved in 50% aqueous ethanol [27] (3.0 mL) and passed through a sterile Millex LG filter [28] (0.22 μm) into a shielded sterile product vial. Total synthesis time consumed in this automated production was ~32 min. Major ASU-controlled unit reaction parameters are indicated in Table 1.

TABLE 1 Major ASU-Controlled Reaction Parameters

Parameter		Reaction Step	Set Point Value	Set Point Control
K-2.2.2./^{18}F$^-$-complex drying	Total time Temperature and vacuum	Generating anhydrous activated K^{18}F + K-2.2.2 complex	9 min 85°C (6 min, 22 KPa) 95°C (3 min, 5 KPa)	±3.0 s ±0.1°C (±1 KPa) ±0.1°C (±1 KPa)
[^{18}F]-Labeling	Reaction time Temperature	Conversion of **1** to MOM [^{18}F]FES sulphate, **2**	10 min 100°C	±1.0 s ±0.1°C
Hydrolysis of **2**	Reaction time Reaction temperature	Generation of [^{18}F]FES, **3**	12 min 100°C	±1.0 s ±0.1°C

2.4 Purification of [^{18}F]FES

[^{18}F]FES solution was dispensed from the ASU and injected onto a high performance liquid chromatographic (HPLC) system [29] equipped with a 5-mL injection loop, C-18 reversed-phase guard column [30] (Phenomenex; 5 × 1 cm, 10 μm), reversed-phase C-18 column [30] (Whatman 10-ODS3; 30 × 0.9 cm, 5 μm), UV detector [31] (set at 280 nm), and radioactivity detector [32]. The crude [^{18}F]FES was eluted with ethanol : water (52:48, v/v; flow rate = 1.8 mL/min) at a retention time of 16.5 min (±5%) (Fig. 2). The column outlet was connected to a sterilized manifold [33] that contained a sterile LG filter [28] (Millex; pore size = 0.22 μm) and a sterile needle (16 G) assembly attached to a preweighed sterile multidose product vial (Mallinckrodt; 20 mL). The peak elution corresponding to [^{18}F]FES was passed through this sterile setup and was collected in the final product vial preloaded with sterile saline [34] (5 mL). Once the product collection (5 ± 0.5 mL) was complete, the vial was removed and weighed to determine the total weight.

The results for isolated yields of [^{18}F]FES in three standard preparations are given in Table 2.

Note 2: [^{18}F]FES formulated as described above in Section 2.4 will be 15–18% v/v ethanol in SWFI when released. This product must be further diluted with appropriate amounts of sterile saline to bring the final EtOH concentration to <1% prior to patient administration.

3 STABILITY OF [^{18}F]FES RADIOPHARMACEUTICAL

The stability of [^{18}F]FES in the clinical formulation described above was determined for expected normal storage and transportation conditions (22°C) in upright and inverted positions of the product vials over a period of 24 h (normal clinical study period). Evaluation of the appearance, pH, radiochemical identity, and radiochemical purity (RCP) for three batches were included in this study design as reflected in Table 3. [^{18}F]FES was found to be completely stable in its current formulation and met all the specifications required for human administration.

4 QUALITY CONTROL PROCEDURES

CAUTION: Radiopharmaceutical preparations for human use will generally fall under local and national guidelines and be the subject of review and approval by regulatory agencies. Our product adheres to the guidelines of the Biologics, Radiopharmaceuticals and Genetic Therapies Directorate of Health Canada, which match or exceed the guidelines of other jurisdictions. Health Canada regulates radiopharmaceutical manufacture and use. The quality and efficacy of products is established by prerelease and retrospective quality control (QC) testing. All tests and final release authorization must be performed by trained personnel.

A summary of the prerelease, retrospective, and supporting characterization QC results and specifications for [^{18}F]FES, is provided in Tables 4–6 [35] for

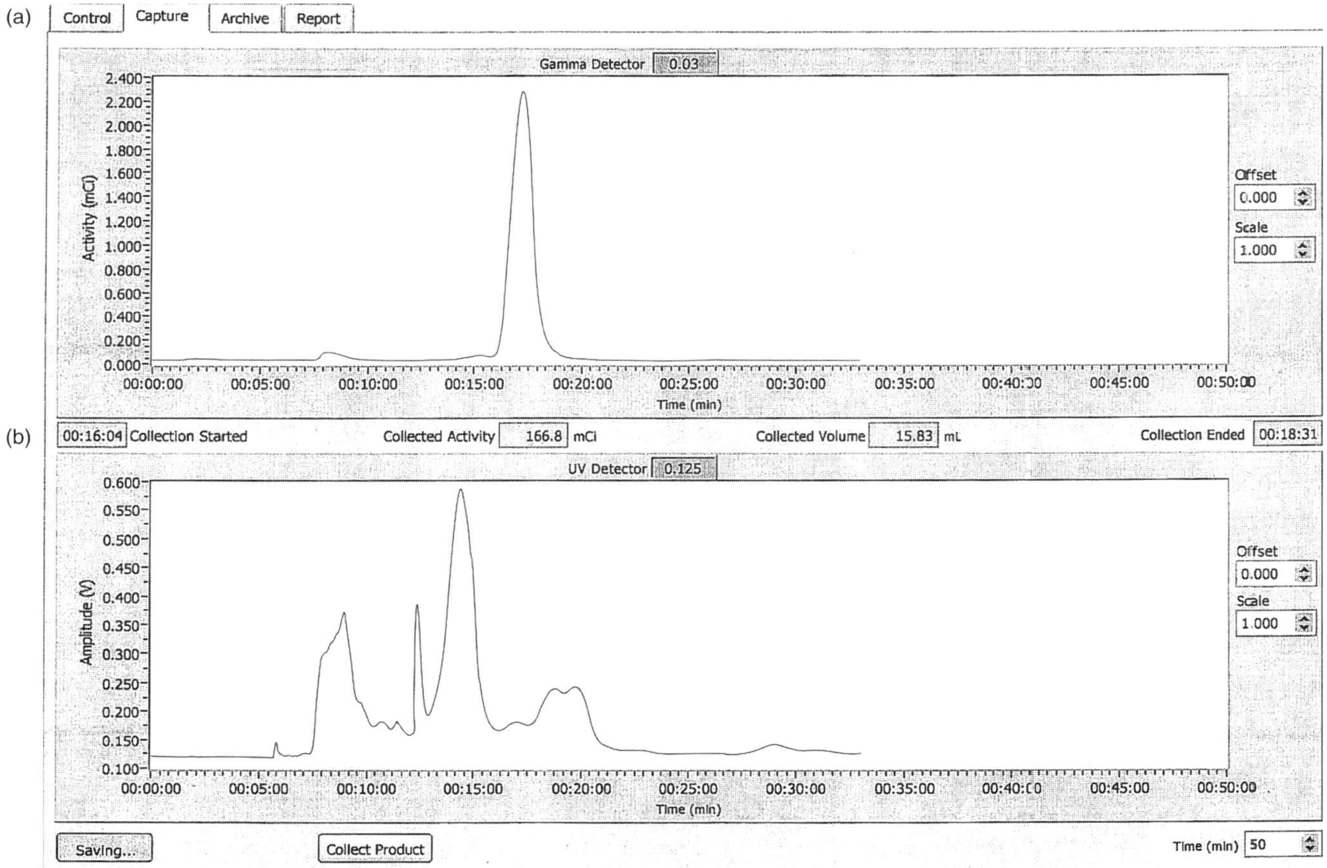

FIGURE 2 HPLC chromatogram of [^{18}F]FES crude reaction mixture. (a) Radioactivity and (b) UV absorption at 280 nm.

TABLE 2 Typical Amounts of [^{18}F]Fluoride and Percentage RCYs of [^{18}F]FES

	^{18}F Activity Ci (GBq)			
Lot No.	$^{18}F^-$ at Start of Synthesis	End of Synthesis as Crude [^{18}F]FES	Postpurification as [^{18}F]FES[a]	Radiochemical yield (RCY, decay corrected) (%)
1	1.5 (54.4)	0.9 (33.9)	0.37 (13.7)	41.0
2	0.8 (30.0)	0.3 (11.5)	0.18 (6.5)	33.2
3	1.4 (53.0)	0.6 (21.0)	0.22 (8.1)	25.8

[a]Postsampling activity of [^{18}F]FES formulated in 15–18% sterile EtOH in SWFI; specific activity of [^{18}F]FES = 250–300 GBq/µmol.

TABLE 3 Stability Lot Data for [^{18}F]FES in High Activity (>2 GBq) Formulation[a]

QC Test	Elapsed Time (h)	Batch 1 Upright	Batch 1 Inverted	Batch 2 Upright	Batch 2 Inverted	Batch 3 Upright	Batch 3 Inverted
Appearance	0	Clear	Clear	Clear	Clear	Clear	Clear
	4	Clear	Clear	Clear	Clear	Clear	Clear
	8	Clear	Clear	Clear	Clear	Clear	Clear
	24	Clear	Clear	Clear	Clear	Clear	Clear
pH	0	6.5	6.5	6.5	6.5	6.5	6.5
	4	6.5	6.5	6.5	6.5	6.5	6.5
	8	6.5	6.5	6.5	6.5	6.5	6.5
	24	6.5	6.5	6.5	6.5	6.5	6.5
Radiochemical identity (RT)[a]	0	N/D	N/D	16.56	16.56	16.15	16.15
	4	16.71	16.36	16.80	16.42	16.68	N/D
	8	16.64	16.71	16.63	16.71	16.63	16.60
	24	Decayed	Decayed	Decayed	Decayed	Decayed	17.0
Radiochemical purity (%)	0	100	100	100	100	98.8	98.8
	4	100	99.9	100	100	98.9	99.9
	8	100	100	100	100	100	100
	24	Decayed	Decayed	Decayed	Decayed	Decayed	100

[a]Radiochemical identity of stability test samples was confirmed using HPLC as described in Section 2.4 by comparing the retention time (RT) of [^{18}F]FES, which is ±5% of the retention time for standard FES (N/D = not done).

three representative lots. These procedures follow the guidelines for F-18 labeled fluorodeoxyglucose ([^{18}F]FDG) from United States Pharmacopeia [36]. All three batches met the QC specificities of the regulatory protocols.

4.1 Prerelease QC Tests

4.1.1 Appearance

The appearance of the [^{18}F]FES product is determined by visual inspection of the product vial, that is placed behind a lead shield and handled from behind this

TABLE 4 Prerelease and Retrospective Quality Control Tests and Data on Three Lots of Clinical [^{18}F]FES Syntheses

QC Tests	Specifications	Lot 1	Lot 2	Lot 3
Appearance	Clear, colorless, no particulate	Pass	Pass	Pass
Radioactivity assay (MBq/mL)	>50	935	442	473
Radiochemical identity (TLC)	R_f of [^{18}F]FES is ±0.05 of FES_{STD} (UV)	$R_{f[18F]FES}$ = 0.60 $R_{f\ STD}$ = 0.60	$R_{f[18F]FES}$ = 0.61 $R_{f\ STD}$ = 0.61	$R_{f[18F]FES}$ = 0.63 $R_{f\ STD}$ = 0.66
Radiochemical purity (TLC) (%)	>90% [^{18}F]FES <5% [^{18}F]F$^-$	[^{18}F]FES = 97% [^{18}F]F$^-$ = 2.1%	[^{18}F]FES = 98% [^{18}F]F$^-$ = 1.4%	[^{18}F]FES − 98% [^{18}F]F$^-$ = 2.0%
pH	5.5–7.5	5.5	5.5	5.5
Filter integrity (KPa)	>30	>30	>30	>30
Radionuclidic identity ($t_{1/2}$) (min)	105–115	111.5	109.5	110
Residual K-2.2.2 (μg/mL)	<50	Pass	Pass	Pass
Residual CH_3CN (% w/v)	<0.04	Pass	Pass	Pass

TABLE 5 Summary of Retrospective QC Test Specifications for [^{18}F]FES

Test	Method	Specification	Lot 1	Lot 2	Lot 3
Endotoxin Test	Bacterial endotoxin test USP on pool	<175 EU/6 mL	Pass	Pass	Pass
Sterility	Sterility test USP on pool	Sterile	Pass	Pass	Pass

TABLE 6 Summary of Supporting Characterization Tests

Test	Method	Preliminary Specification
Radionuclidic purity	γ-Spectrometry	>99.5% 0.511 Mev+1.02 Mev+associated Compton
Chemical purity:		
Estratriol	HPLC	<3.6 μg/dose
FES	HPLC	<3.6 μg/dose

shield using tongs. The product solution is swirled to distribute any settled particles and release entrained bubbles prior to viewing. The product solution must be clear, colorless and not have any visual particulates present to pass the test.

4.1.2 Radioactivity Assay

Radioactivity in the product solution is determined by assay in a standardized radioisotope dose calibrator, and the radioactive concentration is calculated from the total volume in the product vial [37]. The specifications for this assay will vary by users, and our product must contain 0.05 GBq/mL of [^{18}F]FES (at least one patient dose) to pass the test.

4.1.3 Radiochemical Purity (RCP) and Identity

RCP of [^{18}F]FES is determined by micro TLC [38] (Whatman® silica gel, 7.5 × 2.5 cm; 250 μm thickness). RCP check by TLC has an advantage over HPLC test for the purified material since postdevelopment scan of the TLC plate offers a complete analytical profile of the "prerelease" material, while HPLC demonstrates only an elution profile. Any chemical or radiochemical impurity, if present in the purified material and retained by the guard and HPLC columns, would never be noticed using an HPLC check.

Two spots of standard FES (3–5 μg) are applied at the origin. Purified [^{18}F]FES solution is also applied at the same position as one of the spots to validate the development of standard FES and the corresponding radiopharmaceutical [^{18}F]FES under the same chromatographic development conditions. The TLC plate is air dried (nearly 2 min) and then developed using $CHCl_3$: MeOH (8:1 v/v). After drying, the plate is analyzed for radioactive components in the purified product using a BioScan AR 2000 TLC scanner [39]. Radioactive regions are integrated for fluoride, FES, and any other component that may be present in the purified [^{18}F]FES sample.

Two well-defined peaks typically contain radioactivity: the peak representing radiofluoride appears at $R_f = 0.15$ (±0.03) and [^{18}F]FES appears at $R_f = 0.65$ (±0.05; Fig. 3) (the distance traveled is counted from the bottom of the plate by an AR 2000 scanner, and not from the base line). The chromatogram must

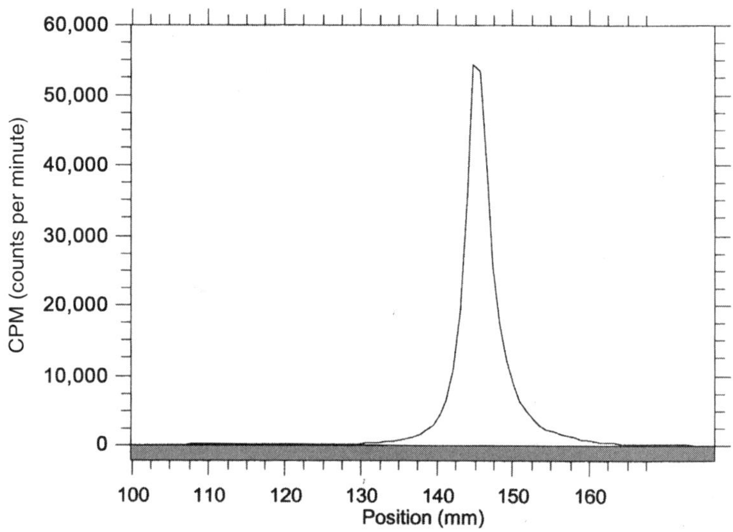

FIGURE 3 TLC chromatogram of pure [^{18}F]FES.

contain ≥90% of the total plate radioactivity as [^{18}F]FES, and the origin peak ([^{18}F]fluoride) area must contain ≤5% of the total plate radioactivity to pass this test. The position of nonradioactive FES on the plate is determined using a UV lamp, and the R_f is compared with the radioactive peak position. The identity of the radioactive peak is confirmed as [^{18}F]FES if the R_f values for radioactive and nonradioactive standards are the same, ±0.05 units.

4.1.4 pH Determination

The purified [^{18}F]FES product solution (2 μL) is applied to a small section of commercially available narrow-range pH (ColorpHast™) paper [38], and the color of the wetted area is immediately compared to the reference color indicator chart supplied with each lot of pH paper. The product pH must be in the range of pH 5.0–7.5 to pass this test.

4.1.5 Filter Integrity

The terminal Millex LG sterile filter originally connected to the product vial assembly is flushed with SWFI (15 mL) from a 20-mL disposable syringe and then attached to the filter integrity test gas supply/regulator/pressure gauge assembly. The gas pressure is slowly increased to a flow of 90 mL/min by adjusting the gas regulator control until a steady stream of gas bubbles emanate from the filter outlet tubing immersed in the beaker of SWFI [21]. The pressure gauge reading at this point must be greater than the specific bubble point (30 KPa) to pass this test.

4.1.6 Radionuclidic Identity

Radionuclidic identity is confirmed by the measurement of decay half-life. Radioactivity in the product solution is determined at two periods by assay in a standardized radioisotope dose calibrator [40], and the times and activities were used to calculate the half-life. The determined half-life value must be between 105 and 115 min to pass the test.

4.1.7 Residual Kryptofix-2.2.2

Residual K-2.2.2 in the product is determined by comparative TLC using a spot test procedure. An analytical TLC plate (Whatman silica gel, 7.5 × 2.5 cm, 250 μm thickness) is spotted at the origin with purified [^{18}F]FES product sample in one lane and an equal volume of standard K-2.2.2 solution (50 μg/mL) in the other lane. After the development with 30% ammonium hydroxide : methanol (9:1 v/v) or acetonitrile : water (95:5 v/v), the plate is removed from the chamber, dried with air flow, and transferred to a closed glass chamber containing iodine vapor (over iodine crystals), which reacts with K-2.2.2 in a concentration-dependent manner to form a colored complex. The visual position and intensity of the standard K-2.2.2 sample lane is compared to that of the product sample [40]. The product sample lane color must be of lesser intensity than that of the standard K-2.2.2 lane to pass the test.

4.2 Retrospective Quality Control (QC) Tests

4.2.1 Sterility

A pooled* sample of radiation-decayed [^{18}F]FES product solution is prepared by sequentially withdrawing 0.05 mL from each individual lot sample into a single 1-mL syringe and redistributing the total volume equally into two separate, sterile, pyrogen-free, septum-sealed, capped, labeled vials. One of the vials is sent to an approved contract testing laboratory [41] for the conduct of the USP sterility test. The second vial is retained as a backup for future reference. The product must not show any growth of anaerobes, aerobes, microaerophiles, and fastidious and nonfastidious microorganisms over a period of two weeks to pass the USP sterility test. All FES samples have passed this test.

**Note 3: The use of pooled samples in sterility testing may not be permitted in all jurisdictions. Country and local regulations should be consulted to ensure appropriate testing procedures are used.*

4.2.2 Bacterial Endotoxin Test (BET)

The bacterial endotoxin content of the samples is determined by testing a pool of radiation-decayed samples using the USP BET methodology. The pools are prepared by withdrawing 0.1 mL of [^{18}F]FES product solution from each individual lot sample and redistributing the total volume equally into 2 separate, sterile, pyrogen-free, septum-sealed, capped, labeled vials. The samples were sent to a commercial laboratory [41] for the conduct of the USP BET. The product must contain $\leq$175 endotoxin units (EU)/6 mL to pass the USP BET test.

Note 4: The use of pooled samples in bacterial endotoxin testing may not be permitted in all jurisdictions. Country and local regulations should be consulted to ensure that appropriate testing procedures are used.

4.3 Supporting Characterization Tests

4.3.1 Radionuclidic Purity

The radionuclidic purity of the cyclotron-produced [^{18}F]fluoride is determined periodically by γ-spectrometry. Potential radionuclides from impurities in the target $H_2{}^{18}O$ water, and released from the metal targets and foils, can be detected by their characteristic γ-ray emissions. A test sample of $^{18}F^-$ in $H_2{}^{18}O$ is placed in a calibrated Na(Tl) or solid-state crystal-based detector, and the γ-ray emissions from the sample are detected and analyzed to produce a spectrum of γ-ray energies and associated intensities [42]. The sample γ-ray spectrum must have γ-rays (>99.5%) from ^{18}F decay known to be 0.511 MeV (and from methodological inherent sum peaks at 1.02 MeV and associated Compton peaks) to pass the test.

4.3.2 Chemical Test for Residual Acetonitrile

Residual acetonitrile in the [^{18}F]FES synthesis product for each batch is determined retrospectively by standard methodology with a gas chromatograph (GC)

[43]. An aliquot of the product sample is analyzed by injection onto a gas GC containing an EC-WAX open tubular column [44] (30 m) operated at 40–250°C (ramp) over 3 min using a helium carrier gas (mobile phase). The column eluent contents are detected using a hydrogen-ion-fed flame ionization detector. The signals are recorded and analyzed by the GC system to produce a quantitative chromatographic profile. Based on the calibration curves prepared from known concentration standards, the sample amount associated with acetonitrile is quantified.

4.3.3 Chemical Purity Test for Estratriol and FES

Estratriol may be present as a result of the hydrolysis of excess unreacted precursor, while "cold" FES is formed during [^{18}F]FES synthesis as a result of the reaction with cold ^{19}F, which is always present as an isotopic impurity and a diluent in ^{18}F samples. These products were determined by analysis using an HPLC separation method with UV detection. The corresponding sample peak areas were compared to a reference calibration curve for each compound for quantitative assessment of sample content [38]. These chemical components are acceptable provided the concentrations are below 3.6 μg per patient dose. The manufacture process is validated by collecting data for three consecutive manufacture lots that meet required specifications. These tests are performed every six months to maintain the consistency of the manufacturing quality (approved by Health Canada).

CHEMICAL ABSTRACTS NOMENCLATURE (REGISTRY NUMBER)

Acetonitrile (75-05-8)
Ammonium hydroxide (1336-21-6)
Carbonic acid, potassium salt (1:2)(584-08-7)
Estra-1,3,5(10)-triene-3,17-diol, 16-fluoro-, (16β,17β)- (9CI) (84693-92-5)
Ethanol (64-17-5)
4,7,13,16,21,24-Hexaoxa-1,10-diazabicyclo[8.8.8]hexacosane (23978-09-8)
Hydrochloric acid (7647-01-0)
Methanol (67-56-1)

REFERENCES AND NOTES

1. Michaelis LC, Ratain MJ. Nat Rev Cancer 2006;6:409–414.
2. Wachter, K.OB/GYN News. Oct 2003.
3. Dehdashti F, Flanagan FL, Mortimer JE, Katzenellenbogen JA, Welch MJ, Siegel BA. Eur. J. Nucl. Med. Mol. Imaging. 1999;26(1):51–56.
4. Kumar R. Eur J Nucl Med Mol Imaging 2007;34(3):346–353.
5. Tonkin K, Sponsor, Alberta Health Services, Study to determine the Utility of FES-PET and FDG-PET in the Prediction of Response to Hormone Therapy in Women with Estrogen Positive Metastatic Breast Cancer. http://clinicaltrials.gov/ct2/show/NCT00358098.

6. Kumar P, Mercer JR, Doerkson C, Tonkin K, McEwan AJB. J Pharm Pharm Sci 2007;10(2): 256–265.
7. Kieswetter DO, Kilbourne MR, Landvatter SW, Heiman DF, Katzenellenbogen JA, Welch MJ. J Nucl Med 1984;25:1212–1221.
8. Kieswetter DO, Katzenellenbogen JA, Kilbourne MR, Welch MJ. J Org Chem 1984;49: 4900–4905.
9. Romer J, Fuchtner F, Steinbach J, Johannsen B. Automated synthesis of 16-α-[^{18}F]Fluoroestradiol for breast cancer imaging. Nucl Med Biol 1999;26:473–479.
10. Lim JL, Zheng L, Berridge MS, Tewson TJ. Nucl Med Biol 1996;23:911–915.
11. Berridge MS, Rosenfeld P, Franceschini MP, Tewson TJ. J Org Chem 1990;55:1211–1216.
12. Tewson TJ, Mankoff DA, Peterson LM, Woo I, Petra P. Nucl Med Biol 1999;26:905–913.
13. Romer J, Steinbach J, Kasch H. Appl Radiat Isot 1996;47:395–399.
14. Virgin [^{18}O]H_2O was purchased from Huyai Isotopes, Inc. Canada, and used as received.
15. Cyclotron was purchased from Advanced Cyclotron Systems, Inc., Burnaby, Canada.
16. Niobium target was purchased from Advanced Cyclotron Systems, Inc., Burnaby, Canada.
17. Double train TRACERlab FX_{FDG}™ synthesis unit was purchased from GE International, USA.
18. Class 100 hot cells were purchased from Comecer, Italy.
19. Anion exchange QMA cartridges were purchased from Macherey-Nagel, Germany.
20. Potassium carbonate was purchased from Sigma-Aldrich, USA, and then sterile repacked in required concentrations by Edmonton Radiopharmaceutical Centre, Canada.
21. Sterile water for injection (10-mL sterile vials and 250-mL sterile bags) was purchased from Braun, USA.
22. Nitrogen gas was supplied by Praxair, Canada.
23. Kryptofix-2.2.2 was purchased from Sigma-Aldrich, Canada, and then sterile repacked in specified concentrations as an acetonitrile solution by Edmonton Radiopharmaceutical Centre, Edmonton, Canada.
24. Acetonitrile was purchased from Sigma-Aldrich, Canada, in bulk, and then repackaged by Edmonton Radiopharmaceutical Center in the amounts specific to the reagent.
25. FES precursor was synthesized and supplied in sterile packages by Edmonton Radiopharmaceutical Center, Edmonton, Canada.
26. HCl solution (10 N) was bought from J.T. Baker, USA, diluted to required concentrations in specified experimental solvents, and then sterile packed by Edmonton Radiopharmaceutical Centre, Edmonton, Canada.
27. EtOH for human use was purchased from Sabex Inc., Canada.
28. Sterile Millex LG filters were purchased from Millipore, USA.
29. Quark Technologies, Australia.
30. Phenomenex, Inc., USA.
31. Beckman, Inc., USA.
32. Edmonton Radiopharmaceutical Centre, Edmonton, Canada.
33. Twenty-milliliter multidose sterile vials were purchased from Mallinckrodt, USA.
34. Sterile saline was purchased as 10-mL packs from BD, USA.
35. Unpublished results from EPC clinical manufactures.
36. US Pharmacopoeia <823> radiopharmaceuticals for positron emission tomography compounding. USP 32–NF 27 (2009).
37. Unpublished procedures of CCI regulatory documents.
38. Whatman® silica gel as per EPC manufacturing specifications.
39. Bioscan, Inc., USA.
40. Dose calibrator purchased from Capintec, USA.
41. Keystone Labs, Inc., Canada.
42. Ortec multichannel γ-spectrum analyzer, USA.
43. SRI8610C Gas chromatograph was purchased from SRI, Inc., USA.
44. EC-WAX open tubular column was purchased from Alltech, USA.

CHAPTER 9

SYNTHESIS OF *N*-SUCCINIMIDYL 4-[^{18}F]FLUOROBENZOATE ([^{18}F]SFB)

XIA SHAO

Department of Radiology, University of Michigan Medical School, Ann Arbor, Michigan

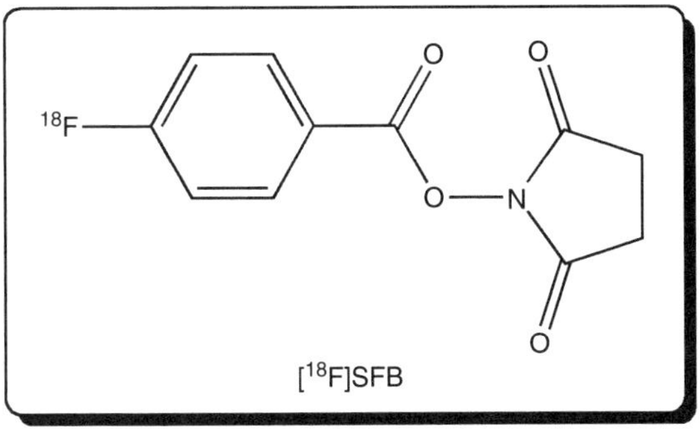

1 INTRODUCTION

N-Succinimidyl 4-[^{18}F]fluorobenzoate ([^{18}F]SFB) is one of the most popular bifunctional prosthetic groups used as acylation agents for radiolabeling of peptides [1], proteins [2], and antibodies [3,4]. Such bioactive compounds when labeled with a 4-[^{18}F]fluorobenzoyl group can be useful for in vivo studies of physiological processes by positron emission tomography (PET) [5].

Initially the preparation of [^{18}F]SFB was achieved by lengthy syntheses requiring multiple reaction vessels and was done manually [6,7]. With the increasing utilization of this compound, several improved methods as well as automated approaches have been developed [5,8–10]. We adapted the three-step, one-pot method for preparing [^{18}F]SFB, reported by Kabalka [8], for use with a General Electric Medical Systems (GEMS) TRACERlab FX_{FN}. This method uses solid-phase extraction (SPE) cartridges, thereby eliminating high performance liquid chromatographic (HPLC) purification, and has proven extremely robust in our hands.

Radiochemical Syntheses, Volume 1: Radiopharmaceuticals for Positron Emission Tomography, First Edition. Edited by Peter J. H. Scott, Brian G. Hockley.

2 SYNTHESIS PROCEDURES

CAUTION: All radiochemical syntheses must be carried out using appropriate equipment in a facility authorized for the use of radioactive materials. Personal protective equipment must be worn and all local radiation safety laws followed.

2.1 Production of [^{18}F]Fluoride

[^{18}O]H_2O (2 mL) [11] was loaded into the [^{18}F]fluoride target of a GEMS PET-Trace cyclotron [12]. The target was bombarded (40-μA beam for 30 min) to generate ~1.5 Ci (55.5 GBq) of [^{18}F]fluoride by the ^{18}O(p,n)^{18}F nuclear reaction.

2.2 Azeotropic Drying of [^{18}F]Fluoride

The [^{18}F]fluoride was delivered to a GEMS TRACERlab FX_{FN} synthesis module [13]as a solution in [^{18}O]H_2O (2 mL). This solution was passed through a QMA Light Sep-Pak® cartridge [14] to trap the [^{18}F]fluoride and recycle the [^{18}O]H_2O. The [^{18}F]fluoride was then eluted into the TRACERlab FX_{FN} glassy-carbon reaction vessel using a solution of aqueous potassium carbonate (3.5 mg in 0.5 mL H_2O) [15]. A solution of Kryptofix-[2.2.2] (15 mg) [16] in MeCN (1 mL) was added, and the reaction mixture was azeotropically dried, initially at 80°C under vacuum for 4 min and subsequently at 60°C with both vacuum and argon flow for an additional 4 min.

2.3 Synthesis of [^{18}F]SFB

Following drying, the trimethylbenzenaminium triflate precursor (**1**) (5 mg) [17] in 0.5 mL DMSO [18] was added to the reaction vessel, and the fluorination reaction was heated at 90°C for 10 min to provide [^{18}F]ethyl 4-fluorobenzoate (**2**). After this time, tetrapropylammonium hydroxide [19] (TPAH, 1.0 M solution in water, 20 μL in 0.5 mL MeCN) was added to saponify the ester. The saponification reaction was heated at 120°C for 3 min to yield [^{18}F]4-fluorobenzoic acid (**3**) as the corresponding tetrapropylammonium salt. On completion of the saponification, MeCN (1 mL) was added and evaporated to remove excess water from the addition of the aqueous solution of TPAH. *O*-(*N*-Succinimidyl)-*N,N,N′,N′*-tetramethyluronium tetrafluoroborate [20] (TSTU, 10 mg in 0.6 mL MeCN) was then added, and the reaction vessel was heated at 90°C for 5 min to provide [^{18}F]SFB (**4**) (Fig. 1).

2.4 Purification of [^{18}F]SFB

The crude reaction mixture was cooled to 40°C and transferred from the reactor to a dilution flask containing 1.5% acetic acid (20 mL) [21]. The resulting solution was transferred through a Waters C18 Plus Sep-Pak cartridge [22] to trap the crude reaction components. Following trapping, the C18 Sep-Pak was washed with 10% MeCN (10 mL) to elute unwanted hydrophilic impurities including unreacted fluoride-18 to waste. Subsequently, [^{18}F]SFB was eluted off into a collection vial with MeCN (2 mL). The [^{18}F]SFB was either used as the obtained solution in MeCN or evaporated to dryness and redissolved in a solvent more

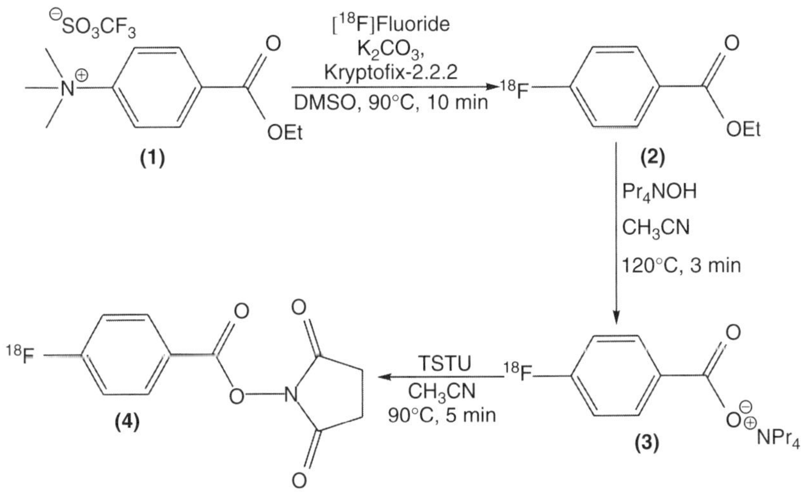

FIGURE 1 Synthesis of [^{18}F]SFB.

appropriate for the subsequent coupling reactions (e.g., DMSO). Typical yields of [^{18}F]SFB are 38% non-decay corrected.

3 QUALITY CONTROL PROCEDURES

CAUTION: SFB is not being used directly for clinical imaging protocol. Quality control procedures for the intermediate must be carried out by trained personnel, and each dose must meet all established quality control criteria before release as prosthetic group.

3.1 Radiochemical Identity

Radiochemical identity was determined using a Shimadzu LC-2010 HPLC equipped with a Bioscan FC3300 radioactivity detector and UV detector under the following conditions: column: Chromolith RP18, 100 × 4.6 mm [23]; mobile phase: 0.1% TFA in 50% MeOH; flow rate: 0.8 mL/min; UV: 254 nm. The retention time of [^{18}F]SFB was compared to that of nonradioactive [^{19}F]SFB reference standard [24]. Our internal specification for release stipulates that the difference in retention times between the radioactive and UV signals had to be within ±10%.

3.2 Radiochemical Purity

Radiochemical purity was analyzed using the HPLC system described in Section 3.1. Radiochemical purity had to be greater than 95%. Representative analytical HPLC tracers are displayed in Figure 2.

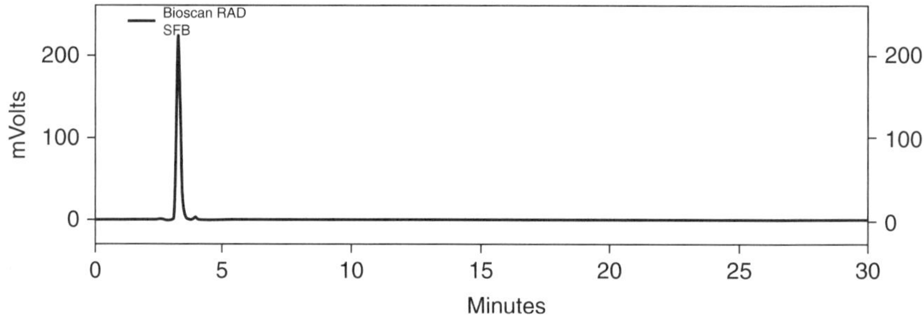

FIGURE 2 Typical radiochemical HPLC trace of SPE-purified [^{18}F]SFB.

3.3 Specific Activity

The concentration of [^{18}F]SFB can be determined using the HPLC system described in Section 3.1. Specific activity had to be greater than 1 Ci/μmol.

3.4 Radionuclidic Identity

Activities were measured using a Capintec CRC®-15R Radioisotope Dose Calibrator [25], and half-life was calculated using Equation 1. Calculated half-life must be 105–115 min.

$$t_{1/2} = -\ln 2(\text{time difference}/(\ln(\text{ending activity}/\text{starting activity}))) \quad (1)$$

CHEMICAL ABSTRACTS NOMENCLATURE (REGISTRY NUMBER)

Acetonitrile (75-05-8)

Benzenaminium, 4-(ethoxycarbonyl)-*N,N,N*-trimethyl-, 1,1,1-trifluoromethanesulfonate (1:1) (124915-06-6)

Benzoic acid, 4-(fluoro-^{18}F)-, 2,5-dioxo-1-pyrrolidinyl ester (141-762-27-8)

Benzoic acid, 4-(fluoro-^{18}F)-, ethyl ester(124915-09-9)

Carbonic acid, potassium salt (1:2)(584-08-7)

Ethanol (64-17-5)

4-Fluorobenzoate (1:1) (2365-27-7)

4,7,13,16,21,24-Hexaoxa-1,10-diazabicyclo[8.8.8]hexacosane(23978-09-8)

Methanaminium, *N*-[(dimethylamino)[(2,5-dioxo-1-pyrrolidinyl)oxy]methylene]-*N*-methyl-, tetrafluoroborate(1-) (1:1) (105832-38-0)

Methane, 1,1′-sulfinylbis- (67-68-5)

1-Propanaminium, *N,N,N*-tripropyl-, hydroxide (1:1) (4499-86-9)

Sodium hydroxide (1310-73-2)

REFERENCES AND NOTES

1. Wust F, Hultsch C, Bergmann R, Johannsen B, Henle T. Radiolabeling of isopeptide N-(r-glutamyl)-L-lysine by conjugation with N-succinimidyl-4-[18F]fluorobenzoate. Appl Radiat Isot 2003;59:43–48.
2. Wester HJ, Hamacher K, Stocklin G. A comparative study of n. c. a. fluorine-18 labeling of proteins via acylation and photochemical conjugation. Nucl Med Biol 1996;23:365–372.
3. Vaidyanathan E, Bigner DD, Zalutsky MR. Fluorine-18 labeled monoclonal antibody fragments: a potential approach for combining radioimmunoscintigraphy and positron emission tomography. J Nucl Med 1992;33:1535–1541.
4. Zijlstra S, Gunawan J, Burchert W. Synthesis and evaluation of a ^{18}F-labelled recombinant annexin-V derivative, for identification and quantification of apoptotic cells with PET. Appl Radiat Isot 2003;58:201–207.
5. Mading P, Fuchtner F, Wust F. Module-assisted synthesis of the bifunctional labelling agent N-succinimidyl 4-[^{18}F]fluorobenzoate ([^{18}F]SFB). Appl Radiat Isot 2005;63:329–332.
6. Vaidyanathan G, Zalutsky MR. Synthesis of N-succinimidyl 4-[^{18}F]fluorobenzoate, an agent for labeling proteins and peptides with ^{18}F. Nat Protoc 2006;1(4):1655–1661.
7. Okarvi SM. Recent progress in fluorine-18 labelled peptide radiopharmaceuticals. Eur J Nucl Med 2001;28:929–938.
8. Tang G, Zeng W, Meixiang Y, Kabalka G. Facile synthesis of N-succinimidyl 4-[18F]fluorobenzoate ([18F]SFB) for protein labeling. J Labelled Comp Radiopharm 2008;51: 68–71.
9. Johnstrom P, Clark JC, Pickard JD, Davenport AP. Automated synthesis of the generic peptide labelling agent N-succinimidyl 4-[^{18}F]fluorobenzoate and application to ^{18}F-label the vasoactive transmitter urotensin-II as a ligand for positron emission tomography. Nucl Med Biol 2008;35:725–731.
10. Glaser M, Årstad E, Luthra SK, Robins EG. Two-step radiosynthesis of [18F]N-succinimidyl-4-fluorobenzoate ([18F]SFB). J Labelled Comp Radiopharm 2009;52:327–330.

11. Enriched [^{18}O]H_2O was purchased from ABX, Germany, or Medical Isotopes, USA, and used as received.
12. GEMS silver high yield [^{18}F]fluoride target.
13. GE Healthcare, USA.
14. QMA-Light Sep-Pak cartridges were purchased from Waters (part no. WAT023525) and conditioned with 10 mL ethanol, 10 mL water, 10 mL 0.5 M sodium bicarbonate, and a further 10 mL water before use.
15. Potassium carbonate was purchased from Aldrich (part no. 209619) and used as received. Sterile water was purchased from Hospira (part no. 0409-4887-50) and used as received.
16. Kryptofix-[2.2.2] was purchased from Acros (part no. 29195-0010) and used as received. Anhydrous acetonitrile was purchased from Acros (part no. 61096-1000) and used as received.
17. Trimethylbenzeneaminium triflate precursor was purchased from ABX, Germany (product no. 4391.0100).
18. Anhydrous DMSO was purchased from Acros (part no. 61042-10000) and used as received. We noticed a gradual decline in yield throughout the lifetime of a bottle of DMSO and stress the importance of using anhydrous solvents in this synthesis.
19. Tetrapropylammonium hydroxide was purchased from Sigma-Aldrich, Switzerland (part no. 254533), and used as received.
20. *O*-(*N*-Succinimidyl)-*N,N,N′,N′*-tetramethyluronium tetrafluoroborate was purchased from Sigma-Aldrich, USA (part no. 385530), and used as received.
21. Acetic acid was purchased from Sigma-Aldrich, USA (part no. 242853) and diluted to 1.5% (v/v) with water.
22. Waters C18 Plus Sep-Pak cartridges were purchased from Waters (part no. WAT020515) and conditioned with 10 mL ethanol and 10 mL water before use.
23. HPLC column was purchased from Merck, Germany (part no. 1021290001).
24. *N*-Succinimidyl 4-fluorobenzoate was purchased from ABX, Germany (product no. 4392.0100).
25. Capintec, Inc., USA.

CHAPTER 10

SYNTHESIS OF 4-(2′-METHOXYPHENYL)-1-[2′-(*N*-2″PYRIDINYL)-*p*-[^{18}F]FLUOROBENZAMIDO]ETHYLPIPERAZINE [^{18}F]MPPF

MARION ALVAREZ AND DIDIER LE BARS

CERMEP-Imagerie du Vivant, Hospices Civils de Lyon, ICBMS UMR, Université Lyon 1, Bron (Lyon), France

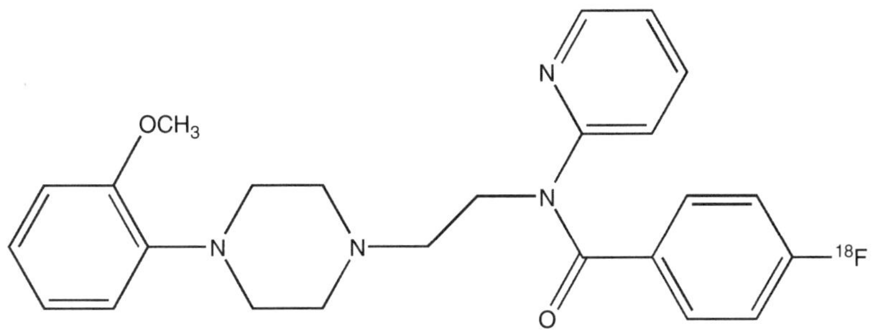

[^{18}F]MPPF

1 INTRODUCTION

4-(2′-Methoxyphenyl)-1-[2′-(*N*-2″-pyridinyl)-*p*-[^{18}F]fluorobenzamido]ethylpiperazine (*p*-[^{18}F]MPPF) is a radioligand developed for positron emission tomographic (PET) imaging of serotonin-1A (5-HT$_{1A}$) receptors in the central nervous system. It is a selective 5-HT$_{1A}$ radiotracer labeled with fluorine-18, a positron emitter with a physical period of 109 min [1].

The serotonin (5-hydroxytryptamine (5-HT)) neurotransmission system is involved in the modulation of various important physiologic functions and behavior, such as thermoregulation, cardiovascular function, aggressive and sexual behavior, mood, appetite, and the sleep–wake cycle. The effects of 5-HT are mediated by as many as seven classes of receptor populations (5-HT$_1$–5-HT$_7$),

Radiochemical Syntheses, Volume 1: Radiopharmaceuticals for Positron Emission Tomography, First Edition. Edited by Peter J. H. Scott, Brian G. Hockley.

many of which also contain several subtypes. The 5-HT$_{1A}$ subtype of the G-protein-coupled 5-HT$_1$ receptor family is located primarily in the limbic forebrain (the hippocampus, entorhinal cortex, and septum). 5-HT$_{1A}$ appears to function both as a presynaptic (somatodendritic) autoreceptor in the raphe nuclei and as a postsynaptic receptor in the terminal fields. This receptor subtype is involved in the modulation of emotion and the function of the hypothalamus, and is implicated in the pathogenesis of anxiety, depression, hallucinogenic behavior, motion sickness, dementia, schizophrenia, and eating disorders.

Many compounds have been radiolabeled and studied for visualization and quantification of these receptors by PET or single photon emission computed tomography (SPECT). WAY 100635 has been developed as a highly selective, silent antagonist of 5-HT$_{1A}$ receptors at both pre- and postsynaptic sites and labeled with carbon-11 for PET [2]. Zhuang et al. [3,4] first developed a series of arylpiperazine-benzamido derivatives and 4-(2′-methoxyphenyl)-1-[2′-[*N*-(2″-pyridinyl)-*p*-iodobenzamido]ethyl]piperazine ([^{125}I]MPPI), which appeared to selectively bind to the 5-HT$_{1A}$ receptors in vitro and in vivo. Zhuang et al. also synthesized a fluoro analog, *p*-MPPF, which displayed a high binding affinity for 5-HT$_{1A}$ receptors and acted as a receptor antagonist in vivo. In 1994, Shiue et al. [5] synthesized *p*-[^{18}F]MPPF for PET studies of 5-HT$_{1A}$ receptors; the synthesis was improved and MPPF evaluated in vitro and in vivo by Le Bars et al. [6,7]). MPPF has since evolved as a major tool for PET exploration of serotonin 5-HT$_{1A}$ system [8].

2 SYNTHESIS PROCEDURES

CAUTION: All radiochemical syntheses must be carried out using appropriate equipment in a facility authorized for the use of radioactive materials. Personal protective equipment must be worn and all local radiation safety laws followed.

2.1 Production of [^{18}F]Fluoride

[^{18}O]H_2O (820 μL) [9] was loaded into the [^{18}F]fluoride target [10] of an IBA Cyclone 18/9 cyclotron. The target was bombarded (16 μA beam for 60 min) to generate ~1 Ci (37 GBq) of [^{18}F]fluoride by the ^{18}O(p,n)^{18}F nuclear reaction.

2.2 Azeotropic Drying of [^{18}F]Fluoride

The [^{18}F]fluoride was delivered to a Coincidence/General Electric Medical Systems (GEMS) TracerLAB MX synthesis module as a solution in [^{18}O]H_2O (820 μL). This solution was passed through a QMA-Light Sep-Pak® cartridge [11] to trap the [^{18}F]fluoride and recycle the [^{18}O]H_2O. The [^{18}F]fluoride was then eluted into the TracerLAB MX reaction vessel with 600 μL of potassium carbonate/K222 eluent [12].

2.3 Synthesis of [^{18}F]MPPF

A standard sterile kit for FDG was used on the GEMS TracerLAB MX, and a second inlet line with a pinch valve was added to the reactor vial. This pinch

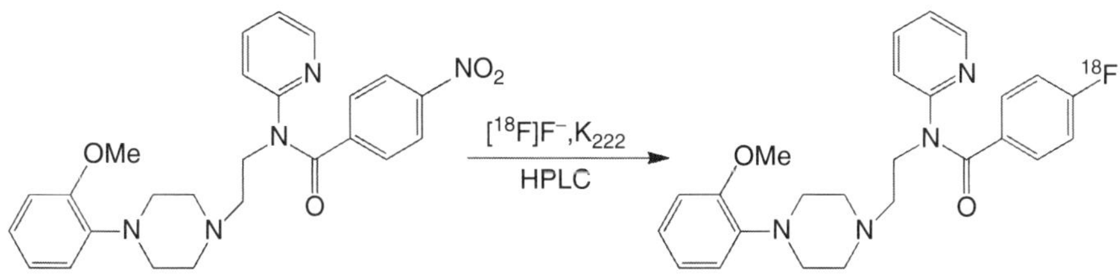

FIGURE 1 Synthesis of [^{18}F]MPPF.

valve was actuated using one of the 24-V logical output signals available on the Synthesizer [7]. After obtaining the dry $^{18}F^{-}$/K222 complex, 10 mg nitro precursor [13] dissolved in DMSO [14] (3 mL) was automatically transferred into the reactor vial through the pinch valve. The labeling reaction was achieved at 175°C for 10 min in the closed reaction vial. The C18 cartridge on the kit (originally intended for alkaline hydrolysis on solid support for FDG) was used for the preliminary purification step after labeling. The reaction mixture was diluted with 20 mL of water [15] and then passed through the preconditioned C18 Sep-Pak cartridge. The labeled precursor was trapped onto the cartridge and rinsed with 10 mL of water to remove impurities such as unreacted $^{18}F^{-}$ and Kryptofix. Elution was achieved with MeOH [16] (1.5 mL) passed through the kit C18 cartridge using the 3 mL syringe originally intended for injection of NaOH into the cartridge for FDG, in a vial containing 750 μL of water, pH 5 (3 mL of acetic acid adjusted with NaOH for 1 L) (Fig. 1).

2.4 Purification and Formulation of [^{18}F]MPPF

After injection into a 6-mL loop, preparative high-performance liquid chromatography (HPLC) was performed on a Waters Symmetry-Prep C18 7 μm, 7.8 × 300 mm [17] column, with water, pH 5, (3 mL acetic acid adjusted with NaOH for 1 L) 725 mL, THF [18] 150 mL, MeOH 250 mL, flow rate = 3 mL/min, and monitored for UV 254 nm and radioactivity, as shown in Fig. 2.

The collected fraction (Rt ∼40 min) is diluted in 40 mL of water, formulated by SPE [19] on absolute ethanol [20] preactivated C18 Sep-Pak, rinsed with 10 mL of water, and eluted with 1 mL of absolute ethanol in a 10 mL sterile vial [21]. Five milliliters of 0.9% saline [22] is finally added before sterile filtration on Millipore 0.22 μm Millex-MP filters [23].

3 QUALITY CONTROL PROCEDURES

CAUTION: All radiopharmaceuticals produced for clinical use must have local regulatory approval before use in humans. Trained personnel must carry out quality control procedures, and each dose must meet all established quality control criteria before release to the clinic (Table 1).

TABLE 1 QC Data for Three Repeat Runs of [^{18}F]MPPF

Quality Control (QC) Test	Release Criteria	Run 1	Run 2	Run 3
Yield mCi (GBq)	NA	65 (2.4)	65 (2.4)	70 (2.6)
Visual inspection	Clear, colorless	Pass	Pass	Pass
Radiochemical identity	0.9 <RRT> 1.1	1.0	1.0	1.0
Radiochemical purity (%)	≥95	≥98	≥98	≥98
Specific activity	>200 mCi/μmol	5.5 Ci/μmol	6.5 Ci/μmol	5.2 Ci/μmol
Residual solvent analysis (ppm)	Methanol	6	7	BDL
	THF	10	27	14
	DMSO	63	0.5	10
	CH_3CN	BDL	BDL	BDL
Dose pH	4.5–8.5	7	7	7
Residual Kryptofix-[2.2.2] (μg/mL)	<50	<10	<10	<10
Sterile filter integrity test (psi)	>50	Pass	Pass	Pass
Radionuclidic identity ($t_{1/2}$) (min)	105–115	106	110	109
Endotoxin analysis (EU/6 mL)	≤175	<0.05	<0.05	<0.05
Sterility testing	No growth after 14 d	Pass	Pass	Pass

NA, not applicable; BDL, beyond detection limit.

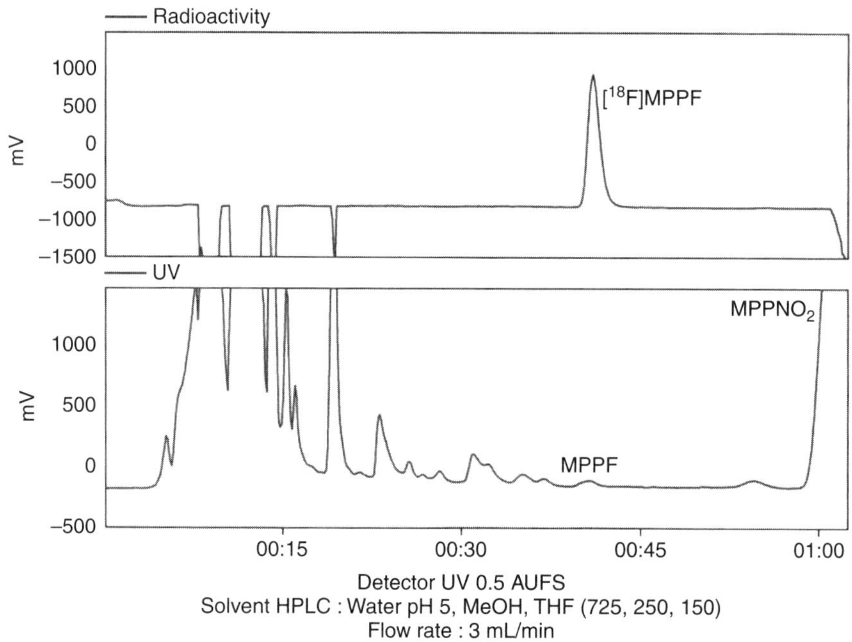

FIGURE 2 Semipreparative UV and radioactive HPLC traces for [^{18}F]MPPF.

3.1 Visual Inspection

The [^{18}F]MPPF dose must be clear, colorless, and free of particulate matter.

3.2 Radiochemical Identity

HPLC analysis of radiochemical identity was conducted using a Beckman Gold 32K Liquid Chromatograph fitted with a 126 LC Pump, 507e injector, 168 UV diode array detector, and Raytest GABI ©-detector (column: Waters Symmetry-Prep C18 5 μm, 4.6 × 30 mm [24]; mobile phase: 50% water pH 5 (3 mL acetic acid adjusted with NaOH), 50% mixture of THF : MeOH 40:60; 0.5 mL/min). [^{18}F]MPPF retention time is compared to the [^{19}F]MPPF reference standard [25] and must be within 10% (relative retention time (RRT) 0.9–1.1).

3.3 Radiochemical Purity

With the same chromatographic system as above, radiochemical purity is determined and must be over 95%. Representative analytical HPLC traces are displayed in Fig. 3.

Although fluorine anion is excluded from the final solution because of preparation conditions, a search for fluorine ions can be performed on thin layer chromatography (TLC) silica plates [26] with developing solvent CH_2Cl_2: MeOH, 93:7; Rf MPPF = 0.4±0.1; and Rf $^{18}F^-$ = 0.

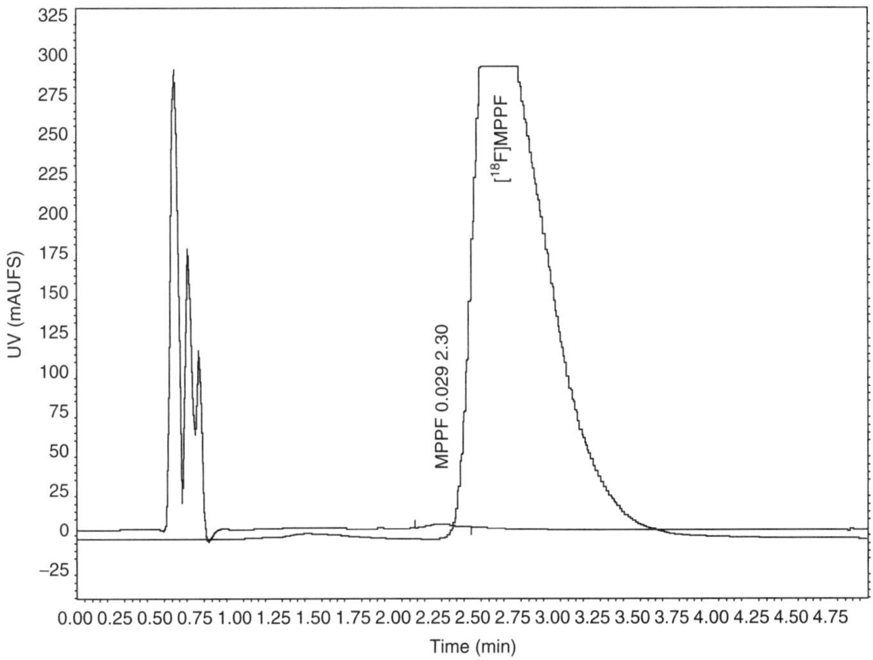

FIGURE 3 Analytical UV and radioactive HPLC trace for [^{18}F]MPPF.

3.4 Specific Activity

A specific activity release limit for [^{18}F]MPPF of 200 mCi/μmol (7.4 GBq/μmol) is in place at CERMEP.

3.5 Residual Solvent Analysis

Analysis of residual solvent levels (DMSO, methanol, THF, acetonitrile) in [^{18}F]MPPF doses was conducted using a Varian 3900 Gas Chromatograph with capillary column CP7424; temperature program: 40°C for 6 min, then 40°C/min up to 200°C for a total of 11 min. Limits imposed by the ICH [27] Harmonised Tripartite Guidelines are <410 ppm/day for class 2 solvents (THF, methanol, acetonitrile) and <5000 ppm/day for class 3 solvents (DMSO).

3.6 Dose pH

The pH of the [^{18}F]MPPF dose was analyzed by applying a small amount of the dose to pH-indicator strips [28] (2.0–9.0) and determined by visual comparison to the scale provided. The pH must be between 4.5 and 8.5.

3.7 Residual Kryptofix-[2.2.2] Analysis

Residual Kryptofix-[2.2.2] level in [^{18}F]MPPF doses is analyzed using the established spot test [29]. Strips of plastic-backed silica gel TLC plates saturated with iodoplatinate reagent are spotted (4 μL) with water (negative control), 50 μg/mL and 10 μg/mL of Kryptofix-[2.2.2] standard (positive controls), and [^{18}F]MPPF dose. If Kryptofix-[2.2.2] is present in a sample, a blue-black spot appears. Spots for the three samples were compared, and a visual determination of residual Kryptofix-[2.2.2] in the [^{18}F]MPPF dose was made. A content <50 μg/mL is acceptable.

3.8 Sterile Filter Integrity Test

The sterile filter from the [^{18}F]MPPF (with needle still attached) was connected to a nitrogen supply via a regulator. The needle was submerged in water, and the nitrogen pressure was gradually increased. If the pressure was raised above the filter acceptance pressure (50 psi) without seeing a stream of bubbles, the filter was considered intact. If a stream of bubble occurs at <50 psi, the test fails.

3.9 Radionuclidic Identity

Activities were measured using a Capintec CRC-15R Radioisotope Dose Calibrator, [30] and half-life was calculated using Equation 1. Calculated half-life must be 105–115 min.

$$t_{1/2} = -\ln 2(\text{time difference}/(\ln(\text{ending activity}/\text{starting activity}))) \quad (1)$$

3.10 Endotoxin Analysis

Endotoxin content in doses of [^{18}F]MPPF was analyzed by independent laboratory according to the European Pharmacopeia [31]. Doses must contain less than 175 Endotoxin Units (EU)/6 mL.

3.11 Sterility Testing

Samples of [^{18}F]MPPF were placed onto fluid thioglycolate (aerobes), trypcase soybean (anaerobes), and Sabouraud (fungi) [32] media. Tubes were incubated with positive and negative controls for 14 days at 37°C according to current European guidelines [33]. Tubes were visually inspected on the third, eighth, and fourteenth days of the test period and compared to the positive and negative standards. Positive standards must show growth (turbidity) on the plates, and [^{18}F]MPPF/negative controls must have no culture growth after 14 days to be indicative of sterility.

CHEMICAL ABSTRACTS NOMENCLATURE (REGISTRY NUMBER)

Acetic acid (64-19-7)

Acetonitrile (75-05-8)

Carbonic acid, potassium salt (584-08-7)

Dimethyl sulfoxide DMSO (67-68-5)

Ethanol (64-17-5)

4-Fluoro-*N*-[2-[4-(2-methoxyphenyl)piperazin-1-yl]ethyl]-*N*-pyridin-2-ylbenzamide (155204-26-5)

4,7,13,16,21,24-Hexaoxa-1,10-diazabicyclo[8.8.8]hexacosane (23978-09-8)

Methanol (67-56-1)

Nitro precursor: 4-nitro-*N*-[2-[4-(2-methoxyphenyl)piperazin-1-yl]ethyl]-*N*-pyridin-2-ylbenzamide (155204-27-6)

THF tetrahydrofurane (29347-89-5)

Sodium hydroxide (1310-73-2)

For detailed supplier information, see Appendix 1 (page 335).

REFERENCES AND NOTES

1. Cheng KT. 4-(2′-Methoxyphenyl)-1-[2′-(N-2″-pyridinyl)-*p*-[^{18}F]fluorobenzamido] ethylpiperazine [*p*-[^{18}F]MPPF]. In: Molecular Imaging and Contrast Agent Database (MICAD) [database online]. Bethesda (MD): National Library of Medicine (US), NCBI; 2004–2010. Available at: http://micad.nih.gov.
2. Pike VW, McCarron JA, Lammertsma AA, Osman S, Hume SP, Sargent PA, Bench CJ, Cliffe IA, Fletcher A, Grasby PM. Eur J Pharmacol 1996;301:5–7.
3. Zhuang ZP, Kung MP, Chumpradit S, Mu M, Kung HF. J Med Chem 1994;37:4572–4575.
4. Zhuang ZP, Kung MP, Kung HF. J Med Chem 1994:37;1406–1407.
5. Shiue CY, Shiue GG, Mozley PD, Kung MP, Zhuang ZP, Kim HJ, Kung HF. Synapse 1997;25:147–154.
6. Le Bars D, Lemaire C, Ginovart N, Plenevaux A, Aerts J, Brihaye C, Hassoun W, Leviel V, Mekhsian P, Weissmann D, Pujol JF, Luxen A, Comar D. Nucl Med Biol 1998;25:343–350.
7. Le Bars D, Bonmarchand G, Alvarez JM, Lemaire C, Mozdzianowski C. J Labelled Comp Radiopharm 2001;44:S1045.
8. Aznavour N, Zimmer L. Neuropharmacology 2007;52:695–707.
9. Enriched water, ROTEM, www.rotemi.co.il.

10. Target IBA Niobium Small Volume, www.iba.be.
11. Sep-Pak Light QMA Waters, ref WAT023525.
12. Eluent: for 10 vials, 70 mg potassium carbonate, Aldrich ref 367877-10G, in 3 mL water + 220 mg Kryptofix, Aldrich ref 29,111-0, in 3 mL acetonitrile SDS, ref 00637G16; 600 μL in each vial.
13. Nitro precursor ABX Radeberg Germany ref 3240, www.abx.de.
14. DMSO: Acros ref 326881000, used as received.
15. Water sterile for injection, Braun ref 014361.
16. MeOH, Sigma Aldrich ref 34860, used as received.
17. Sunfire™ Waters Prep C18 5 μm 10 × 250 mm, part no 186002564.
18. THF, Acros ref 164240025, used as received.
19. Sep-Pak Plus tC18 Waters ref WAT036810: Lemaire C, Plenevaux A, Aerts J, et al. J Labelled Comp Radiopharm 1999;42:63–75.
20. Absolute ethanol: AP-HP Ageps ref 3628442, 10 ml ampoules used as received.
21. Sterile Vial TC-ELU-10 IBA Molecular Imaging.
22. Sodium chloride ProAmp 0.9% solution for injection, Aguettant, Lyon, France.
23. Millex-MP sterile 0.22 μm filters, Millipore (part no. SLMPL25SS).
24. SunFire™ Waters C18 5 μm 4.6 × 30 mm part no 186002556, equilibrated 60 min before use.
25. MPPF standard, ABX ref 3250.
26. Type Polygram Sil G/UV254.
27. International Conference on Harmonisation of Technical Requirements for Registration of Pharmaceuticals for Human Use (ICH).
28. pH-indicator sticks, Macherey-Nagel 92118.
29. Mock BH, Winkle W, Vavrek MT. Nucl Med Biol 1997;24:193–195.
30. Capintec, Inc., USA.
31. European Pharmacopeia Biological tests Endotoxins 2.06.14.
32. Fluid Media ref THIO-T 42074, TSB-T 42100, SAB B-T 42108, BioMérieux, France, www.biomerieux.fr.
33. European Pharmacopeia Biological tests Sterility 2.06.01.

CHAPTER 11

SYNTHESIS OF [^{18}F]-FALLYPRIDE

LAURENT BRICHARD, VALENTINA FERRARI, ROB SMITH, AND FRANKLIN I. AIGBIRHIO

Wolfson Brain Imaging Centre, Department of Clinical Neurosciences, University of Cambridge, Addenbrooke's Hospital, Cambridge, UK

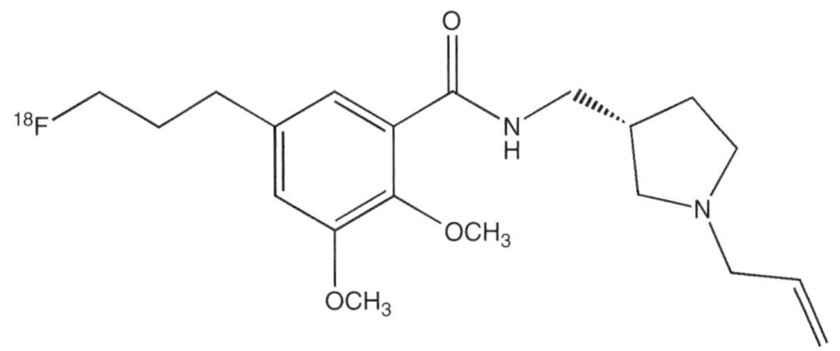

[^{18}F]Fallypride

1 INTRODUCTION

[^{18}F]Fallypride [1] is a widely used positron emission tomographic (PET) radiotracer for imaging the dopaminergic system in the brain. It is a high-affinity antagonist for D2/D3 receptors and has been employed to measure striatal and extrastriatal D2/D3 receptor availability in animals [2–4] and in humans [5–7].

2 SYNTHESIS PROCEDURES

CAUTION: All radiochemical syntheses must be carried out using appropriate equipment in a facility authorized for the use of radioactive materials. Personal protective equipment must be worn and all local radiation safety laws followed.

Radiochemical Syntheses, Volume 1: Radiopharmaceuticals for Positron Emission Tomography, First Edition. Edited by Peter J. H. Scott, Brian G. Hockley.

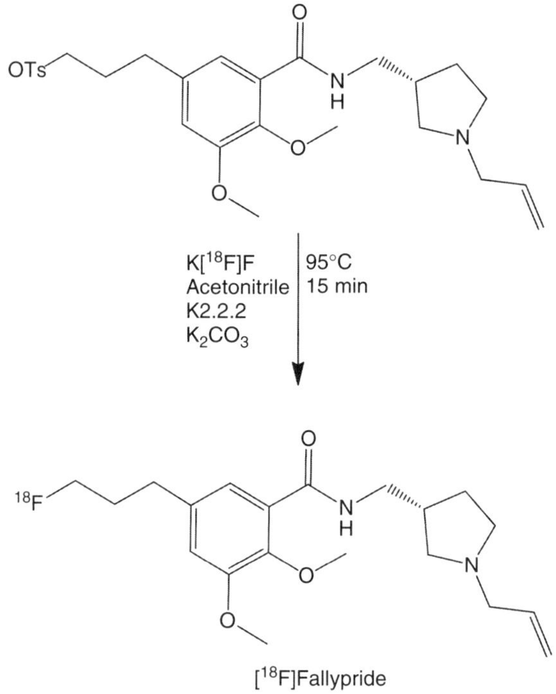

FIGURE 1 Synthesis of [^{18}F]fallypride.

[^{18}F]Fallypride is synthesized in one step, by aliphatic nucleophilic substitution with [^{18}F]fluoride on (*S*)-2,3-dimethoxy-5-[3-[[(4-methylphenyl)-sulfonyl]oxy]-propyl]-*N*-[[1-(2-propenyl)-2-pyrrolidinyl]methyl]-benzamide (tosylate precursor) (Fig. 1). The procedure is performed on a GE Healthcare TRACERlab FX_{FN}.

2.1 Production of [^{18}F]Fluoride

[^{18}F]Fluoride is produced via the ^{18}O(p,n)^{18}F by irradiating a target filled with 98% enriched [^{18}O]water placed on a GE PETtrace cyclotron. For [^{18}F]fallypride preparations, the target is irradiated at 40 μA for 20 min.

2.2 Azeotropic Drying of [^{18}F]Fluoride

After the delivery of the radioactivity from the cyclotron, the [^{18}F]fluoride in [^{18}O]water is trapped on a QMA cartridge (preconditioned to the carbonate form using potassium carbonate solution(1 M, 10 mL) and water 20 mL), while the enriched water is recovered in a separate vial. The [^{18}F]fluoride is then eluted from the cartridge to the reactor with a solution of potassium carbonate (7 mg, 50 μmol) and Kryptofix 2.2.2 (22 mg, 58 μmol) in 0.6 mL acetonitrile : water (50:50 v/v). In order to dry the [^{18}F]fluoride, the temperature of the reactor is set to 95°C, and a stream of helium as well as vacuum is applied. During the

5 min of the drying process, 1 mL of acetonitrile is added to the reactor in three portions to create an azeotrope that facilitates the removal of water.

2.3 Synthesis of [^{18}F]Fallypride

After the drying, 3 mg (5.8 μmol) of tosylfallypride precursor in 1 mL dry acetonitrile is added to the reaction vessel and heated at 95°C for 15 min to produce [^{18}F]fallypride. The resulting reaction mixture is then cooled to 60°C and mixed with 3 mL of the preparative high-performance liquid chromatography (HPLC) mobile phase (acetonitrile : water : triethylamine 56:44:0.1 v/v). The resulting solution is injected on the semipreparative HPLC column.

2.4 Purification and Formulation of [^{18}F]Fallypride

The HPLC purification is performed on a semipreparative Luna column (5 μm, C_{18} 100A 250 × 10 mm, Phenomenex, 00G-4041-N0) at a flow rate of 3 mL/min (acetonitrile : water : triethylamine 56:44:0.1 v/v). Figure 2 shows a typical chromatogram that is obtained for the production of [^{18}F]fallypride, with product starting elution at approximately 26 min. The collected peak is then diluted with 30 mL of water for injection. This solution is passed through an Oasis HLB

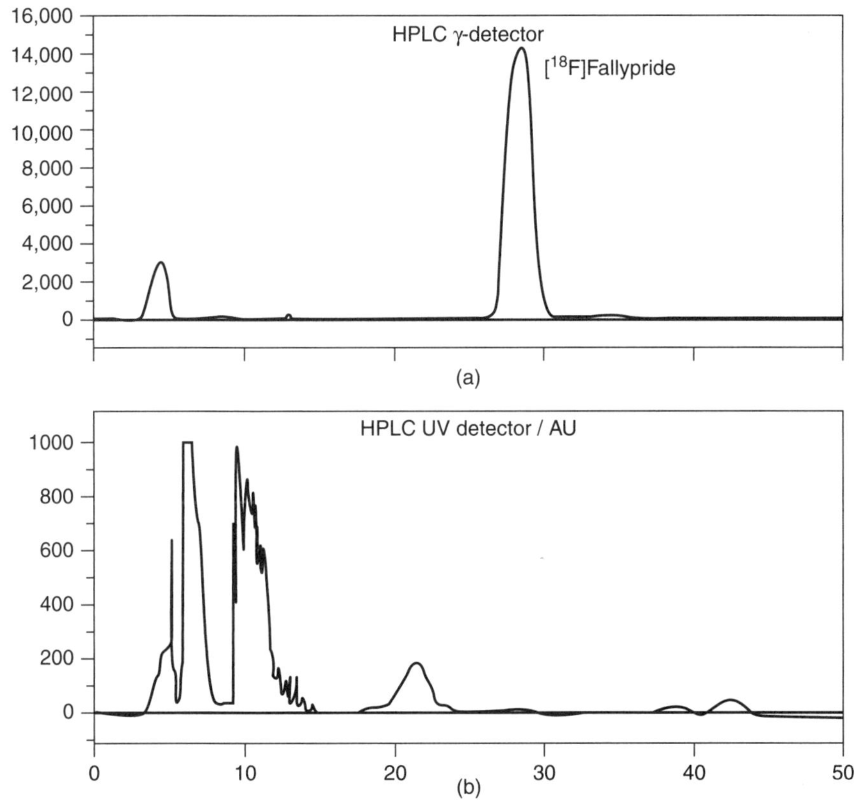

FIGURE 2 Semipreparative radioactive (a) and UV (b) HPLC traces for [^{18}F]fallypride.

30-mg cartridge, which had been previously conditioned with 5 mL of ethanol and 5 mL of water. The cartridge is rinsed with 10 mL of water for injection, and then [^{18}F]fallypride is eluted from the cartridge with 0.8 mL of absolute ethanol and mixed with 10 mL of saline solution. This mixture is subsequently subjected to final filtration through a Millex GP filter. This solution is then split into two sterile vials, one for PET studies and one for quality control (QC) procedures.

3 QUALITY CONTROL PROCEDURES

CAUTION: All radiopharmaceuticals produced for clinical use must have local regulatory approval before use in humans. Trained personnel must carry out QC procedures, and each dose must meet all established QC criteria before release to the clinic.

Owing to the short half-life of ^{18}F (109.7 min) and high levels of radioactivity, QC methods are divided into prerelease and postrelease tests. Prerelease tests have to be performed before release, and all tests have to be passed before release of the production batch to the clinic. These postrelease tests are performed after the release of the product.

3.1 Prerelease Tests

3.1.1 Visual Inspection

The [^{18}F]fallypride solution is examined and must be clear, colorless, and free of particulate matter.

3.1.2 pH

Using a 10-μL pipette, place 6-μL drops of the final formulation on each of the 3 bands of a pH stick (pH range, 4.5–10; Fisher Scientific, EU code: FB33011). The pH of the solution should be between 6 and 8.

3.1.3 Radiochemical Identity

About 5 μL of the QC sample is mixed with 400 μL of a cold fallypride standard solution (3 μg/mL), and 100 μL of the resulting solution is injected on the analytical HPLC column. The HPLC conditions for [^{18}F]fallypride are as follows:

Column	Primesphere 5 μm C18-HC, 250 × 4.6 mm (Phenomenex, 00G-3034-E0)
Injection loop	100 μL
Injection volume	150 μL
Eluent	Acetonitrile/water (56/44 v/v) + 0.1% triethylamine
Pump settings:	
Flow rate	1.5 mL/min
Radioactivity detector:	
Bioscan flow-count with NaI detector	
UVDetector:	
Gilson 118 UV/Vis	λ = 254 nm; AUFS = 0.001

TABLE 1 QC Data for Three Repeat Runs of [^{18}F]Fallypride

QC Test	Release Criteria	Run 1	Run 2	Run 3
Yield mCi (GBq)	NA	162 (5.99)	168 (6.20)	192 (7.12)
Visual inspection	Clear and free from particulates	Pass	Pass	Pass
Radiochemical identity	Difference in RT between 7 and 10 s	1	1	1
Radiochemical purity (%)	≥95	99.67	99.89	100
Chemical amount (μg/mL)	<3	1.03	0.47	0.13
Specific activity (GBq/μmol)	>25	155.5	69.2	103.6
Residual solvent analysis (ppm)	Acetonitrile <410	ND	ND	ND
Dose pH	6–8	7.0	7.0	7.0
Residual Kryptofix-[2.2.2] (μg/mL)	<220	<10	<10	<10
Sterile filter integrity test (bar)	≥3.86	5.3	5.3	5.4
Endotoxin analysis (EU/mL)	<17.5	0.5	0.5	1
Sterility testing	Sterile	Pass	Pass	Pass

NA, not applicable; ND, not detected.

The retention time of [^{18}F]fallypride should correspond to the reference fallypride standard, corrected for the dead volume between UV and radiodetector (the difference between retention times should be 7–10 s).

3.1.4 Radiochemical Purity

The QC sample solution (100 μL) is injected on the same analytical HPLC as described above, and the radiochromatogram is analyzed for potential radioactive impurity. The radiochemical purity of [^{18}F]fallypride must be higher than 95%.

3.1.5 Specific Activity

The specific activity is then calculated using a validated QC Excel Sheet.

3.1.6 Endotoxin Analysis

The endotoxin analysis is performed by testing the presence of pyrogen component in the QC sample using the limulus amebocyte lysate (LAL) method. For this purpose, the equipment used is the Endosafe Portable Test System (Charles River). The specification for [^{18}F]fallypride is to have <17.5 EU/mL in the QC sample.

3.2 Postrelease Tests

3.2.1 Filter Integrity Test

The integrity of the filter is tested using the bubble point test; to pass the test the bubble point has to be ≥3.86 bar (56 psi) [8].

3.2.2 *Sterility Test*

The test for sterility is performed by an external contractor on decay of the radioactivity in the product. The sterility of the final product is tested by the membrane filtration method [9].

3.2.3 *Bacterial Endotoxins*

Endotoxin testing is performed using a chromogenic technique with Endosafe limulus amebocyte lysate (LAL) system in a Portable Test System (PTS) Kinetic Reader (Charles River Laboratories, Charleston, SC). The LAL test is performed immediately after release of the product, and the acceptance criteria for release is that the endotoxin concentration is below 17.5 EU/mL [10].

3.2.4 *Residual Kryptofix 2.2.2*

The analytical method used to detect the level of Kryptofix 2.2.2 in the formulation is a color spot test developed by Mock et al. [11,12]. The procedure is as follows:

- Immerse strips of silica gel TLC in the iodoplatinate spray reagent solution (Aldrich) for few seconds and allow to dry overnight.
- Prepare four standard solutions of Kryptofix 2.2.2 in water of the following concentrations: 50, 25, 20, and 10 μg/mL.
- Apply 2 μL of each Kryptofix 2.2.2 standard solution and 2 μL of the QC sample on the pretreated silica gel strip, and let the spots dry at room temperature.
- Compare the sample spot visually with the standard spots for color intensity. The spots will appear as a blue-black circle in the center surrounded by a white circle of silica gel and an outer brown ring from unreacted reagent washed aside by the aqueous solvent. At low concentrations (<20 μg/mL), the central spot might be slightly brown colored, but it remains clearly outlined with a blue-black border. The higher the concentration of Kryptofix 2.2.2 the more intense the blue-black color is.
- The specification for [^{18}F]fallypride production is to have a concentration of Kryptofix 2.2.2 <220 μg/mL.

3.2.5 *Residual Solvent Analysis (Postrelease Test)*

Residual solvent analysis is performed on a gas chromatography instrument using the following conditions:

Column: Alltech Gas Chrom 254 80/100, 6′ × 1/8″ × 0.85″ (Alltech, 2486P0)

Detector: flame ionization detector (FID)

Gas flows:

Helium (carrier gas): ~25 mL/min

Hydrogen (FID): ~45 mL/min

Air (FID): ~350 mL/min

Injector temperature: 180°C

Detector temperature: 200°C

Column temperature: 130°C

Manual injection (1 μL)

Specification limit: acetonitrile <410 ppm and formulation solvent ethanol ≤10%

The QC results for 3 consecutive runs of [^{18}F] Fallypride are summarized in Table 1.

CHEMICAL ABSTRACTS NOMENCLATURE (REGISTRY NUMBER)

Acetonitrile (75-05-8)

(*S*)-2,3-Dimethoxy-5-[3-[[(4-methylphenyl)-sulfonyl]oxy]-propyl]-*N*-[[1-(2-propenyl)-2-pyrrolidinyl]methyl]-benzamide [166173-74-6]

Ethanol (64-17-5)

(*S*)-5-(3-Fluoropropyl)-2,3-dimethoxy-*N*-[[(2S)-1-(2-propenyl)-2-pyrrolidinyl]methyl]-benzamide [166173-78-0]

4,7,13,16,21,24-Hexaoxa-1,10-diazabicyclo[8.8.8]hexacosane [23978-09-8]

Potassium carbonate [584-08-7]

Triethylamine [121-44-8]

For detailed supplier information, see appendix (below).

APPENDIX Reagents and Consumables Used for the Production of [^{18}F]Fallypride

Item	Specification	Supplier/Catalogue No.	Storage
Tosylfallypride	>95%	ABX/155.0003	Under argon at -20°C
Kryptofix 2.2.2	99.5+%	Sigma/29,111-0	Store at 4–8°C
Acetonitrile	Anhydrous, over molecular sieve, >99.5%	Fluka/00709	Store at room temperature (RT), use within 6 mo of opening
Potassium carbonate	Analytical grade	Fisher Scientific P/4120/50	Store at RT
Triethylamine	99+%	Aldrich/23,962-3	Store at RT
0.9% w/w NaCl solution injection BP	Sterile	Fannin PL24598/0002	Store at RT
Absolute ethanol	Analytical grade	Fisher E/0650DF/17	Store at RT
Ethanol Ph Eur	Sterile	J M Loveridge Ltd 077–2558	Store at RT
Acetonitrile for HPLC	HPLC grade	Fisher A/0626/17	Store at RT
Water for injection BP	Sterile	Hameln Pharma 0007546500	Store at RT
BP Grade ultrapure water (for HPLC)	Sterile	Severn Biotech 20-9101-10	Store at RT
11 mL Evacuated vial	Sterile	Mallinckrodt DRN 4357	Store at RT
QMA cartridge	N/A	Waters WAT023525	Store at RT
Alumina cartridge	N/A	Waters WAT023561	Store at RT
Oasis HLB cartridge	N/A	Waters WAT094225	Store at RT
Millex GP	Sterile	Millipore/SLGP033RS	Store at RT

REFERENCES AND NOTES

1. Mukherjee J, et al. Nucl Med Biol 1995;22:283–296.
2. Dalley JW, et al. Science 2007;315(5816):1267–1270.
3. Mukherjee J, et al. Brain Res 2005;1032:77–84.
4. Rominger A, et al. Synapse 2010;64(8):654–657.
5. Landvogt C, et al. Epilepsia 2010;51(9):1699–1706.
6. Zald DH, et al. NeuroImage 2010;51(1):53–62.
7. Slifstein M, et al. Synapse 2010;64(5):350–362.
8. Council of Europe. European Pharmacopoeia. 7th ed. Strasbourg; 5.1.1. Methods of preparation of sterile products; 01/2008:50101.
9. Council of Europe. European Pharmacopoeia. 7th ed. Strasbourg; 2.6.1 Sterility; 07/2010:20601.
10. Council of Europe. European Pharmacopoeia. 7th ed. Strasbourg; 2.6.14. Bacterial endotoxins; 01/2010:20614.
11. Mock BH, et al. Nucl Med Biol 1997;24:193–195.
12. Council of Europe. European Pharmacopoeia. 7th ed. Strasbourg; Fludeoxyglucose (18F) injection 07/2008:1325.

CHAPTER 12

SYNTHESIS OF [^{18}F]FLUOROETHYLTYROSINE (^{18}F-FET)

YI ZHAO, AMAR ALFTEIMI, AND MAAZ ZUHAYRA

Department of Nuclear Medicine, Molecular Imaging and Therapy, UKSH, University Hospital Schleswig-Holstein Campus Kiel, Kiel, Germany

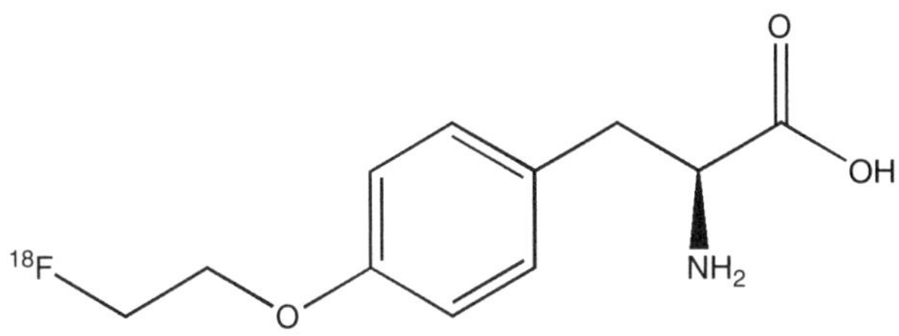

[^{18}F]FET

1 INTRODUCTION

Radiolabeled amino acids and their analogs are established tracers in diagnostic oncology [1]. Their use in single-photon emission tomography (SPECT) and positron emission tomography (PET) is based on their enhanced accumulation into malignant transformed cells by increased expression of amino acid transporters. This is assumed to reflect an enhanced amino acid metabolism and protein synthesis in growing tumors [2]. *O*-(2-[^{18}F]fluoroethyl)-L-tyrosine ([^{18}F]FET) is one of the first ^{18}F-labeled amino acids used for imaging amino acid metabolism in tumors. This tracer overcomes the disadvantages of [^{18}F]fluorodeoxyglucose, [^{18}F]FDG, and [^{11}C]methionine, [^{11}C]MET.

Radiochemical Syntheses, Volume 1: Radiopharmaceuticals for Positron Emission Tomography, First Edition. Edited by Peter J. H. Scott, Brian G. Hockley.

2 SYNTHESIS PROCEDURE

CAUTION: All radiochemical syntheses must be carried out using appropriate equipment in a facility authorized for the use of radioactive materials. Personal protective equipment must be worn and all local radiation safety laws followed.

2.1 Synthesis of 2-Bromoethyl Triflate (BE-OTf 1)

BE-OTf **1** was synthesized as described in the literature [3]. Briefly, 2-bromoethanol [4] (1 g, 8.0 mmol) was dissolved in 2,6-lutidine [5] (2.05 mL, 17.6 mmol), diluted with CH_2Cl_2 [6] (10 mL), and cooled to 0°C. Trifluoromethanesulfonic acid anhydride [7] (2.83 mL, 16.8 mmol) was added dropwise. After being stirred for 30 min, the solvent was removed in vacuo, and the product was distilled (50°C at 0.5 mm Hg) to give 1.1 g (54%) of **1** as a colorless liquid. Analysis by gas chromatography revealed a single peak with Rt = 7.82 min (column Rtx® 200, 0.25 mm ID, 30 m, carrier gas N_2, 0.95 mL/min, injector 250°C, program 50°C (5-min hold, than temperature gradient with 10°C/min)).

2.2 Production of [^{18}F]Fluoride

[^{18}F]Fluoride was produced via the $^{18}O(p,n)^{18}F$ reaction with a CTI RDS 111 cyclotron (Berlin) [8].

2.3 Azeotropic Drying of [^{18}F]Fluoride

The [^{18}F]fluoride ion (5–10 GBq) was delivered to a nuclear interface synthesis module [9] as a solution in [^{18}O]H_2O (2 mL). This solution was passed through a PS HCO_3^- cartridge [10] to trap the [^{18}F] fluoride ion and recycle the [^{18}O]H_2O. The [^{18}F]fluoride ion was then eluted into the nuclear interface glassy-carbon reaction vessel using a solution of aqueous potassium carbonate (11 mg in 0.8 mL H_2O) [11]. A solution of Kryptofix-[2.2.2] (30 mg in 1.5 mL MeCN) [12] was added, and the reaction mixture was azeotropically dried, initially at 90°C with both vacuum and helium flow for 2 min and subsequently with vacuum only for two additional minutes.

2.4 Synthesis of bromo-[^{18}F]fluoromethane ([^{18}F]BFE 2) and Distillation

o-DCB (*o-dichlorobenzene*) [13] (0.5 mL) was added to the dried [^{18}F]KF–K_{222} mixture. After 1 min of stirring at 90°C, BE-OTf **1** (15 μL) dissolved in *o*-DCB (0.5 mL) was added. [^{18}F]BFE **2** was transferred within 8 min into the second reactor containing (*S*)-(−)-tyrosine [14] (4.5 mg, 24.8 μmol), aqueous NaOH [15] (14.5 μL, 4 M), NaI [16] (5 mg, 33 μmol), and dimethyl sulfoxide (DMSO) [17] (2000 μL).

2.5 Synthesis of [^{18}F]FET 3

Fig. 1 shows the synthesis steps of [^{18}F]FET **3**. After [^{18}F]BFE was transferred by distillation into the second reactor, the reaction was heated at 110°C for 15 min.

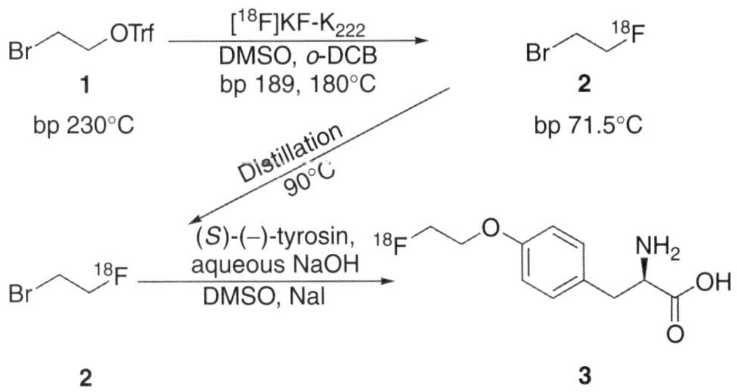

FIGURE 1 Synthesis of [^{18}F]FET.

2.6 Purification and Formulation of [^{18}F]FET 3

After the formation of [^{18}F]FET, the reactor temperature was decreased to 60°C and 8 mL of NH_4Ac (0.1 mol/L) [18] was added, and the reaction mixture was passed through four consecutive C18 Chromafix cartridges (Macherey-Nagel®) [19]. The cartridges were then washed with NH_4Ac (5 mL) and 10% aqueous ethanol (5 mL). [^{18}F]FET **3** was then eluted with 20% ethanol (10 mL). After sterile filtration, [^{18}F]FET **3** was obtained (2–4.1±5% GBq, 41±5% yield, decay corrected) with a radiochemical purity of >99.9% and a specific activity of >80 GBq/μmol at the end of synthesis (EOS). The synthesis time was 35 min from end of bombardment (EOB).

3 QUALITY CONTROL PROCEDURES

CAUTION: All radiopharmaceuticals produced for clinical use must have local regulatory approval before use in humans. Trained personnel must carry out quality control (QC) procedures, and each dose must meet all established QC criteria before release to the clinic.

QC procedures for [^{18}F]FET, based on the current requirements for radiopharmaceuticals laid out in the US Pharmacopeia [20], are summarized below. Complete QC data for three repeat batches of [^{18}F]FET produced using the method described herein are summarized in Table 1. Each of the three doses met all of the established QC criteria.

3.1 Visual Inspection

The [^{18}F]FET dose is examined behind a PET L-block and must be clear, colorless, and free of particulate matter. All three doses met this specification.

3.2 Radiochemical Identity and Purity

High-performance liquid chromatography (HPLC) analysis of radiochemical identity was conducted using an Agilent 1100 LC-Liquid Chromatograph [21]

TABLE 1 QC Data for Three Repeat Validation Runs of [^{18}F]FET

QC Test	Release Criteria	Run 1	Run 2	Run 3
Yield mCi (GBq)	N/A	270 (10)	257 (9.5)	297 (11.0)
Visual inspection	Clear, colorless	Clear, colorless	Clear, colorless	Clear, colorless
Radiochemical identity	RRT = 0.9–1.1	1.0	1.0	1.0
Radiochemical purity (%)	>95	>99.9	>99.9	>99.9
Specific activity (GBq/μmol)	No limit established	>80	>80	>80
Residual solvent analysis	Acetone <5000 ppm	1.29 ppm	1.5 ppm	1.2 ppm
	Ethanol no limit established	20% V	20% V	20% V
	o-DCB no limit established	<0.002 ppm	<0.002 ppm	<0.002 ppm
	DMSO <5000 ppm	<0.002 ppm	<0.002 ppm	<0.002 ppm
	MeCN <410 ppm	<0.02 ppm	<0.02 ppm	<0.02 ppm
Dose pH	4.5–7.5	6.8	6.5	6.0
Residual Kryptofix-[2.2.2] (μ g/mL)	≤50	≤50	≤50	≤50
Sterile filter integrity test (psi)	>50	>50	>50	>50
Radionuclidic identity ($t_{1/2}$) (min)	105–115	109	108	109
Endotoxin analysis (EU/mL)	≤17.5	<0.25	<0.25	<0.25
Sterility testing	No colony growth out to 14 d	Pass	Pass	Pass

fitted with diode array detector and Gina γ-detector [22] (column: Nucleosil 100-5 HD 4·250 mm [23]; solvents: A, NH_4Ac (0.1 M) [18]; B, acetonitrile; gradient: 0–5 min 100% A, 5–30 min 0–100% B; flow rate: 1.0 mL/min)

The retention time of [^{18}F]FET is compared to that of the [^{19}F]FET reference standard [24] and must be within ±10% of the standard (relative retention time (RRT) must be 0.9–1.1).

Radiochemical purity must be >95%. Representative analytical HPLC traces are displayed in Fig. 2. All doses met these specifications.

3.3 Specific Activity

There is no specific activity release limit for [^{18}F]FET in place at our department. The injection volume is based on the injectable mass limit for [^{19}F]FET of 5.5 μg per human patient and is calculated on a dose-by-dose basis by the administering study team.

3.4 Residual Solvent and Precursor Analysis

The final formulation is 20% aqueous EtOH. If desired, the [^{18}F]FET solution can be diluted with saline to be 10% aqueous EtOH.

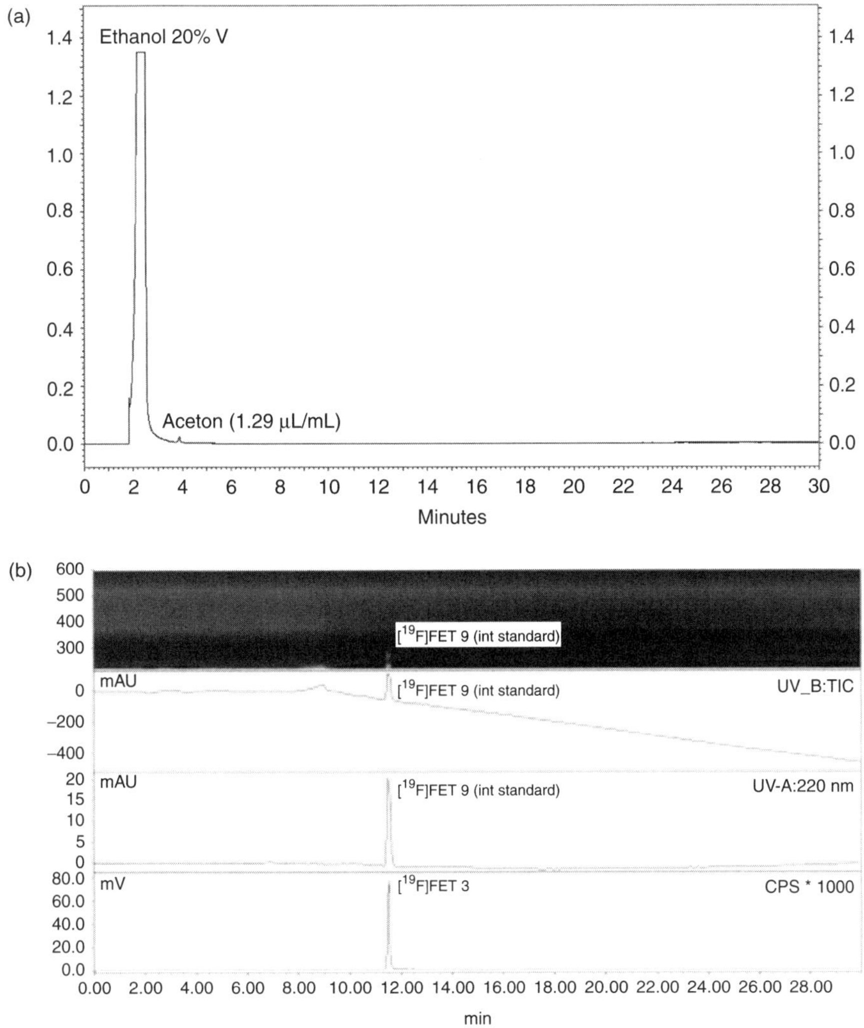

FIGURE 2 Chemical and radiochemical purity detection with (a) gas chromatography and (b) radio HPLC with diode array detection.

Analysis of residual solvent levels of acetone (from the module drying procedure), DMSO (Rt = 16.5 min), *o*-DCB (Rt = 14.2 min), ethanol (Rt = 2.1 min), and acetonitrile (Rt = 4.5 min) in [^{18}F]FET doses was conducted using a Varian 3500 GC Gas Chromatograph [25]. Limits imposed by the ICH Harmonised Tripartite Guidelines are <410 ppm per day for class 2 solvents (acetonitrile) and <5000 ppm per day for class 3 solvents (DMSO, acetone) [26]. Column: Rtx 200, 0.25 mm ID, 30 m [27], carrier gas N_2, 0.95 mL/min, injector 250°C, program 50°C (5-min hold, than 10°C/min). A typical chromatogram is shown in Fig. 2a. All doses met these specifications.

3.5 Dose pH

The pH of [^{18}F]FET dose was analyzed by applying a small amount of the dose to Macherey-Nagel pH 2.0–9.0 nonbleeding pH-indicator strips [28] and determined by visual comparison to the scale provided. Dose pH must be 4.5–7.5. All three doses had pH values that met this specification.

3.6 Residual Kryptofix-[2.2.2] Analysis

Residual Kryptofix-[2.2.2] levels in [^{18}F]FET doses were analyzed using the established spot test [29]. Strips of plastic-backed silica gel thin layer chromatographic (TLC) plates saturated with iodoplatinate spray reagent [30] were spotted with water (negative control), 50 μg/mL Kryptofix-[2.2.2] standard (positive control), and [^{18}F]FET dose. If Kryptofix-[2.2.2] is present in a sample, a blue-black spot appears. Spots for the three samples were compared, and a visual determination of residual Kryptofix-[2.2.2] in the [^{18}F]FET dose was made. Samples containing $<$50 μg/mL were deemed acceptable. All three samples met the specifications for this test.

3.7 Sterile Filter Integrity Test

The sterile filter from the [^{18}F]FET (with needle still attached) was connected to a nitrogen supply via a regulator. The needle was submerged in water, and the nitrogen pressure was gradually increased. If the pressure was raised above the filter acceptance pressure (50 psi) without seeing a stream of bubbles, the filter was considered intact. If a stream of bubble occurs $<$50 psi, the test fails. All filters met the specification for this test.

3.8 Radionuclidic Identity

Activities were measured using a MED®-Radioisotope dose calibrator [31], and half-life was calculated using Equation (1). Calculated half-life must be within a 105–115 min range.

$$t_{1/2} = \frac{\ln(\text{time difference})}{\ln\left[\frac{\text{ending activity}}{\text{starting activity}}\right]} \tag{1}$$

3.9 Endotoxin Analysis

Endotoxin content in the three doses of [^{18}F]FET was analyzed by the Laboratories of the University Hospital Testing System [32] and according to the United States Pharmacopeia (USP) [33]. Doses must contain $\leq$17.5 endotoxin units (EU)/mL to be deemed acceptable. All three doses met this specification.

3.10 Sterility Testing

Samples of [^{18}F]FET were placed onto fluid thioglycolate media (FTM) plates and soybean casein digest media (SCDM) plates. FTM plates are used to test for anaerobes, aerobes, and microaerophiles, while SCDM plates are used to test for nonfastidious and fastidious microorganisms [34]. [^{18}F]FET plates were incubated with positive and negative controls for 14 days. FTM plates were incubated at 32°C, and SCDM plates were incubated at 22°C according to current USP guidelines [35]. Plates were visually inspected on the third, eighth, and fourteenth days of the test period and compared to the positive and negative standards. Positive standards must show growth (turbidity) on thc plates, and [^{18}F]FET negative controls must have no culture growth after 14 days to be indicative of sterility. All samples met the sterility specifications.

CHEMICAL ABSTRACTS NOMENCLATURE (REGISTRY NUMBER)

Acetonitrile (75-05-8)
Carbonic acid, potassium salt (1:2) (584-08-7)
Ethane, 1-bromo-2-(fluoro-18^{18}F)- (9CI) (85725-77-5)
Ethanol (64-17-5)
Ethanol, 2-bromo- (540-51-2)
^{18}F-FET; *O*-(2-fluoroethyl)-L-tyrosine-^{18}F; *O*-(2-[^{18}F]fluoroethyl)-L-tyrosine (178433-03-09)
4,7,13,16,21,24-Hexaoxa-1,10-diazabicyclo[8.8.8]hexacosane (23978-09-8)
Methane, 1,1′-sulfinylbis- (67-68-5)
Methanesulfonic acid, 1,1,1-trifluoro-, 2-bromoethyl ester (103935-47-3)
Sodium hydroxide (1310-73-2)
Sodium iodide (7681-82-5)
Tyrosine (556-03-6)

REFERENCES AND NOTES

1. Vaalburg W, Coenen HH, Crouzel C, Elsinga PH, Långström B, Lemaire C, et al. Nucl Med Biol 1992;19:227–237.
2. Langena KJ, Hamacher K, Weckesser M, Floeth F, Stoffelsa G, Bauer D, Coenen HH, Pauleit D. Nucl Med Biol 2006;33:287–294.
3. Chi DY, Kilbourn MR, Katzenellenbogen JA, Welch MJ. J Org Chem 1987;52:658–664.
4. 2-Bromoethanol was purchased from Aldrich (CAS no. 540-51-2) and used as received.
5. 2-6-Lutidine was purchased from Aldrich (CAS no. 108-48-5) and used as received.
6. Dichloromethane was purchased from J.T.Baker (CAS no. 75-09-2) and used as received.
7. Trifluoromethanesulfonic acid anhydride was purchased from Fluka (CAS no. 358-23-6) and used as received.
8. [^{18}F]Fluoride ion was purchased from Eckert & Ziegler F-Con Deutschland GmbH, and used as received.
9. GE Healthcare, Germany. No hardware modifications are required for the Nina Box to perform the synthesis.
10. Ion exchange cartridges were purchased from Macherey-Nagel, Germany (WAT023525) and conditioned with ethanol (1 mL) and water (1 mL) before use.
11. Potassium carbonate purchased from Aldrich (CAS no. 584-08-7) and used as received. Sterile water purchased from B. Braun (part no. 2351744) and used as received.
12. Kryptofix-[2.2.2] was purchased from Acros (CAS no. 23978-09-8) and used as received. Anhydrous acetonitrile was purchased from Acros (part no. 610961000) and used as received.

13. 1,2-Dichlorobenzene was purchased from Aldrich (CAS no. 95-50-1) and used as received.
14. (*S*)-(−)-Tyrosine was purchased from Merck (CAS no. 60-18-4) and used as received.
15. Sodium hydroxide standard solution (4 mol/L) was purchased from Fluka (CAS no. 1310-73-2) and used as received.
16. Sodium iodide was purchased from Merck (CAS no. 7681-82-5) and used as received.
17. Dimethylsulfoxide was purchased from Fluka (CAS no. 67-68-5) and used as received.
18. Ammonium acetate was purchased from Merck (CAS no.631-61-8) and used as received.
19. C18 Chromafix cartridges were purchased from Macherey-Nagel (Germany). Every cartridge was activated with ethanol (10 mL) and washed with aqueous NH_4Ac (30 mL, 0.1 M).
20. U S Pharmacopeia <823> Radiopharmaceuticals for Positron Emission Tomography-Compounding. USP 32–NF 27. 2009
21. Agilent Corporation, Germany.
22. Ray Test Corporation, Germany.
23. Macherey-Nagel (Germany), part no 721662.46.
24. [^{18}F]FET reference standard was purchased from ABX advanced biochemical compounds, Germany.
25. Varian Corporation, Germany.
26. International Conference on Harmonisation of Technical Requirements for Registration of Pharmaceuticals for Human Use (ICH).
27. Restek Corporation, Germany.
28. Macherey-Nagel, Germany.
29. Mock BH, Winkle W, Vavrek MT. Nucl Med Biol 1997;24:193–195.
30. Sigma-Aldrich, (part no I9157).
31. MED Nuklear-Medizintechnik Dresden GmbH, Germany.
32. University Hospital Schleswig-Holstein Campus Kiel, UKSH.
33. U S Pharmacopeia <85> Bacterial Endotoxins Test. USP 32–NF *27*. 2009.
34. Becton, Dickinson and Company, USA.
35. U S Pharmacopeia <71> Sterility Tests. USP 32–NF *27*. 2009.

CHAPTER 13

SYNTHESIS OF [^{18}F]FLUMAZENIL ([^{18}F]FZ)

RALF SCHIRRMACHER

McConnell Brain Imaging Centre, Department of Neurology and Neurosurgery, Montreal Neurological Institute, McGill University, Montreal, Quebec, Canada
Lady Davis Institute for Medical Research, Department of Neurology and Neurosurgery, McGill University, Montreal, Quebec, Canada

ALEXEY KOSTIKOV

McConnell Brain Imaging Centre, Montreal Neurological Institute, McGill University, Montreal, Quebec, Canada

GASSAN MASSAWEH

McConnell Brain Imaging Centre, Montreal Neurological Institute, McGill University, Montreal, Quebec, Canada
The Molecular Imaging & Research Centre, Department of Radiology, Queen Elizabeth II Health Sciences Centre, Capital Health/Dalhousie University, Halifax, Nova Scotia, Canada

MIRIAM KOVACEVIC AND CARMEN WÄNGLER

McConnell Brain Imaging Centre, Montreal Neurological Institute, McGill University, Montreal, Quebec, Canada

ALEXANDER THIEL

Lady Davis Institute for Medical Research, Montreal, Canada
Department of Neurology and Neurosurgery, McGill University, Montreal, Canada

Radiochemical Syntheses, Volume 1: Radiopharmaceuticals for Positron Emission Tomography, First Edition. Edited by Peter J. H. Scott, Brian G. Hockley.

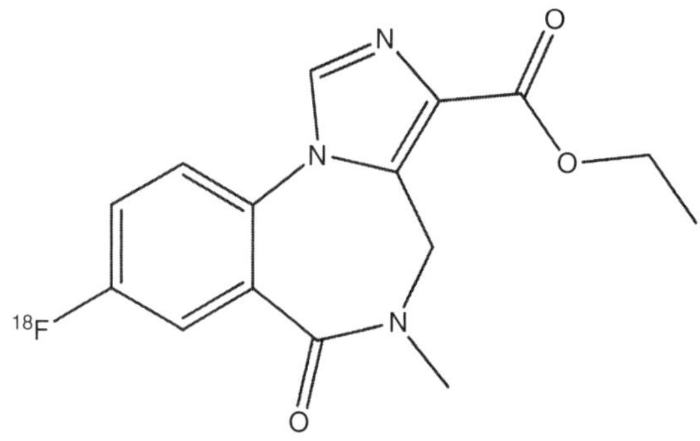

[^{18}F]FZ

1 INTRODUCTION

The central γ-aminobutyric acid (GABA)-A receptor is a chloride ion channel consisting of five subunits. The subunits belong to different classes (e.g., α 1–6, β 1–3, γ1–3), which determine the pharmacological properties of a certain GABA-A receptor type.

Flumazenil (FZ) is a ligand for the benzodiazepine binding site (BZBS) of the central GABA-A receptor. In the human cortex, GABA-A receptors are located in axosomatic and axodendritic synapses on neurons that are innervated by inhibiting interneurons and are thus cortical neuronal markers [1]. FZ is an antagonist at the BZBS of the GABA-A receptor. While GABA itself binds in between the α and β subunits [2], benzodiazepines such as FZ bind in between the α and γ subunits, which is the BZBS [3]. In vitro, FZ binds to the α1, α2, α3, and α5 subunits [4] with equal affinity of $K_i = 1-2$ nM [2,5]. These four subunits are present in more than 90% of all cortical GABA-A receptors [3].

As a marker of neuronal density, [^{11}C]FZ has widely been used to detect epileptic foci in neocortical epilepsy [6] and neuronal loss in the hippocampus of patients with temporal lobe epilepsy [7] and was found to be of diagnostic value in presurgical diagnostics for epilepsy [8]. Furthermore, [^{11}C]FZ has been used as a marker of early and delayed neuronal loss after stroke [9] and to validate diffusion-weighted magnetic resonance imaging (MRI) in acute stroke [10].

The short half-life of [^{11}C]FZ, which necessitates the presence of a cyclotron at the positron emission tomography (PET) site, has imposed limitations to a broader clinical application of this tracer. For research, fluorine-labeled compounds are desirable in order to make use of the full resolution potential of the newest generation of high resolution PET scanners. The synthesis of [^{18}F]FZ has been described by us and others [11–14]. Here we describe the synthesis and a detailed workup procedure for [^{18}F]FZ for human application using a straightforward and reliable high performance liquid chromatographic (HPLC) workup procedure preventing any contamination of the [^{18}F]FZ with the nitro-precursor.

The chemical/radiochemical integrity of the final injectable [^{18}F]FZ solution for human application has been proved by radio-HPLC and radio-thin layer chromatography (TLC). Following Health Canada recommendations of PET studies governed by local ethic committees, three consecutive [^{18}F]FZ batches were produced before starting a human imaging protocol involving multiple subjects, and were subjected to residual solvent testing, residual Kryptofix 2.2.2® (K222) testing, bacterial endotoxin and sterility testing.

2 SYNTHESIS PROCEDURES

CAUTION: All radiochemical syntheses must be carried out using appropriate equipment in a facility authorized for the use of radioactive materials. Personal protective equipment must be worn and all local radiation safety laws followed.

2.1 Production of [^{18}F]Fluoride

Aqueous [^{18}F]fluoride (300–450 mCi, 11.1–16.6 GBq) was prepared by the ^{18}O(p,n)^{18}F nuclear reaction on an enriched [^{18}O]water target (97%, 2.2 mL), which was irradiated with a proton beam (35 μA) using an IBA cyclotron (Cyclon 18/9) for 15 min. After the irradiation, the [^{18}F]F$^-$/H$_2$[^{18}O]O was transferred via a stream of He into the corresponding vessel of the automated synthesis module (HotboxIII, Scintomics, Germany). For a detailed description of the automated synthesis see Section 2.5. A procedure using the GE TRACERlab FX$_{FN}$ synthesis module for [^{18}F]FZ production has been described elsewhere [14].

2.2 Azeotropic Drying of [^{18}F]Fluoride

No-carrier-added (NCA) aqueous [^{18}F]fluoride/H$_2$[^{18}O]O (11.1–16.6 GBq, ~2.2 mL) was passed through a preconditioned ((i) rinse with 10 mL 0.5 M K_2CO_3 and (ii) rinse with 10 mL deionized water) QMA Sep-Pak Light cartridge (Waters). The H$_2$[^{18}O]O is collected, and the fixed ^{18}F$^-$ is eluted from the QMA cartridge with a solution of 30 μL of 0.5 M K_2CO_3 and Kryptofix 2.2.2 (10 mg) in CH_3CN/H_2O (1.5 mL CH_3CN + 30 μL H_2O). The residual water is removed by coevaporation to dryness with CH_3CN (this step is repeated two times with 0.5 mL CH_3CN) using a stream of argon at 105°C and a gentle vacuum of about 300 mbar. After about 4 min, the argon flow is stopped, the vacuum is set to maximum (~10 mbar), and the final drying is continued for 2 min.

2.3 Synthesis of [^{18}F]FZ

The synthesis was performed semiautomatically in a GE TRACERlab FX$_{FN}$ synthesis module or fully automated in a Scintomics HotboxIII synthesis module (Scintomis, Germany), including the injection of the crude reaction mixture onto the HPLC system (see below). To the "dried" [^{18}F]F$^-$/Kryptofix 2.2.2/K$^+$ complex, the labeling precursor Ro-15 2344 (7 mg, 0.02 mmol), dissolved in 0.6 mL of anhydrous *N,N*-dimethyl formamide (DMF), is added and heated to 150–160°C for 30 min (stirring is not necessary). The reaction mixture turns

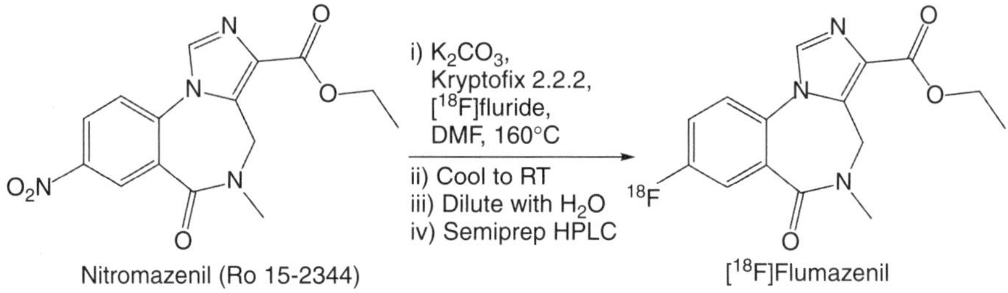

FIGURE 1 Synthesis of [^{18}F]flumazenil ([^{18}F]FZ).

dark red. After 30 min, the reaction mixture is cooled by a stream of air to room temperature and diluted with 1.5 mL of water or HPLC solvent for the direct injection onto the described HPLC system. The cooling step before adding the water or HPLC solvent appears to be crucial. If the temperature is too high during the addition, the yields of [^{18}F]FZ after HPLC purification drop considerably (Fig. 1). If necessary, a small sample of the crude reaction mixture can be taken before injecting the whole solution onto the HPLC system to measure the radiochemical yields (RCYs) by radio-TLC to estimate final RCYs.

2.4 Purification and Formulation of [^{18}F]FZ Using the Scintomics HotboxIII Module

2.4.1 Purification

The injection of the mixture onto HPLC is done automatically using a syringe pump connected to the injection vessel containing the mixture to be injected (Fig. 2). The radioactivity peak corresponding to the [^{18}F]FZ ($R_f = 15$ min, HPLC conditions: HPLC column, Waters Symmetry Prep 7 µm C18 (7.8 × 300 mm); HPLC solvent, 0.05 M NaOAc pH 5/tetrahydrofuran (THF)/MeOH (80/10/10); flow, 2.5 mL/min, isocratic) was collected in a 50-mL round bottom flask (Fig. 4).

2.4.2 Solvent Evaporation and Final Formulation

The HPLC solvent was removed by evaporation using a water bath at 100°C and a gentle vacuum (~400–700 mbar) (Fig. 3). This evaporation step was repeated two times after the addition of 4 mL of ethanol (100%) to entirely remove all traces of HPLC solvent. *Note: The whole procedure does not need more than 4–5 min and needs less space than a rotary evaporator and is easier to operate*. The purified [^{18}F]FZ was dissolved in ethanol (0.5 mL) and diluted with physiological saline (2.5 mL) and sterile phosphate buffer (2.5 mL) for human application.

An injectable solution was obtained by filtering the obtained solution through a 0.22-µm sterile filter (Millex®). The average RCY (uncorrected for radioactive decay) of 7–9% seems to be highly dependent not only on the quality of the

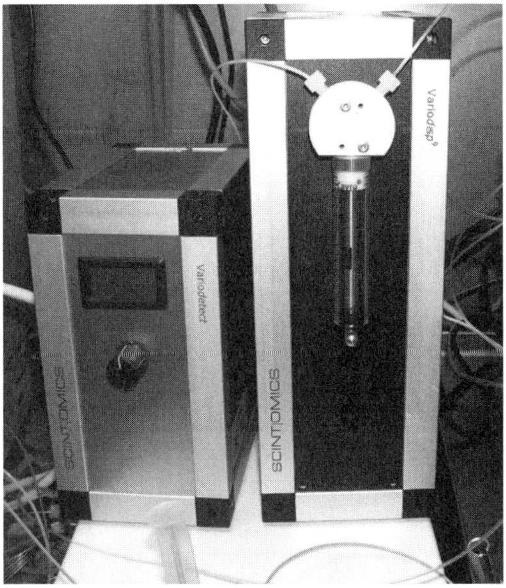

FIGURE 2 Injection of the crude reaction mixture is done by a syringe pump controlled by the HotboxIII module.

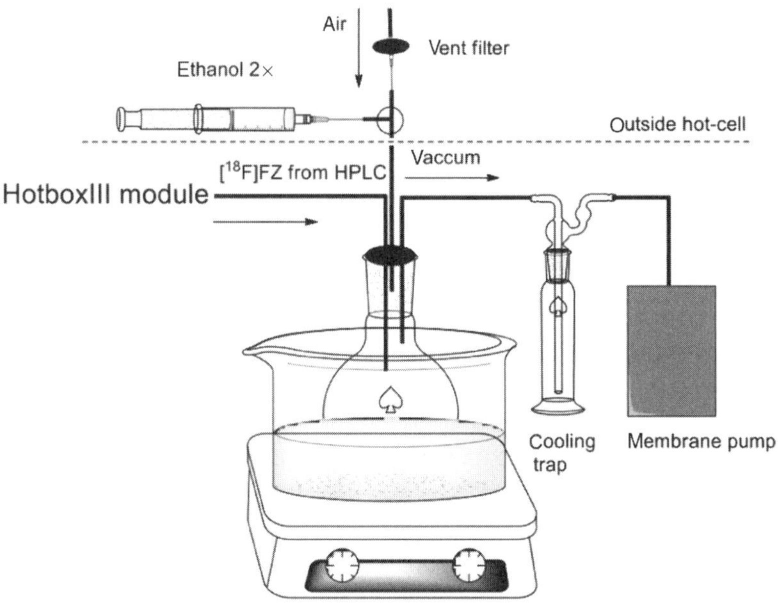

FIGURE 3 Simple but efficient setup to evaporate the HPLC solvent from the [^{18}F]FZ solution. Ethanol for coevaporation is supplied through a T-valve fitting, which is then flipped to a vent filter air supply. The air that is constantly streaming through the solution helps to quickly evaporate the solvent. This step is repeated two times.

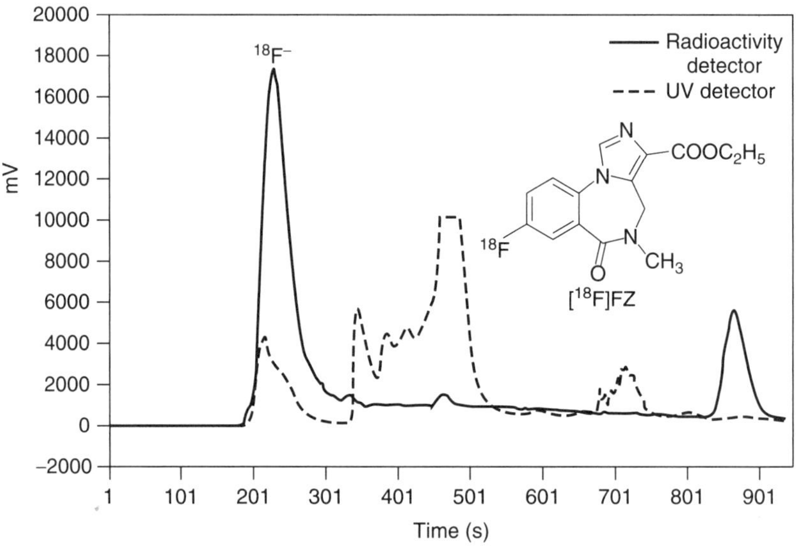

FIGURE 4 Semipreparative UV and radioactive HPLC traces for [^{18}F]FZ. All UV active components of the reaction mixture elute well separated from the [^{18}F]FZ.

precursor material but also on the individual HPLC solvent batch as well as the maturity of the DMF used for the reaction. From our experience, the synthesis of [^{18}F]FZ is one of the most demanding and sensitive syntheses in radiochemistry and requires some trial and error before working properly.

2.5 Step-By-Step Explanation of the Fully Automated Synthesis of [^{18}F]FZ Using a Scintomics HotboxIII Module

The detailed scheme shown below using the Scintomics HotboxIII synthesis unit explains all the necessary automated steps for the routine remotely controlled production of [^{18}F]FZ. Figure 5 shows a schematic overview of the Scintomics synthesis module, followed by a detailed description of operation. This protocol can be easily adapted to any other commercially available module.

1. $^{18}F^-/[^{18}O]H_2O$ is delivered from the cyclotron into vial labeled F-18.
2. This solution is passed through the QMA cartridge into vial labeled O-18 under vacuum by opening valves V3 and V4; the reactor is heated to 105°C. The $^{18}F^-$ gets fixed on the QMA cartridge.
3. After 50 s, valves V3 and V4 are closed and the $^{18}F^-$ is eluted from QMA with the solution from R1 (Kryptofix2.2.2/K_2CO_3/CH_3CN) into the reactor by opening valves V6, V9, V14, V15, and V39.
4. After 30 s, valves V6, V9, V14, and V39 are closed and valves V11, V14, and V40 are opened. The solvents are evaporated under vacuum and a stream of inert gas (nitrogen or argon).

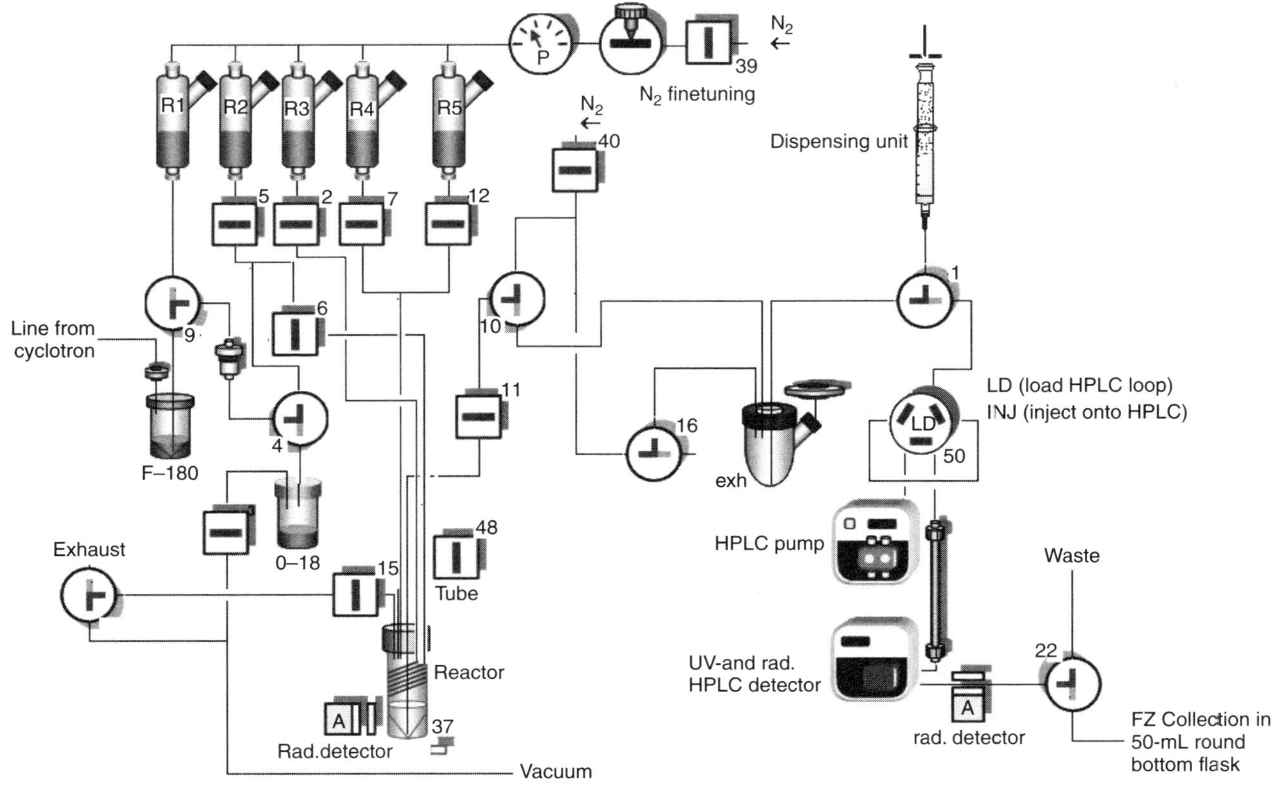

FIGURE 5 Schematic overview of the HotboxIII Scintomics synthesis module used in the synthesis of [^{18}F]FZ. R1 = [K222 (10 mg)/0.5 M K_2CO_3 (30 μL), H_2O (30 μL), CH_3CN (1.5 mL)], [R2 = R5 = MeCN (0.5 mL)], [R3 = nitromazenil (7 mg, 0.02 mol) in DMF (0.7 mL)], [R4 = HPLC eluent (see preparative HPLC conditions) or water (1.5 mL)].

5. After 4 min, valve V11 is closed, and CH_3CN is added from vial R2 by opening valves V5, V6, and V39.
6. After 30 s, valves V5, V6, and V39 are closed again and valve V11 is opened to allow the stream of inert gas into the reactor. CH_3CN is evaporated along with traces of water.
7. After 3 min, CH_3CN is added from vial R5 by opening valves V12 and V39 for 30 s and then closing them back again. Again, CH_3CN is evaporated along with traces of water.
8. After 2.5 min, valves V11 and V40 are closed to allow final drying under vacuum (~5 mbar) without a stream of nitrogen.
9. After 1.5 min, the K^+/K222/$^{18}F^-$ complex is considered completely dry, and the precursor is added from vial R3 into the reactor by opening valves V2 and V39 and flipping valve V14 to the atmosphere.
10. After 30 s, valves V2, V15, and V39 are closed and the temperature of the reactor is brought up to 150°C. Nucleophilic ^{18}F-fluorination takes place.
11. After 30 min reaction time, the reactor is cooled down to 25°C by opening valve V37, which provides a stream of pressurized air to the outside of the reactor. Crucial step: if the reaction mixture is too hot when the solvent is added, the yields drop considerably.
12. After 5 min, valve V37 is closed and the HPLC eluent (or water) is added to the reactor from vial R4 by opening valves V7, V15, and V39.
13. After 10 s, the resulting solution is transferred from the reactor by the stream of inert gas (nitrogen or argon) into the mixing vessel through valve V48 (pushing the central line into the reactor), closing valve V15 and opening valves V10, V11, and V16.
14. HPLC solvent flow is brought up to 2.5 mL/min; valves V7, V10, and V39 are closed; and the crude reaction mixture is transferred to a dispensing vial through valve V1.
15. After 1 min, valve V1 is changed toward HPLC, valve V50 (injection valve) is put to load position "LD," and the mixture is injected on HPLC.
16. After 1 min, valves V1 and V16 are closed and V50 is put into position "INJ" (the solution is brought onto HPLC).
17. Once the product peak appears on the chromatogram (approximately after 14–15 min at our individual HPLC setup), valve V22 is switched and the peak is collected into the receiving flask for further workup.
18. For solvent evaporation, further workup, and final formulation of the injectable [^{18}F]FZ solution, see Section 2.4.2. Quality control procedures are outlined in Section 3 and Table 2.

3 QUALITY CONTROL PROCEDURES

CAUTION: All radiopharmaceuticals produced for clinical use must have local regulatory approval before use in humans. Trained personnel must carry out

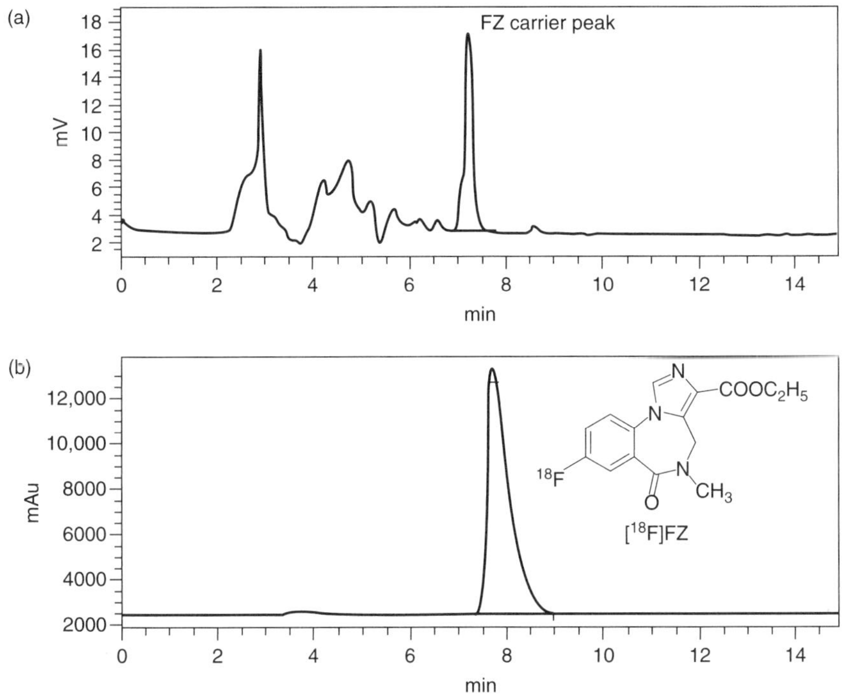

FIGURE 6 UV (a) and radio-HPLC (b) chromatogram of the purified [^{18}F]FZ for quality control.

quality control procedures, and each dose must meet all established quality control criteria before release to the clinic.

3.1 Visual Inspection

The [^{18}F]FZ dose is examined behind a lead glass window and must be clear, colorless, and free of particulate matter. Even a slight yellow color indicates contamination with the nitro-precursor or other unwanted side products, and the solution has to be discarded.

3.2 Radiochemical Identity and Specific Activity (SA)

The radiochemical identity of the final injectable solution is proved by HPLC and radio-TLC (Fig. 6). The specific activity (SA) of the final [^{18}F]FZ solution is determined by HPLC via a UV calibration curve. The radiochemical identity was further confirmed by injecting a mixture of nonradioactive FZ and the purified [^{18}F]FZ fraction from the preparative HPLC purification. The relative retention time (RRT) was 0.94.

3.2.1 Radio-TLC (silica gel-60 plates)

Solvent: Ethyl acetate/MeOH/Et_3N (75/20/5); R_f[^{18}F]FZ = 0.5

3.2.2 *Analytical HPLC*

Column: Phenomenex Prodigy ODS 5 μm, 250 × 4.6 mm
Solvent: acetonitrile/KH_2PO_4 (20 mM, pH 6–7) 40/60; R_f[^{18}F]FZ = 8 min
Flow: 0.7 mL/min; UV at 254 nm.

3.2.3 *Determination of Specific Activity*

For SA determination, a UV calibration curve was prepared by injecting different amounts of nonradioactive [^{19}F]FZ dissolved in 40 μL of water onto an HPLC system (see Section 3.2.2, analytical HPLC conditions). Solutions of different concentrations are prepared from a stock solution by dilution. In our particular setup, we injected the following amounts of [^{19}F]FZ (Table 1) in 40 μL of 99+% water (+ traces of CH_3CN, see UV calibration curve) into the HPLC and created a UV calibration curve (Fig. 7). From this calibration curve, the HPLC software creates a linear equation, which serves to determine the amount of the [^{18}F]FZ radiotracer and nonradioactive carrier that has to be put into relationship to the

TABLE 1 Injection of Variable Amounts of [^{19}F]FZ into HPLC to Generate a UV Calibration Curve

Amount of [^{19}F]FZ (ng) in 40 μL of 99+% Water (μL)[a]	UV Area (mAU) (Three Injections)
40	(131.16, 129.02, 129.29) **129.8**
80	(249.91, 252.31, 252.88) **251.7**
200	(640.76, 641.34, 643.80) **641.9**
400	(1281.40, 1274.40, 1290.60) **1282.1**

[a]The average over three different measurements (represented in bold) for each concentration has been used for UV area determination. HPLC conditions: Phenomenex Prodigy ODS, 5 μm, 250 × 4.6 mm; acetonitrile/KH_2PO4 (20 mM, pH 6–7) 40/60; flow 0.7 mL/min; UV at 254 nm.

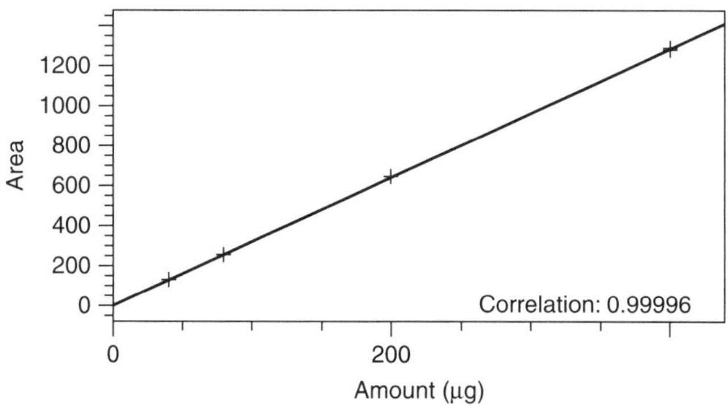

FIGURE 7 UV calibration curve for the determination of the SA of [^{18}F]FZ showing a high correlation of 0.99.

injected radioactivity (the radioactivity amount present in the 40-μL injection volume). After conversion of the obtained FZ amount in nanogram into micromoles or millimoles, the SA is usually expressed in gigabecquerel per micromole or Curie per millimole.

Preparation of the UV Calibration Curve (Table 1 and Figure 7) FZ (1.3 mg, 4.3 μmol) was dissolved in CH_3CN (1 mL). An aliquot of this solution (7.7 μL) was diluted with water (992 μL) to bring the total volume to 1 mL. The concentration of the FZ in the resulting solution was 10 μg/mL). The other three solutions (1, 2, and 5 μg/mL) were prepared by dilution of the 10 μg/mL solution. Therefore, the solvent in all of these solutions is 99+% water.

3.3 Quality Assurance

TABLE 2 QC Data for Three Repeat Runs of [^{18}F]FZ

QC Test	Release Criteria	Run 1	Run 2	Run 3
Yield [^{18}F]FZ mCi (MBq)	Yield ≥20 mCi (dose: 5–7 mCi for HRRT PET)	20 (740)[a]	23 (851)[a]	21 (777)[a]
Visual inspection	Must be a c.l[b]	c.l[b]	c.l[b]	c.l[b]
Radiochemical identity (min)	R_t-HPLC[c] = 8 (R_f − TLC = 0.5[d])	8 (0.5)	8 (0.5)	8 (0.5)
Radiochemical purity (%)	>95	97.5	97.9	97.8
Specific activity (Ci/mmol)	>800	1860	2370	2660
Residual solvent analysis (ppm)	Methanol <3000 THF <720	<30 <7	<30 <7	<30 <7
Dose pH	pH = 6–7	7.0	7.0	7.0
Residual Kryptofix 2.2.2 (mg/mL)	<0.05	Pass	Pass	Pass
Sterile filter integrity test (psi)	>50	>50	>50	>50
Osmolality[e] (mOsm/kg)	280–320	Pass	Pass	Pass
Radionuclidic identity ($t_{1/2}$) (min)	$t_{1/2}$ = 110 ± 4	107.4	106.8	112.1
Endotoxin analysis	≤175 EU/V[e]	<25 EU/mL[f]	<25 EU/mL[f]	<25 EU/mL[f]
Sterility testing	Sterile[g]	Sterile[g]	Sterile[g]	Sterile[g]

[a]These are final RCYs of [^{18}F]FZ as a sterile injectable solution for a two human (1 h scanning time each) subject imaging protocol.
[b]c.l = clear solution.
[c]Analytical HPLC conditions: column: Phenomenex Prodigy ODS 5 μm, 250 × 4.6 mm; solvent: acetonitrile/KH_2PO_4 (20 mM, pH 6–7) 40/60, flow rate 0.7 mL/min, UV at 254 nm.
[d]TLC condition: ethyl acetate/MeOH/Et_3N (75/20/5).
[e]Since ethanol impairs osmolatity determination, osmolatity was determined without the addition of ethanol (note that only 40% of [^{18}F]FZ is dissolved without the addition of ethanol).
[f]Max injection volume at 7 mL, thus bacterial endotoxin limit is <25 EU/mL.
[g]Aliquots were taken from each run (~0.25 mL) and sent to an external laboratory (Sterigen, Quebec, Canada) for sterility and endotoxin testing following United States Pharmacopeia (USP) 32 <71> (no growth after 14 d) and USP 32 <85>, respectively.

CHEMICAL ABSTRACTS NOMENCLATURE (REGISTRY NUMBER)

Acetonitrile (75-05-8)
Dimethylformamide (68-12-2)
Ethanol (64-17-5)
Ethyl acetate (141-78-6)
4,7,13,16,21,24-Hexaoxa-1,10-diazabicyclo[8.8.8]hexacosane (23978-09-8)
4H-Imidazo[1,5-a][1,4]benzodiazepine-3-carboxylic acid, 8-fluoro-5,6-dihydro-6-oxo-, ethyl ester (Ro 15–2344) (79089-72-8)
Methanol (67-56-1)
Potassium carbonate (584-08-7)
Sodium acetate (127-09-3)
Tetrahydrofuran (109-99-9)

For detailed supplier information, see appendix (below).

APPENDIX

Precursor (Ro 15–2344); product # 1690.0000, ABX Germany
[^{18}O]H_2O; ABX, Germany
QMA Sep-Pak; part # WAT023525, Waters
4,7,13,16,21,24-Hexaoxa-1,10-diazabicyclo[8.8.8]hexacosane; product # 291110, Sigma-Aldrich
Sodium acetate; product # S-2889, Sigma-Aldrich
Potassium carbonate; product # 209619, Sigma-Aldrich
Acetonitrile; product # 271004, Sigma-Aldrich
Ethyl acetate; product # E1451, Fisher
Dimethylformamide (DMF); product # 227006, Sigma-Aldrich
Methanol; product # A-452, Fisher
Tetrahydrofuran (THF); product # T-425, Fisher
Filter Millex sterile 22 μ GV; product # SLGV033RS, Fisher
HPLC columns; *for preparation*: Symmetry Prep 7 μm C 18; Part # WAT 066235, Waters, *for analysis*: Phenomenex Prodigy ODS 5 μm; Part # 00G-4097-EO, Phenomenex
HPLC system: Agilent Technologies 1200 series
Interface; Agilent Interface 35900E
Radiation detector; Raytest Gabi
Software; Agilent Chem Station Rev. B.04.01 SP1
GC: SRI 8610C gas chromatograph
Osmometer; Advanced Instruments, Inc. Model 3320 Osmometer

REFERENCES AND NOTES

1. Richards JG, Schoch P, Haring P, Takacs B, Mohler H. J Neurosci 1987;7:1866–1886.
2. McKernan RM, Quirk K, Prince R, Cox PA, Gillard NP, Ragan CI, Whiting P. Neuron 1991;7:667–676.
3. Mohler H. Cell Tissue Res 2006;326:505–516.
4. Atack JR, Smith AJ, Emms F, McKernan RM. Neuropsychopharmacology 1999;20:255–262.
5. Smith AJ, Alder L, Silk J, Adkins C, Fletcher AE, Scales T, Kerby J, Marshall G, Wafford KA, McKernan RM, Atack JR. Mol Pharmacol 2001;59:1108–1118.
6. Hammers A, Koepp MJ, Labbe C, Brooks DJ, Thom M, Cunningham VJ, Duncan JS. Neurology 2001;56:897–906.
7. Szelies B, Weber-Luxenburger G, Mielke R, Pawlik G, Kessler J, Pietrzyk U, Bauer B, Heiss WD. Eur J Neurol 2000;7:393–400.
8. Sood S, Chugani HT. Childs Nerv Syst 2006;22:810–820.
9. Heiss WD, Grond M, Thiel A, Ghaemi M, Sobesky J, Rudolf J, Bauer B, Wienhard K. Stroke 1998;29:454–461.
10. Heiss WD, Sobesky J, Smekal V, Kracht LW, Lehnhardt FG, Thiel A, Jacobs AH, Lackner K. Stroke 2004;35:1892–1898.
11. Rhyzhikov NN, Gomzina NA, Fedorova OS, Vasil'ev DA, Kostikov AP, Krasikova RN. Radiochemistry 2004;46:290–292.
12. Rhyzhikov NN, Seneca N, Krasikova RN, Gomzina NA, Shchukin E, Fedorova OS, Vassiliev DA, Gulya's B, Hall H, Savic I, Halldin C. Nucl Med Biol 2005;32:109–116.
13. Mandap KS, Ido T, Kiyono Y, Kobayashi M, Lohith TG, Mori T, Kasamatsu S, Kudo T, Okazawa H, Fujibayashi Y. Nucl Med Biol 2009;36:403–409.
14. Massaweh G, Schirrmacher E, la Fougere C, Kovacevic M, Wängler C, Jolly D, Gravel P, Reader AJ, Thiel A, Schirrmacher R. Nucl Med Biol 2009;36:721–727.

CHAPTER 14

SYNTHESIS OF 6-[^{18}F]FLUORODOPAMINE (6-[^{18}F]FDA)

MICHAEL A. CHANNING, JOHN L. MUSACHIO, AND JOZEF J. KUSMIERZ

National Institutes of Health, PET Department, Clinical Center, Bethesda, Maryland

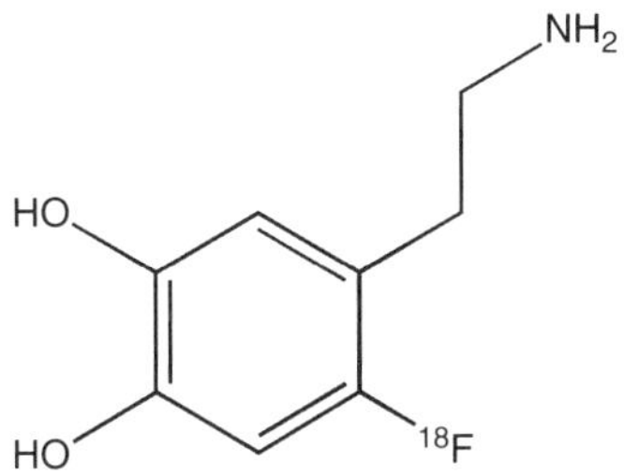

6-[^{18}F]FDA

1 INTRODUCTION

Since the pioneering work of Kirk et al. [1] on the synthesis and biological properties of ring fluorinated biogenic amines much has been invested in the development of 6-[^{18}F]fluorodopamine (6-[^{18}F]FDA) as a positron emission tomographic (PET) tracer. 6-[^{18}F]FDA as an analog of dopamine is readily taken up into vesicles and stored in sympathetic neurons as 6-[^{18}F]FDA, where it is converted by dopamine β-hydroxylase into 6-[^{18}F]fluoronorepinephrine [2]. Early PET studies in animals focused on the metabolic fate of 6-[^{18}F]FDA and ultimately the tracer's distribution within the sympathetic nerve endings of the myocardium [3,4]. This early work laid the foundation for later studies aimed at the assessment of the sympathetic innervation in the heart of patients with Parkinson disease (PD). Patients with PD show little or no uptake of 6-[^{18}F]FDA in the myocardium (Fig. 1) [5].

More recently, 6-[^{18}F]FDA has been used for functional imaging of sympathetic endocrine tumors (pheochromocytoma and paraganglioma) [6]. These

Radiochemical Syntheses, Volume 1: Radiopharmaceuticals for Positron Emission Tomography, First Edition. Edited by Peter J. H. Scott, Brian G. Hockley.

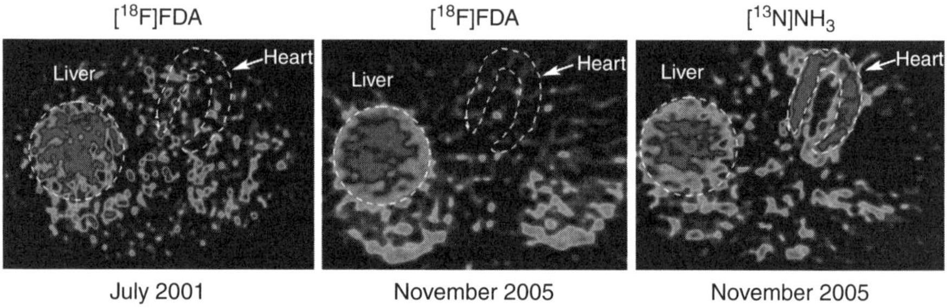

FIGURE 1 Thoracic 6-[^{18}F]FDA and [^{13}N]ammonia images in July 2001 and November 2005 in a patient who first developed symptoms of PD in about May 2005. Note the left ventricle myocardial 6-[^{18}F]FDA-derived radioactivity in both 2001 and 2005, indicating cardiac sympathetic denervation. Myocardial perfusion, as indicated by [^{13}N]ammonia-derived radioactivity, was normal. *Source*: Reprinted with permission from Ref. [5].

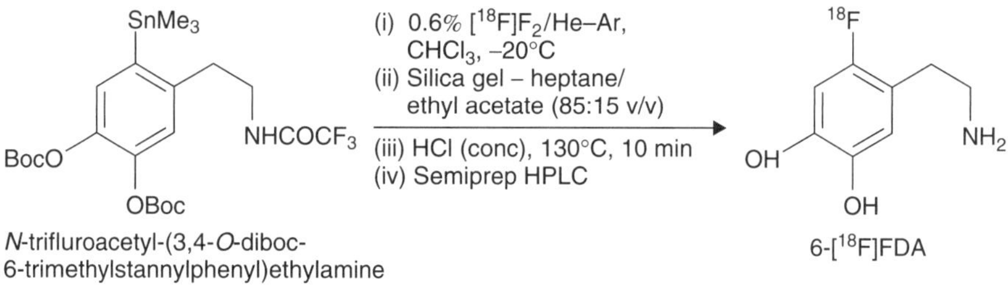

FIGURE 2 Radiosynthesis of 6-[^{18}F]FDA.

tumors are derived from chromaffin cells that produce and secrete catecholamines. The tumors express both the membrane and vesicular transporters, and as a consequence, 6-[^{18}F]FDA is readily translocated and stored in the sympathetic vesicles. 6-[^{18}F]FDA is proving to be a valuable diagnostic tool for the localization and characterization of pheochromocytoma [7,8].

6-[^{18}F]FDA was initially prepared from 6-[^{18}F]fluoroDOPA by enzymatic decarboxylation using aromatic L-amino-acid-decarboxylase (AADC, EC 4.1.1.28) [4]. Later, direct radiosynthetic methods were developed involving the fluoro demetalation of either the appropriately protected 6-aryl-mercury(II) trifluoroacetate or trimethylstannyl derivatives [9,10]. Alternatively, high specific activity 6-[^{18}F]FDA has been prepared via a no-carrier-added nucleophilic ^{18}F-fluorination in a three-step process [11]. This chapter details the automated synthesis of 6-[^{18}F]FDA using the 6-aryl trimethylstannyl precursor (Fig. 2).

2 SYNTHESIS PROCEDURES

CAUTION: All radiochemical syntheses must be carried out using appropriate equipment in a facility authorized for the use of radioactive materials. Personal protective equipment must be worn and all local radiation safety laws followed.

2.1 Production of [^{18}F]Fluorine

An evacuated $^{18}F_2$ proton target [12] of a General Electric Medical Systems (GEMS) PETtrace cyclotron [13] was pressurized ($\Delta P \sim 150$ psig) with [^{18}O]oxygen gas (97% ^{18}O) [14]. The first irradiation sequence continued, and the cyclotron was tuned to deliver protons (15 MeV) at a beam current of 40 μA for 30 min ($^{18}O(p,n)^{18}F$ nuclear reaction) to produce ~600 mCi (22.2 GBq) of fluorine-18 activity. The [^{18}O]oxygen gas was cryogenically transferred back to its supply reservoir. All the fluorine-18 activity remained adsorbed on the walls of the aluminum target.

A second irradiation sequence was initiated, and the target was automatically filled to a final ΔP 154 psig by first slowly adding 3% F_2 in helium [15] (ΔP 29 psig) followed by argon [16] to give a mixture of helium and argon that was 0.6% in F_2. The fluorine target was irradiated a second time at 35 μA for 20 min, exchanging the fluorine-18 activity from the target walls with "cold" fluorine gas (~200 μmol F_2). The target gases ($^{18}O_2$, 3% F_2/He, and Ar), vacuum pump, and cryogenic transfer were all maintained utilizing the GEMS F_2 Proton Target Rig located within a hot cell.

2.2 Fluorodestannylation

The $^{18}F_2$ target outlet valve was opened, and the target contents delivered via an in-line stainless steel pressure regulator (30 psig) and needle valve assembly to the Eckert & Ziegler Modular Lab [^{18}F]F_2-fluoroDOPA/fluorodopamine synthesis module [17] in the hot cell. The progress of the radiosynthesis was continually monitored by way of the control screen (Fig. 3). The gas was allowed to flow (150 mL/min) into a chloroform (10 mL) solution containing the 6-[^{18}F]FDA precursor [18] (120 mg; 196 μmol) at −20°C. The transfer was completed within 20 min. Residual fluorine-18 activity exiting the vent of the reaction vessel was trapped in a 22-L gas sampling bag [19].

2.3 Normal Phase Flash Chromatography

The reaction vessel was pressurized (argon), and the reaction mixture was loaded onto the head of a silica flash cartridge (15 g) [20]. The cartridge was eluted with heptane/ethyl acetate (85:15 v/v) [21] while monitoring the effluent for γ-radioactivity. After ~110 mL of the flash cartridge effluent had been delivered to the waste vessel, the desired protected 6-[^{18}F]FDA was collected (~50 mL eluate) into a glass 100-mL conical flask at 90°C and the solvent evaporated to dryness under vacuum (30 mbar) [22].

2.4 Hydrolysis

To the residual material in the evaporation flask was added 2 mL of concentrated HCl (37%), and the vessel was heated to 130°C. After 10 min, the HCl was evaporated to dryness. Water (2 mL) was added and evaporated to dryness once more. Finally, 4 mL of the high performance liquid chromatographic (HPLC) eluent, 98:2 50 mM H_3PO_4: ethanol [23], was added and allowed to mix.

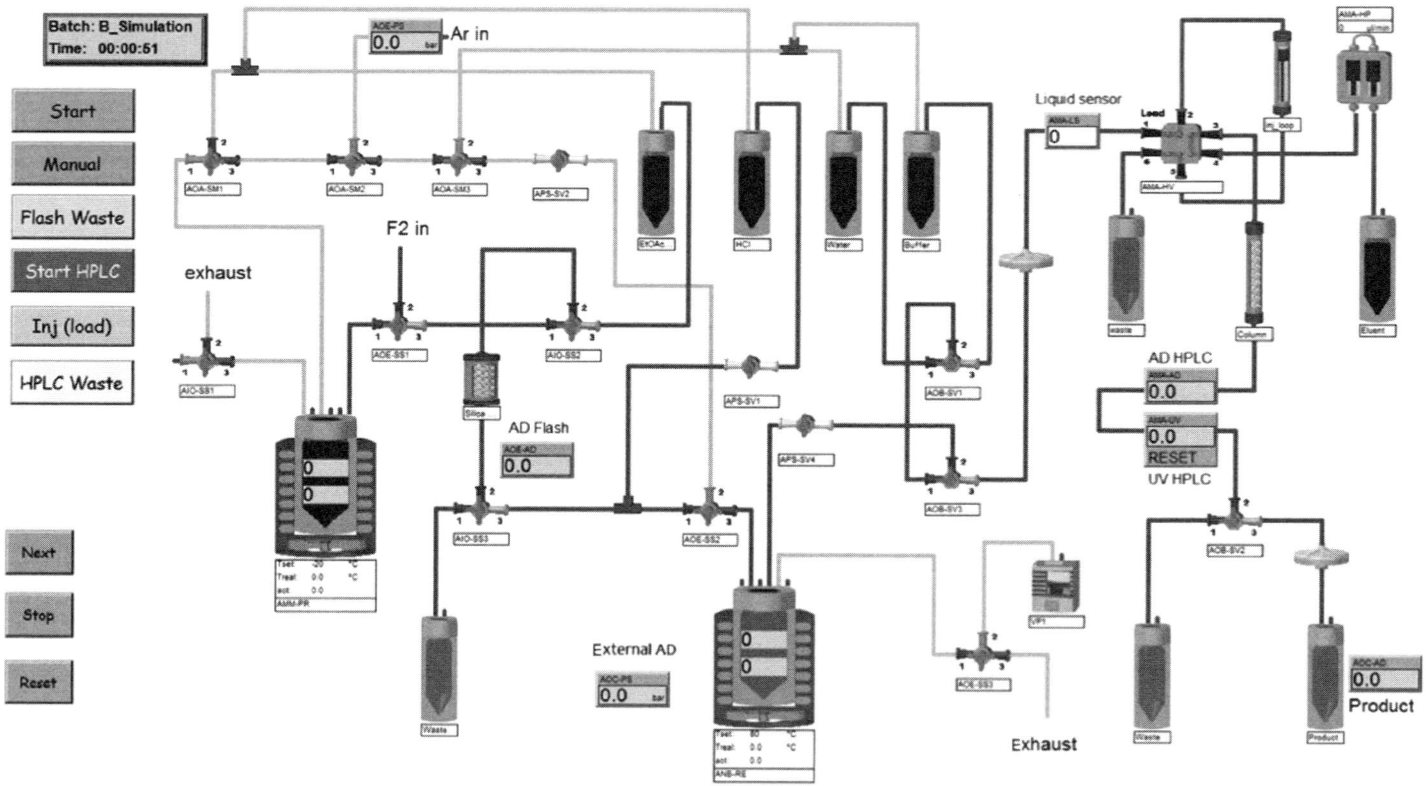

FIGURE 3 Eckert & Ziegler module control screen.

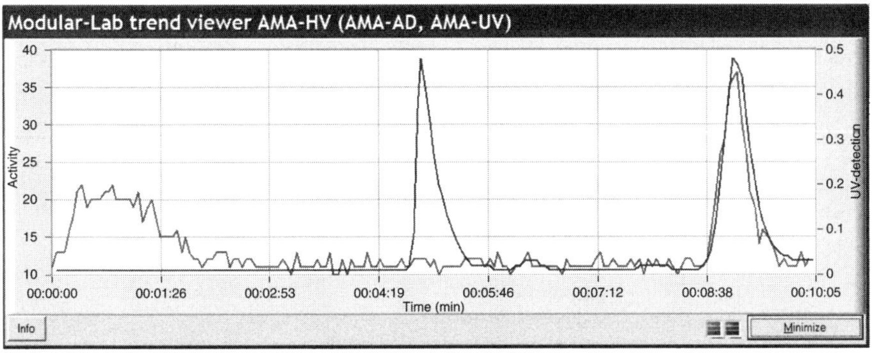

FIGURE 4 Semipreparative HPLC trace of 6-[^{18}F]FDA purification. Smoother trace is UV detection at 254 nm; noisier trace is radioactivity detection.

2.5 HPLC Purification

The contents of the evaporation flask were transferred through a 25-mm (diameter) 1.2-μm porosity filter [24] to the injection loop (5 mL) of the semipreparative Knauer HPLC system [25]. The loop valve was switched to the inject position. The material eluted at 3 mL/min with 98:2 50 mM H_3PO_4: ethanol through the semipreparative HPLC column [26]. The effluent was monitored for UV (254 nm) and γ-radioactivity (Fig. 4).

The desired product, which eluted at 8 min, was directed through a Millex GS 0.22-μm sterile filter into a 20-mL sterile amber serum vial containing FDA ascorbate/buffer (12 mL) [27] and equipped with a second Millex GS 0.22 sterile vent filter [28]. Typically 57 mCi (2.11 GBq) of 6-[^{18}F]FDA is obtained with an uncorrected radiochemical yield of 20%.

Note: If ascorbic acid is not added, the final product will initially pass all quality control (QC) procedures but fail to meet the full 5-h expiry.

3 QUALITY CONTROL PROCEDURES

CAUTION: All radiopharmaceuticals produced for clinical use must have local regulatory approval before use in humans. Trained personnel must carry out QC procedures, and each dose must meet all established QC criteria before release to the clinic.

QC procedures for 6-[^{18}F]FDA, based on the current requirements for compounding PET radiopharmaceuticals described in the US Pharmacopeia [29], are summarized below. Complete QC data for three consecutive batches of 6-[^{18}F]FDA produced using the radiosynthetic procedure are summarized in Table 1.

Each of the three batches met all of the established QC criteria. The specific activities are reported at end-of-synthesis (EOS) for these qualifying samples

TABLE 1 Quality Control Data for Three 6-[^{18}F]FDA Qualifying Preparations

Test	Acceptance Criteria	528 VR3	530 VR4	532 VR5
Radionuclidic identity	The measured half-life is between 105 and 115 min	Pass	Pass	Pass
Radiochemical identity	The retention time of 6-[^{18}F]FDA for injection (±10%) to that of fluorodopamine reference standard	Pass	Pass	Pass
Radiochemical purity (%)	≥90.0% 6-[^{18}F]FDA for injection	>99	>99	>99
Dose mass limit (mg)	≤6.0 mg per injected dose	3.43	5.51	4.53
Specific activity	≥100mCi/mmol (3.7 GBq/mmol) at time-of-injection	1402 (EOS)	1344 (EOS)	1331 (EOS)
Chemical impurity mass limit (%)	Sum of integrated UV impurities ≤10%	4.0	4.9	3.1
Chemical impurity: tin (ppm)	≤5ppm total Sn	0.76	0.25	<0.10 (LOQ)
Residual solvent:				
Chloroform (ppm)	≤60	<LOD	<LOD	<LOD
Ethyl acetate (ppm)	≤5000	<LOD	<LOD	<LOD
Heptane (ppm)	≤5000	<LOD	<LOD	<LOD
Ethyl alcohol (ppm)	≤20,000	3266	2987	1875
pH	5.0–8.0	Pass (6.1)	Pass (6.1)	Pass (7.0)
Sterility testing	No growth observed in 7 d (BACTEC™)	Pass	Pass	Pass
Membrane filter integrity	Sterile 0.22-μm filters are single use. Each membrane tested at 50 psig	Pass	Pass	Pass
Bacterial endotoxins Limulus amebocyte lysate (LAL) assay (EU/mL)	≤2.5Endotoxin units (EU) per kilogram body weight and not >175 EU per subject	<1.6	<1.6	<1.6

LOQ, limit of quantization; LOD, limit of detection.

since there was no patient data. The values given for dose mass limit are the total mass in milligrams in the vial (15 mL formulation buffer).

3.1 Radionuclidic Identity

An aliquot (0.2 mL) was removed and dispensed into the official 5-mL assay ampoule [30] containing 4.8 mL water diluent. The ampoule was sealed with Parafilm®, placed in a dose calibrator [31], and the reading (highest sensitivity range) and initial time entered into the half-life calculator. After >5 min, a second reading was entered with the corresponding final time. The calculated half-life must be within ±5% of the half-life for fluorine-18 (109.8 min).

3.2 Radiochemical Identity

Radiochemical identity was determined on an Agilent 1100 Series HPLC [32] equipped with a UV diode array detector (DAD, 220 nm) and a Bioscan Flow Count 3200 NaI/PMT γ-radiation scintillation detector [33]. The data were captured and processed using Agilent ChemStation for LC 3D systems acquisition software. For the analysis of 6-[^{18}F]FDA, an Agilent Zorbax SB-Aq column [34] was employed (3.5 μm, 4.6 × 100 mm). The injection volume was set to 5 μL. The column was eluted at 1 mL/min with 10/90 acetonitrile/F2 buffer [35]. Before analysis, the HPLC system was equilibrated and qualified as being suitable for use. The system was qualified using three injections of system suitability standard (300 μg/mL of fluorodopamine) followed by blank injection. After sample analysis, the bracketing standard (300 μg/mL of fluorodopamine) was injected. The calculated concentration of the bracketing standard should be within (±10%) of the expected value. The calculated relative standard deviation (RSD) for the system suitability peak retention time and response area must be ≤2%. The retention time of the 6-[^{18}F]FDA sample peak must be within the retention time window (±10%) of the qualifying reference standards [36].

3.3 Radiochemical Purity

Radiochemical purity was determined on an Agilent 1100 Series HPLC equipped with a UV detector (220 nm) and a Bioscan Flow Count 3200 NaI/PMT γ-radiation scintillation detector. The analysis of 6-[^{18}F]FDA for radiochemical purity utilizes the same sample injection and a data set identical to that described in Sections 3.2–3.6. The integrated area for the 6-[^{18}F]FDA peak must be ≥90% of the total integrated area for the γ-radiation channel (Fig. 5).

3.4 Dose Mass Limit

The mass concentration (μg/mL) was determined on an Agilent 1100 Series HPLC equipped with a UV detector (220 nm) and a Bioscan Flow Count 3200 NaI/PMT γ-radiation scintillation detector. A five-level external standard calibration curve was created for 6-fluorodopamine quantitative analysis. The 6-fluorodopamine concentration was calculated by ChemStation using a built-in standard curve. The analysis of 6-[^{18}F]FDA for mass concentration utilizes the same sample injection and a data set identical to that described in Sections 3.2–3.6 (Fig. 6).

3.5 Specific Activity

Specific activity was determined using an Agilent 1100 Series HPLC equipped with a UV detector (220 nm) and a Bioscan Flow Count 3200 NaI/PMT γ-radiation scintillation detector. The analysis of 6-[^{18}F]FDA for specific activity utilizes the same sample injection and a data set identical to that described in

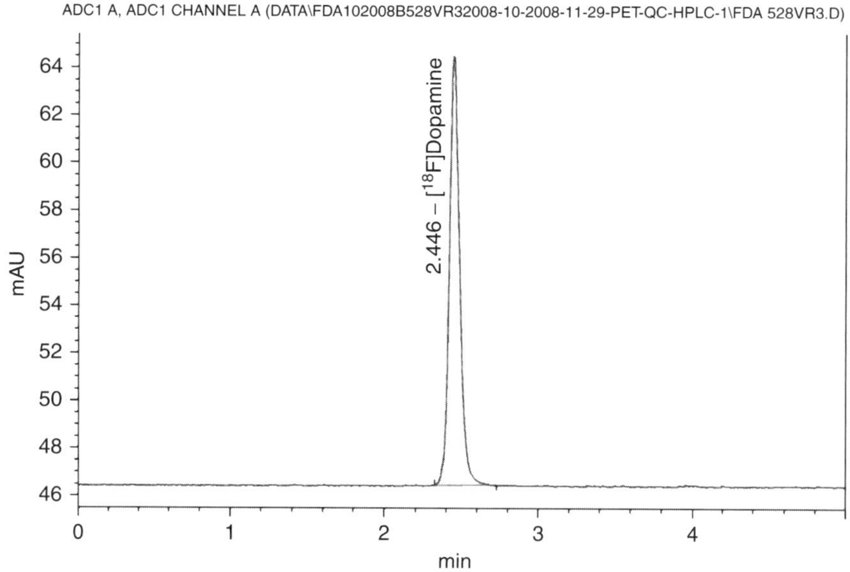

FIGURE 5 Radioactivity analytical HPLC trace for 6-[^{18}F]FDA sample.

Sections 3.2–3.6. The area of the UV peak for 6-[^{18}F]FDA is used to determine the concentration (micrograms per milliliter) of the sample injected onto the HPLC. The activity per unit volume obtained from the official assay is used to calculate the activity in the sample corrected to end-of-bombardment (EOB). The two calculated values are used to calculate the specific activity of the 6-[^{18}F]FDA sample in millicurie per millimole (EOB). The specific activity at the time-of-injection (TOI) is calculated by correction to the actual TOI. The specific activity must not be $\geq$100 mCi/mmol ($\geq$3.7 GBq/mmol).

3.6 Chemical Impurity Mass Limit (UV)

Chemical UV impurity limit was determined on an Agilent 1100 Series HPLC equipped with a UV detector (220 nm) and a Bioscan Flow Count 3200 NaI/PMT γ-radiation scintillation detector. The analysis of 6-[^{18}F]FDA for chemical UV impurity limit utilizes the same sample injection and a data set identical to that described in Sections 3.2–3.6. The sum of the area(s) for impurities (DAD/UV) must be $\leq$10% of the total integrated area (DAD/UV) (Fig. 6).

3.7 Chemical Impurity Limit of Tin

Attributable to use of "*N*-trifluoroacetyl-3,4-di-tert-butoxycarbonyloxy-6-(trimethylstannyl) phenethylamine" as the precursor material in the production of 6-[^{18}F]FDA, the total tin concentration (in parts per million) was determined. An aliquot of every batch of decayed sample was maintained in a -20°C freezer, and a predetermined number of batches were randomly selected quarterly and shipped to the contract laboratory for analysis.

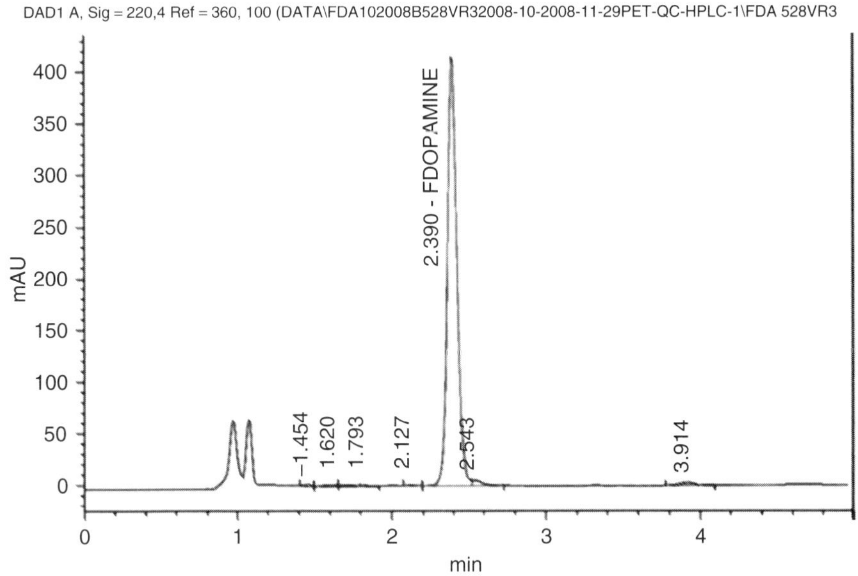

FIGURE 6 UV analytical HPLC trace for 6-[^{18}F]FDA sample.

Samples were processed using a validated microwave-assisted acid digestion method, where a portion of sample was weighed into a microwave vessel and placed in the microwave and heated to a minimum of 180±10°C in 5 min, and maintained at this temperature for 10 min. The vessels were allowed to cool before removal of the digested sample. Concentration of total tin was then determined by inductively coupled plasma mass spectrometry (ICP-MS) and total tin concentration reported to the lowest possible (typically in the range of 0.01–0.5 ppm) [37].

3.8 Residual Solvent Analysis

Analysis of residual solvent levels (chloroform, ethyl acetate, heptane, and ethanol) in 6-[^{18}F]FDA were carried out on an Agilent 6850 gas chromatograph (GC) equipped with a flame ionization detector (FID) and an Agilent 6850 automatic liquid sampler (ALS) [38]. The injection port was configured for split sample injection at a split ratio of 20:1 and operated at 250°C. The column (J & W DBWAX column 30 m × 0.25 mm × 0.25 μm film thickness) [39] was initially operated at 35°C for 2 min, and the temperature was then increased to 85°C (at a rate of 20°C/min) and held at 85°C for 1 min. Helium was used as the carrier gas, and the column flow rate was adjusted to be 2.2 mL/min (average velocity, 43 cm/s). The hydrogen flow rate and air flow rate were set to 40 and 450 mL/min, respectively. The makeup flow of helium was set to a rate of 45 mL/min. The detector was operated at 250°C. A 10-μL syringe was used for the ALS. The injection volume was set to 1.0 μL. Standard solutions

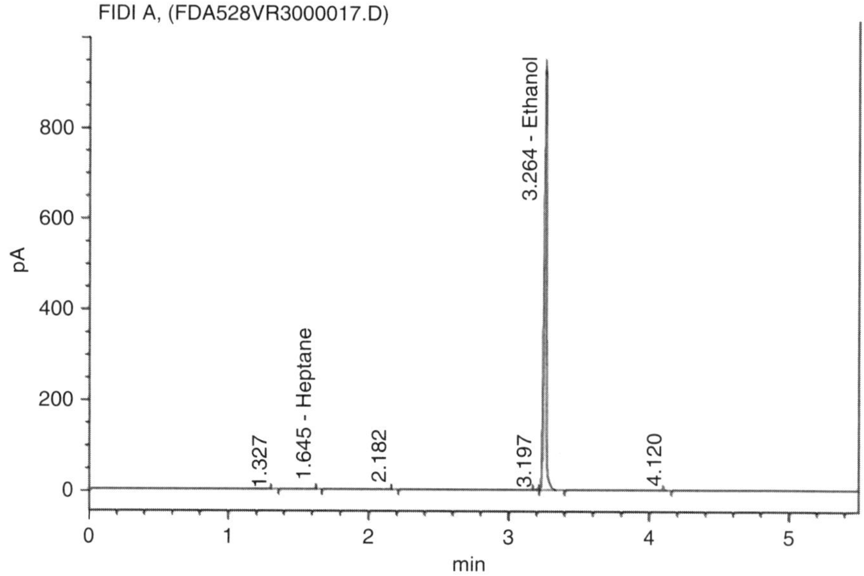

FIGURE 7 Analytical GC trace (FID)—1875 ppm (0.19%) ethanol.

of chloroform, ethyl acetate, heptane, and ethanol were prepared in HPLC grade water for detector calibrations. On the basis of an average of five replicates for each level, five-level external standard calibrations were created for the online quantitative analysis. Limits imposed by the ICH Harmonised Tripartite Guidelines [40] are <60 ppm/day for chloroform and <5000 ppm/day for class 3 solvents (ethyl acetate, heptane, and ethanol) (Fig. 7).

3.9 pH Testing

The pH of the 6-[^{18}F]FDA dose was analyzed by applying a small amount of the dose to ColorpHast® (range 6.5–10) nonbleeding pH-indicator strips [41] and determined by visual comparison to the scale provided. Dose pH must be 5.0–8.0.

3.10 Sterility Testing

Every single batch of the radiopharmaceutical was tested for sterility. Sterility tests were started within 24–72 h after the completion of the production.

An equal aliquot of the decayed PET drug (0.2–10 mL) was added to one of each 40-mL BACTEC standard anaerobic/F and Standard 10 aerobic/F culture bottles using aseptic technique (Becton Dickinson PN 442191 and 442260, respectively). The inoculated medium bottles were transferred to NIH Microbiology Section (10 Center Drive, Clinical Center), where samples were processed no later than 2 h after inoculation. Inoculated BACTEC bottles were incubated in the BACTEC fluorescent series instrument (BACTEC FX) for the qualitative culture and recovery of anaerobic and aerobic microorganisms from the PET radiopharmaceutical. Each vial contained a sensor that was monitored by the instrument every 10 min, which could detect the increase in CO_2 produced by the growth of microorganisms. This allowed positive results to be obtained more rapidly when compared to traditional methods.

3.11 Sterile Filter Integrity Test

The sterile filter from the 6-[^{18}F]FDA was connected to a the bubble test apparatus. The nitrogen pressure was gradually increased. If the pressure was raised above the filter acceptance pressure (50 psi) without seeing a stream of bubbles, the filter was considered intact. If a stream of bubble occurs $<$50 psi, the test fails.

3.12 Endotoxin Analysis

Endotoxin content in doses of 6-[^{18}F]FDA were analyzed by a Charles River Laboratories EndoSafe® Portable Testing System [42]. Doses must contain $\leq$2.5 endotoxin units (EU) per kilogram body weight and no more than 175 EU per subject.

CHEMICAL ABSTRACTS NOMENCLATURE (REGISTRY NUMBER)

Acetonitrile (75-05-8)
Argon (7440-37-1)
Ascorbic acid (50-81-7)
Chloroform (67-66-3)
Ethanol (64-17-5)
Ethyl acetate (141-78-6)
Ethylenediaminetetraacetic acid disodium salt dihydrate (6381-92-6)
Fluorine (7782-41-4)
6-Fluorodopamine (71144-39-3)
Helium (7440-59-7)
Heptane (142-82-5)
Hydrochloric acid (7647-01-0)
Monosodium phosphate monohydrate (7558-80-7)
[^{18}O]Oxygen (labeled, 32767-18-3; unlabeled, 7782-44-7)
N-Trifluoroacetyl-3,4-di-*tert*-butoxycarbonyloxy-6-(trimethylstannyl) phenethylamine (170953-68-1)
Octanesulfonic acid sodium salt (5324-84-5)
Phosphoric acid (7664-38-2)
Sodium phosphate dibasic anhydrous (7558-79-4)

REFERENCES AND NOTES

1. Kirk KL, Cantacuzene D, Nimitkitpaisan Y, McCulloh D, Padgett WL, Daly JW, Creveling CR. Synthesis and biological properties of 2-, 5-, and 6-fluoronorepinephrines. J Med Chem 1979;22(12):1493–1497.
2. Chiueh CC, Zukowska-Grojec Z, Kirk KL, Kopin IJ. 6-Fluorocatecholamines as false adrenergic neurotransmitters. J Pharmacol Exp Ther 1983;225(3):529–533.
3. Eisenhofer G, Hovevey-Sion D, Kopin IJ, Miletich R, Kirk KL, Finn R, Goldstein DS. Neuronal uptake and metabolism of 2- and 6-fluorodopamine: false neurotransmitters for positron emission tomographic imaging of sympathetically innervated tissues. J Pharmacol Exp Ther 1989;248(1):419–427.
4. Goldstein DS, Chang PC, Eisenhofer G, Miletich R, Finn R, Bacher J, Kirk KL, Bacharach S, Kopin IJ. Positron emission tomographic imaging of cardiac sympathetic innervation and function. Circulation 1990;81(5):1606–1621.
5. Goldstein DS. Cardiac denervation in patients with Parkinson disease. Cleve Clin J Med 2007;74(1):S91–S94. Review.
6. Pacak K, Eisenhofer G, Carrasquillo JA, Chen CC, Li S, Goldstein DS. 6-[^{18}F]Fluorodopamine positron emission tomographic (PET) scanning for diagnostic localization of pheochromocytoma. Hypertension 2001;38:6–8.
7. Ilias I, Yu J, Carrasquillo JA, Chen CC, Eisenhofer G, Whatley M, McElroy B, Pacak K. Superiority of 6-[^{18}F]-fluorodopamine positron emission tomography versus [^{131}I]-metaiodobenzylguanidine scintigraphy in the localization of metastatic pheochromocytoma. J Clin Endocrinol Metab 2003;88:4083–4087.
8. Timmers HJ, Chen CC, Carrasquillo JA, Whatley M, Ling A, Havekes B, Eisenhofer G, Martiniova L, Adams KT, Pacak K. Comparison of ^{18}F-fluoro-L-DOPA, $_{18}$F-fluoro-deoxyglucose, and ^{18}F-fluorodopamine PET and ^{123}I-MIBG scintigraphy in the localization of pheochromocytoma and paraganglioma. J Clin Endocrinol Metab 2009;94(12):4757–4767.
9. Chaly T, Dahl JR, Matacchieri R, Bandyopadhyay D, Belakhlef A, Dhawan V, Takikawa S, Robeson W, Margouleff D, Eidelberg D. Synthesis of 6-[^{18}F]fluorodopamine with a synthetic unit made up of primarily sterile disposable components and operated by a master slave manipulator. Appl Radiat Isot 1993;44(5):869–873.
10. Namavari M, Satyamurthy N, Barrio JR. Synthesis of 6-[^{18}F]fluorodopamine, 6-[^{18}F]fluoro-m-tyramine and 4-[^{18}F]fluoro-m-tyramine. J Labelled Comp Radiopharm 1995;36(9):825–833.
11. Ding Y-S, Fowler JS, Gatley J, Dewey SL, Wolf AP, Schyler DJ. Synthesis of high specific activity 6-[^{18}F]fluorodopamine for positron emission tomography studies of sympathetic nervous tissue. J Med Chem 1991;34(2):861–863.
12. GEMS $^{18}F_2$ proton target system, cylindrical aluminum target body (87 mL), Havar foil (0.0025 mm). This included the GEMS $^{18}F_2$ proton rig located in the hot cell. The rig was used for the delivery of all target gases, and vacuum pump (Edwards Model E2M1.5) was used for evacuating the target.
13. GE Healthcare USA. http://www.gehealthcare.com
14. ^{18}O oxygen gas (97% ^{18}O) chemical purity >99.8% PN OLM-212-0 (labeled CAS # 32767-18-3; unlabeled CAS # 7782-44-7) Cambridge Isotopes Laboratories, Inc., 50 Frontage Rd., Andover, MA 01810–5413, USA. The 18-O reservoir was maintained at ≥260 psig. This insured a fill pressure for the first irradiation of ≥140 psig.
15. Certified 2-component gas mixture: 3% F_2 in He, (CGA 679) Matheson Tri-Gas Inc., PN: PZ6953.
16. Argon, UHP Grade, (≥99.996%) (CGA 580) Roberts Oxygen Co., PN: R104C301.
17. Eckert & Ziegler Modular Lab [^{18}F]F_2-fluoroDOPA/fluorodopamine synthesis module composed of : SVM (2), SMM (1), SSM (3), PRM (1), REM (1), and HPLC (1) modules.
18. 6-[^{18}F]FDA precursor: *N*-trifluoroacetyl-3,4-di-*tert*-butoxycarbonyloxy-6-(trimethylstannyl) phenethylamine, ABX Biochemical, ($C_{23}H_{34}F_3NO_7Sn$; FW 612.22) PN: 1371.

19. Cali-5-Bond™ 22 L pillow gas sampling bag (GSB-P/22) Mylar with straight through connector (STC-NPB-F-SR). Calibrated Instruments, Inc., 171 Brady Avenue, Hawthorne, NY 10532, Tel (914) 741–5700 email: gsb@calibrated.com
20. The silica cartridge was a Sepra Si 50, 15 g (Sorbtech PN: AX1248-20), and used in conjunction with the DASi 65 Module Kit (Sorbtech PN: AX1236-1). The cartridge was conditioned before use by wetting with HEP : EtOAc 85:15 (50 mL).
21. Silica cartridge eluent: to the clean 250-mL reservoir labeled "HEP/EtOAc 85:15" was sequentially added 170 mL heptane followed by 30 mL ethyl acetate. The stir bar was added to the reservoir, and the solution was stirred until homogenous. Connect the reservoir to the synthesis module. This solution must be prepared on the day of 6-[^{18}F]FDA production and is discarded to organic waste at clean up.
22. Vacuum pump KNF UN842. During evaporation steps, solvents were trapped using a vacuum trap (KGW Nr. 17377, 20. 5 by 5 cm) submerged in a liquid nitrogen dewar. The vent side of the vacuum pump and evaporator were passed through soda lime (4–8 mesh) to trap HCl vapors.
23. HPLC eluent 6-[^{18}F]FDA: 98:2 50 mM H_3PO_4: ethanol. Dispense ~800 mL of DDI water to a clean 1-L graduated cylinder. Add 3.35 mL of 85% phosphoric acid to the cylinder via a pipette and dilute to the 980-mL mark with DDI water. Measure 20 mL USP grade ethanol in a clean 50-mL graduated cylinder, and add this alcohol to the 1-L cylinder containing the phosphoric acid solution. Add a clean magnetic stir bar to the 1-L cylinder, and stir at room temperature until homogenous. Filter (Pall Ultipor, Nylon filter, 0.45 μm, PN: NX047100) using a vacuum filtration flask, and pour into a 1-L reservoir; label as "98:2 50 mM H3PO4 : ethanol" (label should include: date of preparation, "store at room temperature," and the five-day date of expiry).
24. HPLC injectate filter: Pall Versapor 25-mm filter, 1.2 μm, PN: 4190.
25. The Semi-Prep Knauer HPLC consisted of: WellChrom HPLC-Pump K-120 (PN: A41501), WellChrom Fixed Wavelength Detector K-2501 (A4180), and Knauer 6-Port/2-Channel injection valve (A1369-1).
26. Semi-Prep column: YMC-Pack ODS-AQ, 10 × 250 mm, 5 μm, PN: AQ12S05-2510WT. Column was conditioned with HPLC eluent (98:2 50 mM H_3PO_4: ethanol) at 3 mL/min for ~10 min.
27. Formulation buffer for 6-[^{18}F]FDA: L-ascorbic acid (1.0±0.1 mg) was added to an empty 20-mL vial. The transfer was verified by visual inspection. To this was added 12 mL of dibasic anhydrous sodium phosphate (8.51 mg/mL in sterile water). This solution was prepared on the day of 6-[^{18}F]FDA production. The solution was loaded into a sterile syringe and passed through a sterile 0.22-μm filter (Millipore Millex GS 33 mm, Millipore) into an amber 20-mL sterile dose vial (NIH Pharmacy).
28. Sterilization filter: Sterile 33-mm filter (0.22 μm), Millex GS Millipore, PN: SLGS033SS.
29. United States Pharmacopeia Chapter <823> Radiopharmaceuticals for Positron Emission Tomography—Compounding, USP 32 NF 27 (2009).
30. Official assay glass ampoule size: 5 mL (diameter × height, 16.5 × 84 mm) Wheaton PN: 176779.
31. Dose calibrator: Capintec (6 Arrow Road Ramsey, NJ 07446) CRC15R. The dose calibrator was checked for constancy against traceable NIST standard sources each day.
32. Analytical HPLC system: Agilent 1100 series HPLC was comprised of binary pump G1312A, Autosampler G1313A, diode array UV detector G1315B, thermostated column compartment G1316A, vacuum degasser G1322A, and 35900E A/D dual channel interface with an Agilent ChemStation data acquisition system.
33. Radio-HPLC detector: Bioscan Flow Count 3200. The NaI/PMT assembly was shielded with ~1.75 in of lead to minimize the background.
34. Agilent Zorbax SB-Aq column, (3.5 μm, 4.6 × 100 mm), PN 861953-914.
35. Analytical HPLC eluent: 10:90 acetonitrile : F2 buffer; Monosodium phosphate monohydrate (13.6 g; 0.1 mol), Na_2EDTA dihydrate (100 mg), and octanesulfonic acid (OSA) sodium salt (200 mg) were weighed and transferred into a 1-L volumetric flask; the contents were dissolved in approximately 900 mL of DDI water. The pH of the solution was adjusted with concentrated

phosphoric acid dropwise (~11 drops) to pH 3.5 (meter). The volumetric flask was brought to volume with DDI water and mixed well. The mobile phase was stored at room temperature and used within two months.

36. Reference standard of 6-fluorodopamine hydrobromide was obtained from Isosciences, 386 Maiden Lane, King of Prussia, PA 19406-1803, and used as received. Sample quality control analysis injection scheme: the 6-[^{18}F]FDA samples were analyzed by a single injection (5 μL) without dilution. The following sequence was used: three injections (5 μL each) of FDA reference standard C (300 μg/mL), one injection of analytical blank (eluent), one injection of 6-[^{18}F]FDA sample, and finally one injection of standard C to bracket the injection sequence. The instrument processing method was set up so that the chromatograms were integrated, concentrations were calculated using the standard curve method, and results were printed immediately after each run. The instrument was considered suitable for use as long as the percentage RSD of peak response and retention time for the injection sequence was $\leq 2\%$.
37. Tin analysis was contracted out to Galbraith Laboratories (P.O. Box 51610, Knoxville, TN 37950).
38. Agilent 6850 gas chromatograph equipped with a flame ionization detector (FID) and an Agilent 6850 automatic liquid sampler (ALS).
39. Analytical GC column: J & W DBWAX column 30 m × 0.25 mm × 0.25 mm film thickness; Agilent PN: 122-7032E.
40. ICH Harmonised Tripartite Guidelines. http://www.fda.gov/RegulatoryInformation/Guidances/ucm128223.htm
41. ColorpHast pH-indicator strips pH 6.5–10 nonbleeding (PN: 9583).
42. Charles River Laboratories EndoSafe-PTS™, Charles River Laboratories International, Inc., 251 Ballardvale Street, Wilmington, MA 01887.

CHAPTER 15

SYNTHESIS OF 2-([^{18}F]FLUORO)-3-[(2*S*)-2-AZETIDINYLMETHOXY]PYRIDINE ([^{18}F]2FA)

Hayden T. Ravert, Daniel P. Holt, and Robert F. Dannals

Division of Nuclear Medicine, The Johns Hopkins University School of Medicine, Baltimore, Maryland

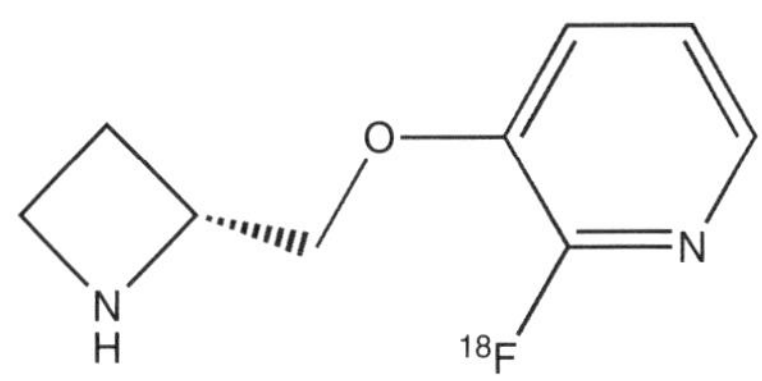

[^{18}F]FA

1 INTRODUCTION

Alterations of the $\alpha 4\beta 2$-subtype of the nicotinic acetylcholine receptor have been implicated in a wide variety of disorders, including Alzheimer's disease, Parkinson's disease, schizophrenia, attention deficit hyperactivity disorder, and depression, as well as in tobacco dependence. A recent review emphasizes the importance of noninvasive imaging of the $\alpha 4\beta 2$-subtype and summarizes the previous radiotracer research in this area [1].

Thus far, only a handful of radioligands have been used to image this binding site in humans. Radiohalogenated analogs of the Abbott Laboratories compound A-85380, a neuronal nicotinic acetylcholine receptor agonist, have historically shown the most promise. 2-([^{18}F]Fluoro)-3-[(2*S*)-2-azetidinylmethoxy]pyridine ([^{18}F]2FA) is one of those analogs, which has been used the most [2–8].

Radiochemical Syntheses, Volume 1: Radiopharmaceuticals for Positron Emission Tomography, First Edition. Edited by Peter J. H. Scott, Brian G. Hockley.

2 SYNTHESIS PROCEDURES

CAUTION: All radiochemical syntheses must be carried out using appropriate equipment in a facility authorized for the use of radioactive materials. Personal protective equipment must be worn and all local radiation safety laws followed.

2.1 Production of [^{18}F]Fluoride

[^{18}O]Water (PETNET Solutions, approximately 1.7 mL) was loaded into the silver body, high yield [^{18}F]fluoride target of a General Electric Medical Systems (GEMS) PETTrace cyclotron. The target was bombarded with a proton beam of 30 μA for 30 min to produce approximately 1 Ci (37 GBq) of [^{18}F]fluoride by the ^{18}O(p,n)^{18}F nuclear reaction.

2.2 Synthesis of [^{18}F]2FA

Multiple radiochemical syntheses of [^{18}F]2FA have been reported in the literature. The two procedures described below are adapted from the published syntheses modified by Horti et al. [9] and Dolle et al. [10], using a GEMS TRACERlab FX_{FN} synthesis module and Advion NanoTek Microfluidic Synthesis System, respectively.

2.3 TRACERlab Synthesis

2.3.1 Preparation of the GEMS TRACERlab FX_{FN}

A Chromafix 30-PS-HCO_3 SPE cartridge [11] was attached between valves 13 and 15 on the GEMS TRACERlab FX_{FN} (as shown schematically in Fig. 2), and C-18 SepPak [12] was attached between valves 14 and 16. Aqueous potassium carbonate [13] (2.3 mg in 300 μL water [14]), Kryptofix-[2.2.2] [15] in acetonitrile [16] (13 mg in 1 mL), the iodo precursor (Boc-2IA [17], 4 mg) in DMSO [18] (400 μL), 600 μL of the first high performance liquid chromatographic (HPLC) mobile phase (described below), methylene chloride [18] (2 mL), trifluoroacetic acid [19] (200 μL), and 600 μL of the second HPLC solvent (described below) were added to the vials attached to valves 1,2,3,4,7,8, and 9, respectively. Compressed air, vacuum, and argon [20] gas were confirmed to be ready, and the appropriate software was loaded.

2.3.2 Azeotropic Drying of [^{18}F]Fluoride

After the preparation of the GEMS TRACERlab FX_{FN}, the [^{18}F]fluoride was delivered to the receiving vessel as a solution of [^{18}O]water. The [^{18}F]fluoride was pulled to the Chromafix 30-PS-HCO_3 SPE cartridge and the [^{18}O]water collected for recycling. The resin cartridge was eluted with the aqueous potassium carbonate into the TRACERlab FX_{FN} glassy-carbon reaction vessel and the Kryptofix-[2.2.2] solution added. The mixture of [^{18}F]fluoride, potassium carbonate, and Kryptofix was azeotropically dried first under vacuum and argon flow at 55±2°C for 2.25 min, then under vacuum alone at 125°C for 2.25 min, and finally under vacuum and argon flow (60–80 mL/min) at 140°C for 1.5 min.

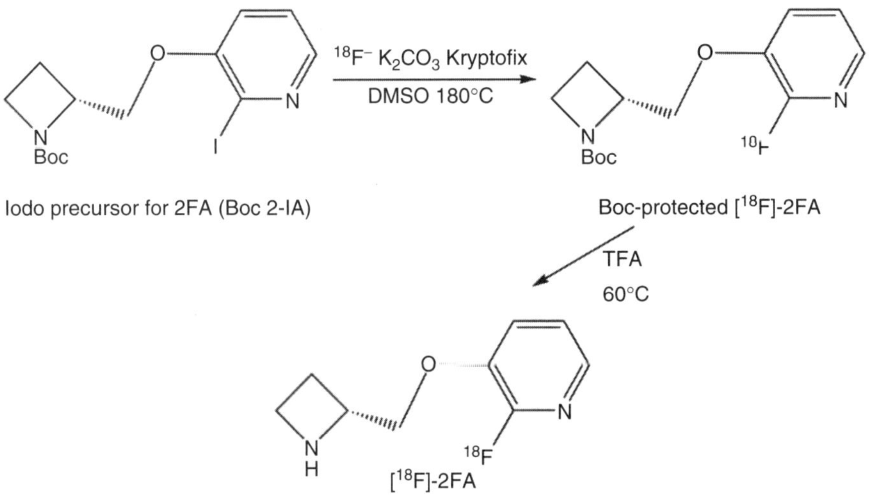

FIGURE 1 Synthesis of [^{18}F]2FA via the iodo precursor.

2.3.3 Synthesis of [^{18}F]2FA

The solution of the iodo precursor (Fig. 1) was added to the reaction vessel containing the [^{18}F]fluoride, potassium carbonate, and Kryptofix, and the reaction mixture was heated at 180°C for 15 min. After cooling to 40°C under vacuum, the aliquot of the first HPLC mobile phase described below was added to the reaction vessel. The TRACERlab system automatically paused operation, and the solution was remotely injected onto a equilibrated (7 mL/min for 20 min) Hamilton PRP-1 C-18 [21] (7.8 × 305 mm) column eluted (48% acetonitrile [22]: 52% water [22]) at a flow rate of 7 mL/min (Fig. 3). The Boc-[^{18}F]2FA intermediate was collected in 30 mL water. The HPLC fraction was pushed by argon (60–80 mL/min) through the C-18 SepPak, dried for 1 min, and eluted from the C-18 SepPak with methylene chloride into the second reaction vessel.

The TRACERlab system was started again, and trifluoroacetic acid was added to the second reaction vessel. The methylene chloride solution was evaporated by heating to 60°C for 2 min. After cooling, an aliquot of the second HPLC solvent described below was added.

2.3.4 Purification and Formulation of [^{18}F]2FA

The crude reaction mixture was purified by HPLC using a Hamilton PRP-1 C-18 (7.8 × 305 mm) column eluted with 6% acetonitrile: 4% methanol [22]: 90% water containing 0.2% trifluoroacetic acid mobile phase at a flow rate of 7 mL/min (Fig. 4, equilibrated at 7 mL/min for 20 min). The [^{18}F]2FA fraction was collected, evaporated to dryness at 80°C under vacuum, dissolved in 7 mL normal sodium chloride [23], and passed through a sterile 0.22-μm filter [24] into a sterile vial containing 3 mL sterile 8.4% sodium bicarbonate [23]. The

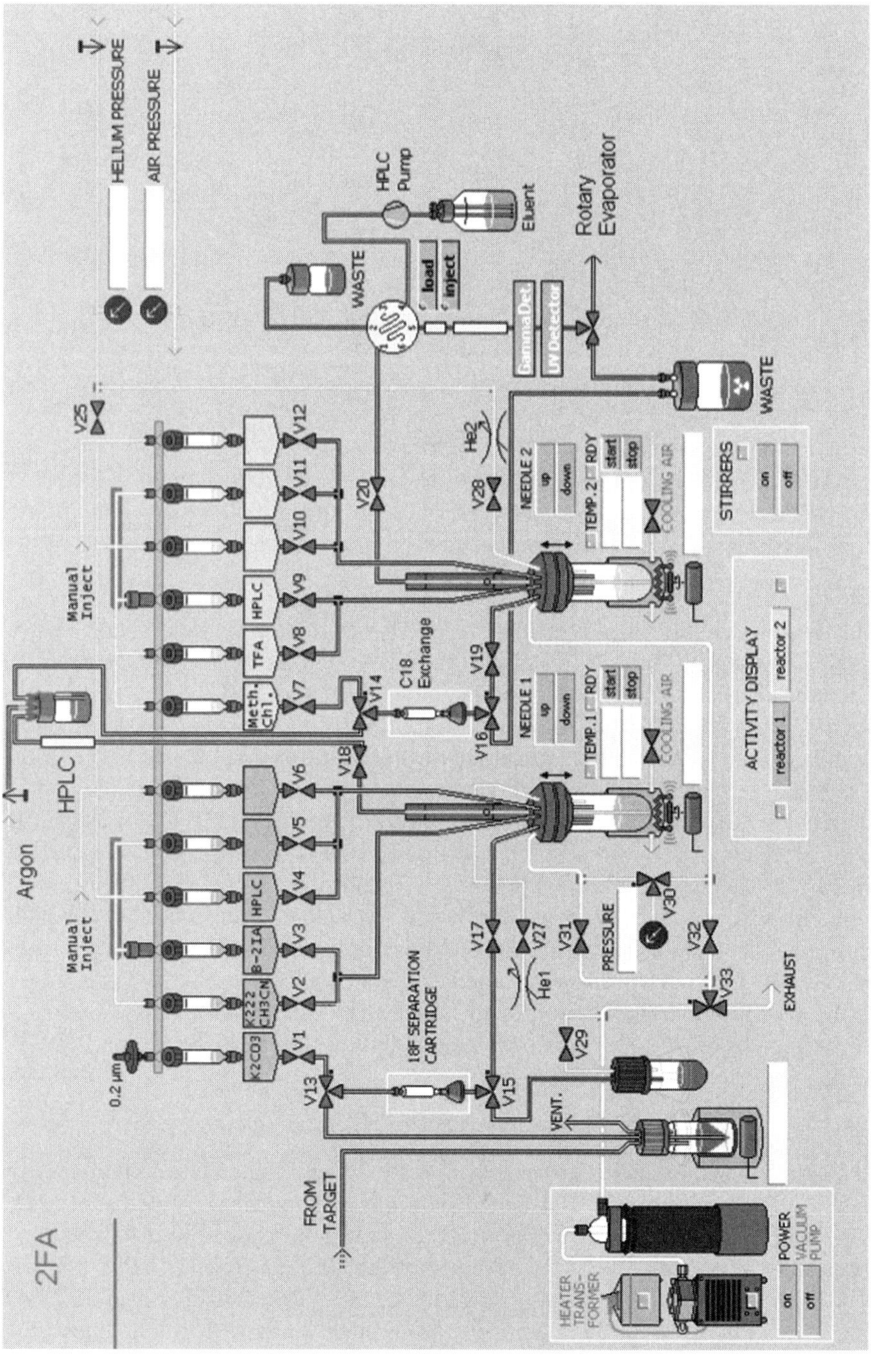

FIGURE 2 TRACERlab setup.

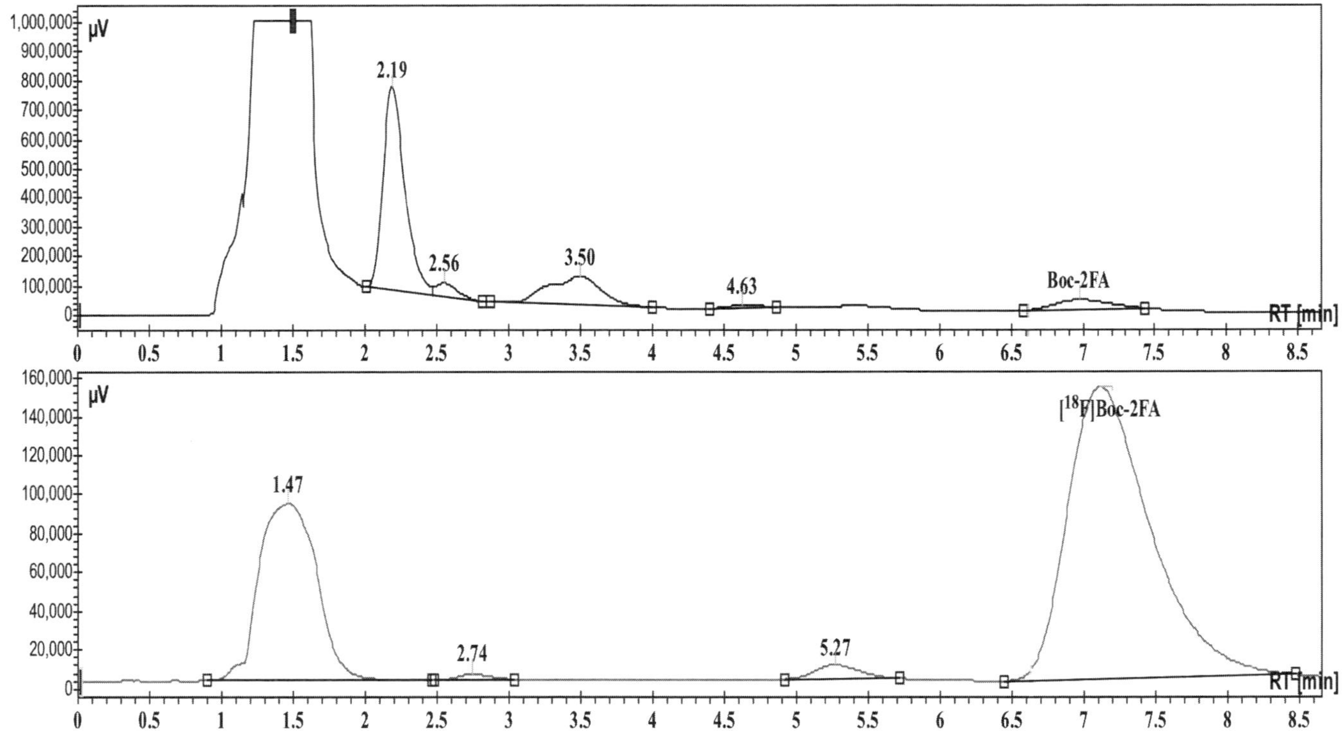

FIGURE 3 Preparative HPLC of Boc-[^{18}F]2FA.

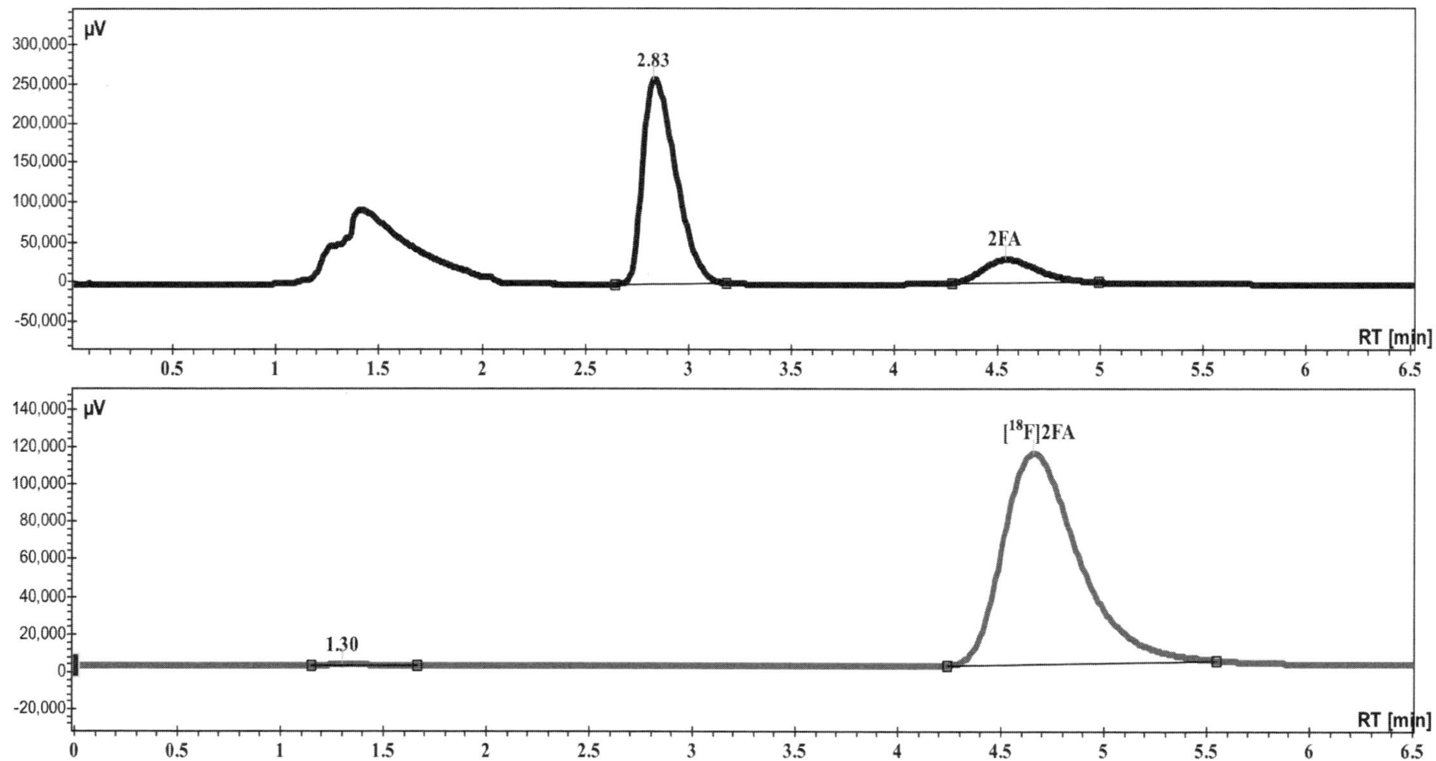

FIGURE 4 Preparative HPLC of [^{18}F]2FA.

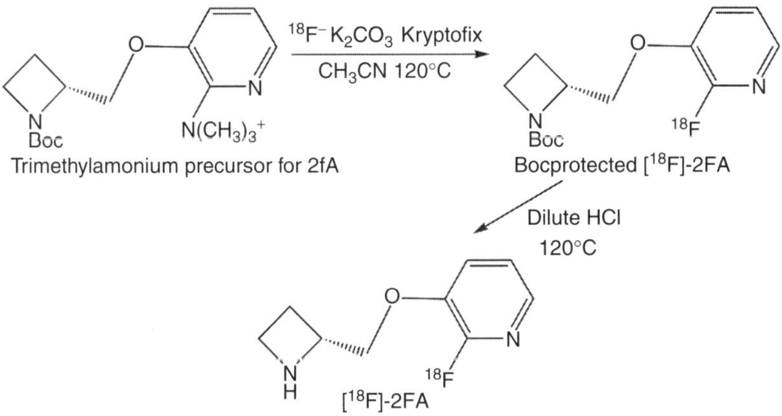

FIGURE 5 Synthesis of [^{18}F]2FA via the trimethylammonium precursor.

[^{18}F]2FA was isolated with an average radiochemical yield of 14±10% (n = 16) in 78±12 min at an average specific radioactivity of 83.9±64.6 GBq/μmole (2267±1747 mCi/μmole).

2.4 Advion NanoTek Microfluidic Synthesis

2.4.1 Preparation of the Advion NanoTek

A Chromafix 30-PS-HCO_3 SPE cartridge was attached to the [^{18}F]fluoride concentrator Trap port (Fig. 6). A dry 5-mL V-vial was attached to the Concentrator cap and placed in the heating block. Acetonitrile and a solution of potassium carbonate/Kryptofix-[2.2.2] (400 μL of 10 mg potassium carbonate and 60 mg of Kryptofix in 2 mL of 10% aqueous acetonitrile) were attached to the appropriate ports on the Concentrator valve. Acetonitrile was attached to the acetonitrile lines of pumps P1 and P3. A solution of the trimethylammonium precursor of 2FA [25] (4 mg in 400 μL acetonitrile) was attached to the precursor line of pump P1. Hydrochloric acid [26] (2 N), aqueous disodium hydrogen phosphate [26] (1 M), and water were attached to the labeled lines of pump P2. A 4-mL vial was attached to the hydrolysis cap and placed in the hydrolysis heater. The inject line of Concentrator 2 was attached to the injector of the preparative HPLC system.

2.4.2 Azeotropic Drying of [^{18}F]Fluoride

After the setup of the Advion NanoTek system, [^{18}F]fluoride was delivered to a V-vial in a solution of [^{18}O]water to a dose calibrator. After a radioactivity measurement was recorded, the drying sequence was started using the Advion NanoTek concentrator module. The sequence consisted of elution of the [^{18}F]fluoride through the Chromafix 30-PS-HCO_3 SPE cartridge, with collection of the [^{18}O] water in the V-vial in the dose calibrator. The solution of potassium carbonate/Kryptofix-[2.2.2] was eluted slowly over the Chromafix column into the 5-mL V-vial in the concentrator heater. The concentrator heater temperature

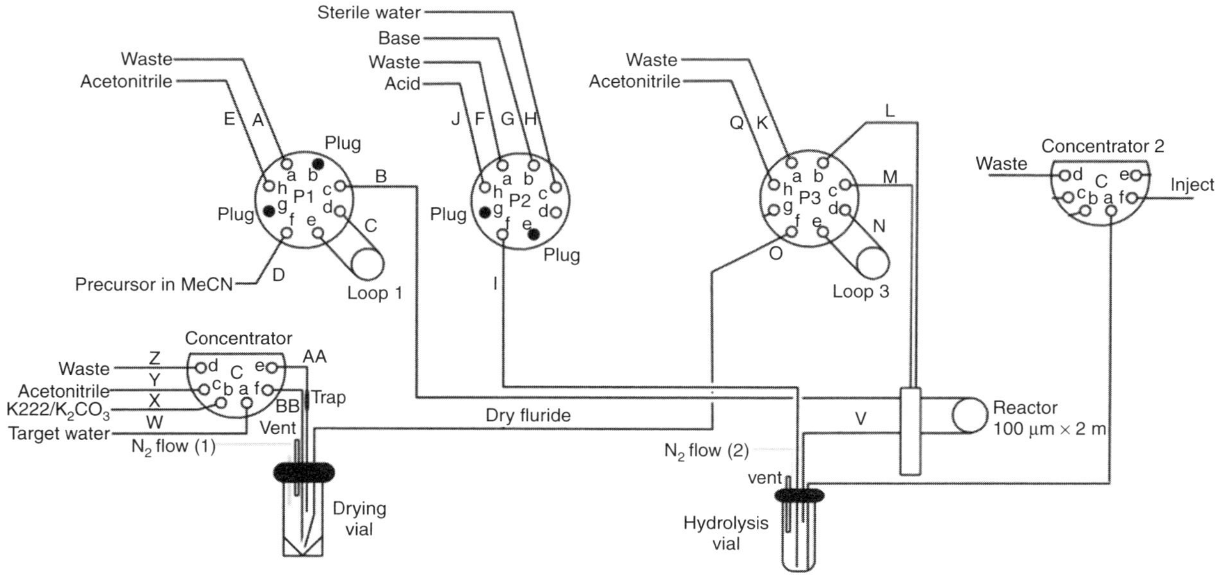

FIGURE 6 Advion NanoTek setup.

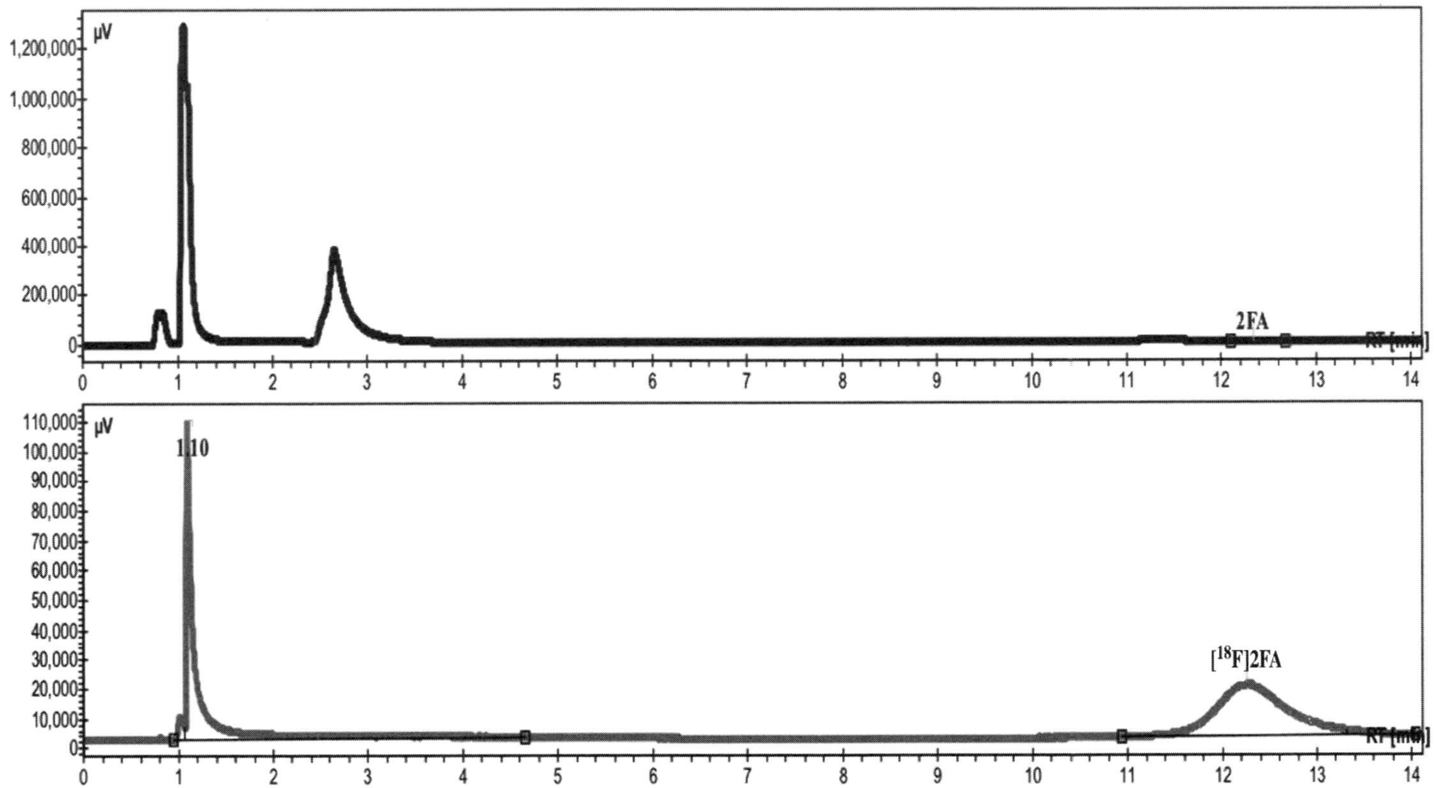

FIGURE 7 Preparative HPLC of [^{18}F]2FA.

was set at 120°C, and a low flow of nitrogen gas flow was established to the concentrator V-vial as the [^{18}F]fluoride was delivered. Aliquots of acetonitrile (1 mL) were added through the Chromafix column to the V-vial, and the contents of the vial were allowed to evaporate to dryness. Two additional aliquots of acetonitrile (500 μL) were added. The contents of the vial were brought to dryness (visual observation), after which the heater and nitrogen gas flow were turned off and the vial was allowed to return to a temperature below 70°C before the addition of a final aliquot of acetonitrile (600 μL) to resolubilize the [^{18}F]fluoride.

2.4.3 *Synthesis of [^{18}F]2FA*

The solution of the trimethylammonium precursor of 2FA was preloaded into a loop extension of the syringe pump on the Advion Nanotek before drying the [^{18}F]fluoride. After drying, the [^{18}F]fluoride in acetonitrile was loaded into a second loop extension of a second syringe pump. The two loops were joined at a T-connector before entering a 100-μm diameter × 2-m long coiled glass tubing reactor (Advion) heated to 120°C. Each loop was simultaneously emptied by their respective syringe pumps driven at a rate of 20 μL/min through the glass reactor. The output of the glass reactor was collected in the hydrolysis V-vial along with a final wash of the reactor with 200 μL acetonitrile. The vial was heated to 85°C under a low nitrogen [20] flow (20–40 mL/min) to remove most of the acetonitrile. While continuing a low nitrogen flow, water (100 μL) was added to the vial as the temperature was increased to 120°C. Hydrochloric acid (100 μL) was added, and the nitrogen flow stopped. After 180 s, the heater was turned off and aqueous disodium hydrogen phosphate (200 μL) was added.

2.4.4 *Purification and Formulation of [^{18}F]2FA*

The reaction mixture from the synthesis above was remotely injected using a Nanotek concentrator syringe pump onto a Waters Atlantis [27] (19 × 100 mm) column with a C-18 guard column equilibrated with 100% 0.1 M NaH_2PO_4 [26]. The column was eluted with a linear gradient: 100% 0.1 M NaH_2PO_4 (held for 5 min) to 100% of 5% ethanol [28]: 95% 0.1 M NaH_2PO_4 (over 20 min) at 18 mL/min. The fraction containing the [^{18}F]2FA (Fig. 7) was collected directly into a sterile vial via a sterilizing 0.22-μm filter [29]. The [^{18}F]2FA was isolated with an average radiochemical yield of 20±8% ($n = 9$) in 87±6 min with an average specific radioactivity of 340.4±187.6 GBq/μm (9201±5072 mCi/μm).

3 QUALITY CONTROL PROCEDURES

CAUTION: All radiopharmaceuticals produced for clinical use must have local regulatory approval before use in humans. Trained personnel must carry out quality control (QC) procedures, and each dose must meet all established QC criteria before release to the clinic.

QC procedures for [^{18}F]2FA, based on the current requirements for radiotracers set forth in the US Pharmacopeia [30], are summarized below. Complete QC data

TABLE 1 QC Data for Three Repeat Runs of [^{18}F]2FA using the NanoTek Method

QC Test	Release Criteria	Run 1	Run 2	Run 3
Yield mCi (GBq)	n/a	174 (6.4)	220 (8.1)	161 (6.0)
Visual inspection	Clear, colorless	Conforms	Conforms	Conforms
Radiochemical identity	Matches standard	Match	Match	Match
Radiochemical purity (%)	≥95	100	100	—
Specific activity (mCi/μmole)	1500	8172	19876	9352
Residual solvent analysis (ppm)	Acetonitrile ≤400	2	0	33
pH	4.5–8.5	5.0	4.5	5.5
Residual Kryptofix-[2.2.2] (μg/mL)	<50	<50	<50	<50
Sterile filter integrity test (psi)	≥13	21	22	21
Radionuclidic Identity ($t_{1/2}$) (min)	105–115	109.3	108.5	107.8
Endotoxin analysis (EU/mL)	≤17.5	<0.5	<0.5	<0.5
Sterility testing	No growth out to 14 d	Pass	Pass	Pass

for three repeat batches of [^{18}F]2FA produced using the methods disclosed herein are summarized in Table 1. Each of the three batches met all of the established QC criteria.

3.1 Visual Inspection

Using remote handling equipment and appropriate radiation shielding (leaded glass or a mirror system), the vial containing the [^{18}F]2FA product was visually inspected under bright light. The product met this acceptance specification if it was clear and colorless with no evidence of foreign matter.

3.2 Radiochemical Identity

The radiochemical identity of [^{18}F]2FA was determined using HPLC (Varian with Galaxie data acquisition and processing software [31]). The following conditions were used: a Waters Atlantis dC-18 5 μm 150 × 4.6 mm [27] eluted with 2.5% acetonitrile: 97.5% 0.1 M triethylamine buffer pH 7.2 at a flow rate of 2 mL/min was monitored by a UV detector (269 nm) in series with radioactivity detector. A reference standard solution (approximately 1 mg/mL of 2FA [25] in absolute ethanol) was injected to establish the suitability of the system conditions. To determine the radiochemical identity, an aliquot of the final injection matrix of [^{18}F]2FA was mixed with an aliquot of the reference standard solution. The product met this acceptance specification if the retention time of the reference material, as determined by a UV detector, was consistent with the retention time of the [^{18}F]2FA, as determined by a radiation detector, with appropriate correction for the offset between the two detector systems.

3.3 Radiochemical Purity

Using the same HPLC system described for the radiochemical identity test, an appropriate volume of [^{18}F]2FA was injected. The quantity injected should avoid uncorrected dead-time loss (for main peak) in the radioactive detection system. The percentage of radiochemical purity of [^{18}F]2FA was determined by dividing the activity found in the [^{18}F]2FA peak by the total activity found in the chromatogram multiplied by 100. The product met this acceptance specification if the radiochemical purity was greater than or equal to 95%.

3.4 Specific Activity

The specific activity of [^{18}F]2FA was calculated by dividing the assayed radioactivity of a calibrated aliquot of [^{18}F]2FA (mCi/mL at end-of-synthesis) by the concentration of [^{18}F]2FA measured by HPLC-UV (micromoles of [^{18}F]2FA per milliliter). The product met this acceptance specification if the specific activity was greater than or equal to 1500 mCi/μmole.

Figure 8 shows typical chromatograms observed during the determination of the radiochemical identity, radiochemical purity, and specific activity. The chromatograms show radiotracer analysis (upper 2 panels) and co-injection with an authentic nonradioactive 2FA standard (lower 2 panels).

3.5 Residual Solvent Analysis

Analysis of residual solvent levels in [^{18}F]2FA batches was conducted using a Varian gas chromatograph and Galaxie software [32] for data acquisition and analysis and a WAX (polyethyleneglycol phase: USP G16, G20) 30-m, 0.25-mm ID, 0.25-μm film column. An aliquot of a standard solution of 10 mL of normal saline (Sodium Chloride Injection, USP) to which was added 63 μL of acetone [18] (5000 ppm), 56 μL of tetrahydrofuran [18] (5000 ppm), 63 μL of absolute ethanol [28] (5000 ppm), 5.1 μL of acetonitrile [18] (400 ppm), 9.3 μL of dimethylformamide [18] (880 ppm), 45 μL of dimethylsulfoxide [18] (5000 ppm), and 5 μL of methylethylketone [18] (400 ppm) was analyzed to determine system suitability. An aliquot of the final [^{18}F]2FA product matrix was injected and the levels of residual solvent calculated by comparison to standards. The product met this acceptance specification if the level of acetonitrile was less than or equal to 410 ppm [33].

3.6 pH

A drop of the [^{18}F]2FA final product matrix was applied to a pH indicator paper (pH 2–9; sensitivity of 0.3–0.5 units) [34]. The strip color was matched to an indicator chart. The product met this acceptance specification if the pH was between 4.5 and 8.5.

3.7 Residual Kryptofix-[2.2.2] Analysis

Residual Kryptofix-[2.2.2] in the [^{18}F]2FA final product matrix was assayed using an established spot test. On a silica thin layer chromatographic (TLC) plate (2.5 × 7.5 cm, EMD [35]), 2-μL aliquots of a 5-mg/mL Kryptofix-[2.2.2] standard solution in absolute ethanol, a standard solution of 50 μg/mL Kryptofix-[2.2.2] in the final matrix, and the [^{18}F]2FA final product matrix were spotted in separate lanes. The strip was developed in a TLC tank using 9:1 (v:v) methanol [18]: 30%

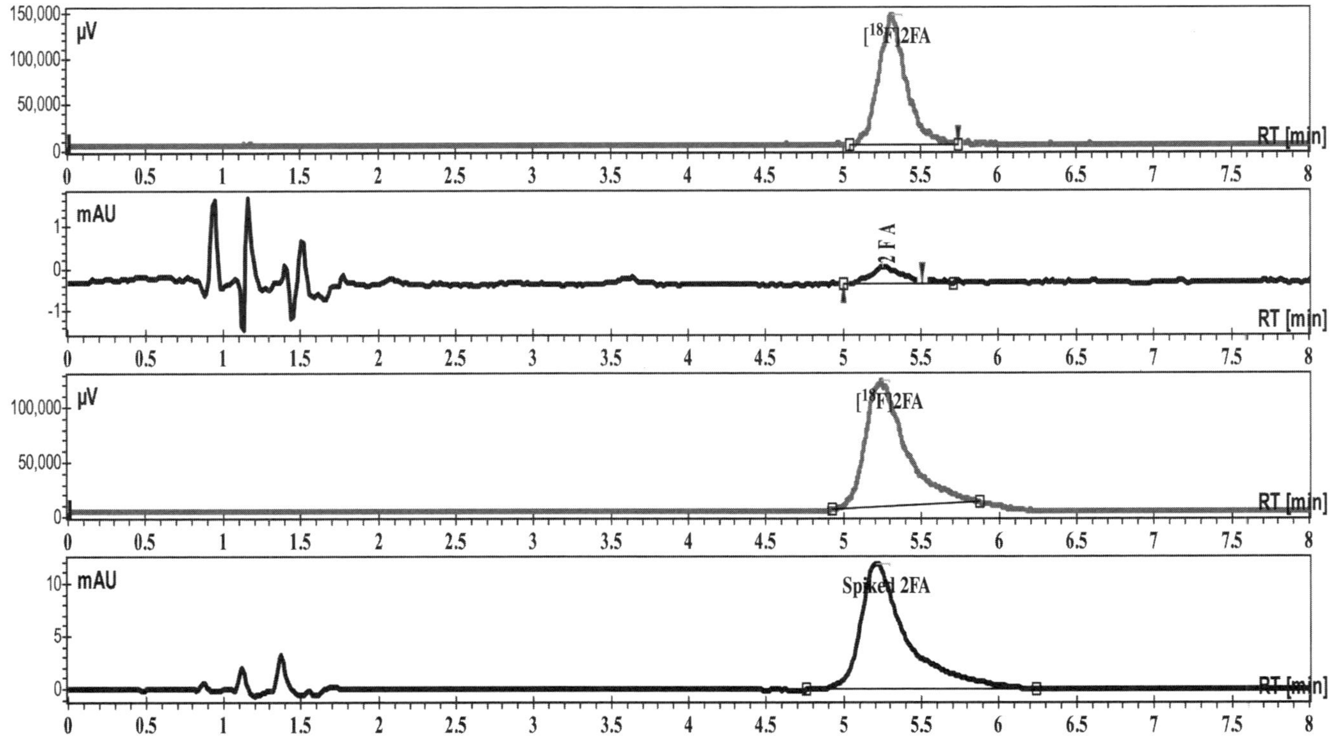

FIGURE 8 Quality control chromatograms of [^{18}F] 2FA.

ammonium hydroxide [26], after which the TLC plate was dried and exposed to iodine vapor to visualize the spots. Spots for the three samples were compared. The product met this acceptance specification if the residual Kryptofix-[2.2.2] level was demonstrated to be less than 50 μg/mL.

3.8 Sterile Filter Integrity Test

The sterile microfilter from the [^{18}F]2FA terminal filtration step was washed with 5 mL of absolute ethanol, left wetted, and attached to a calibrated pressure gauge [36] and air pressure source. The distal end of the filter was placed in a liquid reservoir, and the gas pressure was slowly increased. The product met this acceptance specification for the Millex FG filter if the pressure was greater than or equal to 13 psi without seeing a stream of bubbles.

3.9 Radionuclidic Identity

The radioactivity content (in millicurie) in an aliquot of the [^{18}F]2FA final product matrix was determined in a Capintec CRC-15R Radioisotope Dose Calibrator [37] at time 0 min and again at 15 min. The half-life was calculated using the formula below. The product met this acceptance specification if the calculated half-life was between 105 and 115 min.

$$t_{1/2} = (4.495)/(\log \mathrm{A} - \log \mathrm{B})$$

3.10 Endotoxin Analysis

Endotoxin levels in batches of the [^{18}F]2FA final product matrix was analyzed by a Charles River Laboratories EndoSafe Portable Testing System [38]. The product met this acceptance specification if the endotoxin level was less than or equal to 17.5 USP endotoxin units per milliliter.

3.11 Sterility Testing

In a laminar flow hood, samples of the [^{18}F]2FA final product matrix (approximately 100 μL each) were added to fluid thioglycolate media and soybean casein digest medium [39]. The media were incubated at 32.5±2.5°C and 22.5±2.5°C, respectively, and observed daily for any turbidity indicative of positive growth. The product met this acceptance specification if no growth was observed during the 14-day incubation period.

CHEMICAL ABSTRACTS NOMENCLATURE (REGISTRY NUMBER)

Acetonitrile (75-05-8)

1-Azetidinecarboxylic acid, 2-[[[2-(fluoro-^{18}F)-3-pyridinyl]oxy]methyl]-, 1,1-dimethylethyl ester, (2*S*)- (205594-64-5)

Carbonic acid, potassium salt (1:2)(584-08-7)

Ethanol (64-17-5)

4,7,13,16,21,24-Hexaoxa-1,10-diazabicyclo[8.8.8]hexacosane(23978-09-8)

Hydrochloric acid (7647-01-0)

Methane, 1,1′-sulfinylbis- (67-68-5)

2-Pyridinaminium, 3-[[(2*S*)-1-[(1,1-dimethylethoxy)carbonyl]-2-azetidinyl] methoxy]-*N,N,N*-trimethyl-, 1,1,1-trifluoromethanesulfonate (1:1)(233766-75-1)

Pyridine, 3-[(2*S*)-2-azetidinylmethoxy]-2-(fluoro-^{18}F)- (205594-65-6)

Trifluoroacetic acid (76-05-1)

REFERENCES AND NOTES

1. Horti AG, Gao Y, Kuwabara H, Dannals RF. Life Sci 2010;86:575–584.
2. Mitkovski S, Villemagne VL, Novakovic KE, O'Keefe G, Tochon-Danguy H, Mulligan RS, Dickinson KL, Saunder T, Gregoire MC, Bottlaender M, Dolle F, Rowe CC. Nucl Med Biol 2005;32:585–591.
3. Brody AL, Mandelkern MA, London ED, Olmstead RE, Farahi J, Scheibal D, Jou J, Allen V, Tiongson E, Chefer SI, Koren AO, Mukhin AG. Arch Gen Psychiatry 2006;63:907–915.
4. Picard F, Bruel D, Servent D, Saba W, Fruchart-Gaillard C, Schollhorn-Peyronneau MA, Roumenov D, Brodtkorb E, Zuberi S, Gambardella A, Steinborn B, Hufnagel A, Valette H, Bottlaender M. Brain 2006;129:2047–2060.
5. Ellis JR, Villemagne VL, Nathan PJ, Mulligan RS, Gong SJ, Chan JG, Sachinidis J, O'Keefe GJ, Pathmaraj K, Wesnes KA, Savage G, Rowe CC. Neurobiol Learn Mem 2008;90:404–412.
6. Kimes AS, Chefer SI, Matochik JA, Contoreggi CS, Vaupel DB, Stein EA, Mukhin AG. Neuroimage 2008;39:717–727.
7. Mukhin AG, Kimes AS, Chefer SI, Matochik JA, Contoreggi CS, Horti AG, Vaupel BD, Pavlova O, Stein EA. J Nucl Med 2008;49:1628–1635.
8. Sabri O, Kendziorra K, Wolf H, Gertz HJ, Brust P. Eur J Nucl Med Mol Imag 2008;35:S30–S45.
9. Horti AG, Koren AO, Ravert HT, Musachio JL, Mathews WB, London ED, Dannals RF. J Labelled Comp Radiopharm 1998;41:309–318.
10. Dolle F, Dolci L, Valette H, Hinnen F, Vaufrey F, Guenther I, Fuseau C, Coulon C, Bottlaender M, Crouzel C. J Med Chem 1999;42:2251–2259.
11. ABX GmbH, Germany. The cartridge was washed with 2 mL high purity water and pushed dry with air.
12. SepPak Plus C-18. Waters Corporation, USA.
13. Trace metal basis, 99.995%, Sigma-Aldrich.
14. Sigma-Aldrich Trace Select.
15. Sigma-Aldrich, 98%.
16. Thermo-Fisher, silylation grade.
17. Boc-2IA was synthesized in house as described in Reference 9.
18. HPLC grade.
19. Spectrophotometric grade.
20. Matheson, UHP grade.
21. Hamilton Company, USA.
22. Acetonitrile, methanol, and water in all HPLC methods are HPLC grade.
23. Hospira, Inc., USA.
24. Acrodisc 25 mm, 0.2 μm HT Tuffryn. PALL Life Sciences, USA.
25. ABX GmbH, Germany.
26. ACS grade.
27. Waters Corporation, USA.
28. USP grade.

29. Millex FG 25 mm, 0.2 μm. Millipore Corporation, USA.
30. US Pharmacopeia Chapter <823> Radiopharmaceuticals for positron emission tomography–compounding. USP 32-NF29. 2009.
31. Varian Prostar 210 pump with a Varian Prostar 325 variable wavelength Uv detector and a Bioscan Flowcount PMT detector using Galaxie Chromatogr0061phy Data System Version 1.9 Service Pack 2B. Varian, Inc., USA.
32. Varian 3800 Gas Chromatograph using Galaxie Chromatography Data System Version 1.9, Service Pack 2B. Varian, Inc., USA.
33. International Conference on Harmonisation of Technical Requirements for Registration of Pharmaceuticals for Human Use. ICH Harmonised Tripartite Guideline Impurities: Guideline for Residual Solvents, 1997.
34. ColorPhast pH-indicator strip–pH range 2.0–9.0. EMD Chemicals, Inc., Germany.
35. SilicaGel-60 F254 5 × 7.5 mm Aluminum Back TLC Strips. EMD Chemicals, Inc., Germany.
36. Bubble Point Test Apparatus. Millipore Corporation, USA.
37. Capintec Inc., USA.
38. Endosafe PTS Reader, Integrated Software Version 7.10, Service Pack 2.0, and Printer. Charles River Laboratories, USA.
39. Becton, Dickinson and Company, USA.

CHAPTER 16

SYNTHESIS OF [^{18}F]-SUBSTANCE-P ANTAGONIST-RECEPTOR QUANTIFIER ([^{18}F]SPA-RQ)

SOFIE CELEN

Laboratory for Radiopharmacy, Faculty of Pharmaceutical Sciences, K.U. Leuven, Leuven, Belgium

KIM SERDONS AND TJIBBE DE GROOT

Department of Nuclear Medicine, K.U. Leuven, Leuven, Belgium

TERENCE G. HAMILL

Discovery Imaging, Merck Research Laboratories, West Point, Pennsylvania

GUY BORMANS

Laboratory for Radiopharmacy, Faculty of Pharmaceutical Sciences, K.U. Leuven, Leuven, Belgium

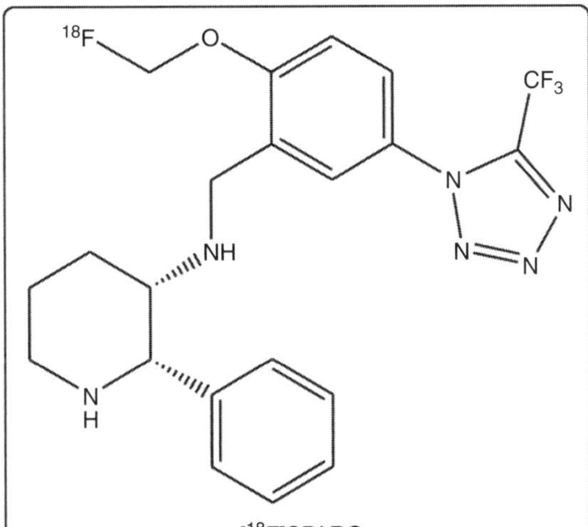

[^{18}F]SPARQ

Radiochemical Syntheses, Volume 1: Radiopharmaceuticals for Positron Emission Tomography, First Edition. Edited by Peter J. H. Scott, Brian G. Hockley.

1 INTRODUCTION

Tachykinins are a class of neuropeptides that are involved in a variety of biological functions in the central and peripheral nervous system. The tachykinin receptors have been divided into three subtypes according to their preferred ligands: NK1 (formerly called the substance P receptor), NK2 (formerly the substance K/substance E receptor), and NK3 (formerly the NK B receptor) [1]. NK1 receptors have been of particular interest because of their potential implication in various neuropsychiatric and other disorders. This receptor has the highest density in the striatum and negligible density in the cerebellum [2]. The mammalian tachykinin substance P (SP) is the best characterized tachykinin. It is the prime agonist of the NK1 receptor, and during the past decade, NK1 antagonists (SP antagonists) have been developed and tested as potential treatments for a variety of disorders including emesis, pain, inflammation, depression, anxiety and in the treatment of chemotherapy-induced nausea and vomiting [3–5]. Noninvasive imaging of the NK1 receptor using positron emission tomography (PET) allows the study of the neuroanatomical distribution of this receptor and of the biological role that this receptor plays in various disease states. Furthermore, an NK1 receptor PET tracer would be useful in the clinical development of NK1 receptor antagonists by giving direct insight into the relationship between receptor occupancy, administered dose, and efficacy of candidate drugs. Reported PET ligands that have been evaluated for NK1 visualization and quantification are [^{11}C]GR205171 [6], [^{11}C]R116301 [7,8], and [^{18}F]SPA-RQ [2,9–11]. SPA-RQ has been shown to be a selective high affinity NK1 receptor antagonist (human $IC_{50} = 67$ pM) with moderate lipophilicity ($\log P = 1.8$) [10], and [^{18}F]SPA-RQ (Fig. 1) has been used in clinical imaging studies [2].

2 SYNTHESIS PROCEDURES

CAUTION: All radiochemical syntheses must be carried out using appropriate equipment in a facility authorized for the use of radioactive materials. Personal protective equipment must be worn and all local radiation safety laws followed.

2.1 Production of [^{18}F]Fluoride

[^{18}F]Fluoride was produced by the [^{18}O(p,n)^{18}F] nuclear reaction by irradiation of 97% enriched [^{18}O]H_2O (2 mL) [12] in a niobium target [13] using 18-MeV protons accelerated in a Cyclone 18/9 cyclotron [14] for 30 min–2 h (35-μA beam), generating ~1–4 Ci (37–148 GBq) of [^{18}F]fluoride.

2.2 Azeotropic Drying of [^{18}F]Fluoride

The [^{18}F]fluoride was delivered to a home-made synthesis module as a solution in [^{18}O]H_2O (2 mL). The irradiated target water was passed over a QMA anion exchange cartridge [15], which retains the [^{18}F]fluoride. The [^{18}F]fluoride

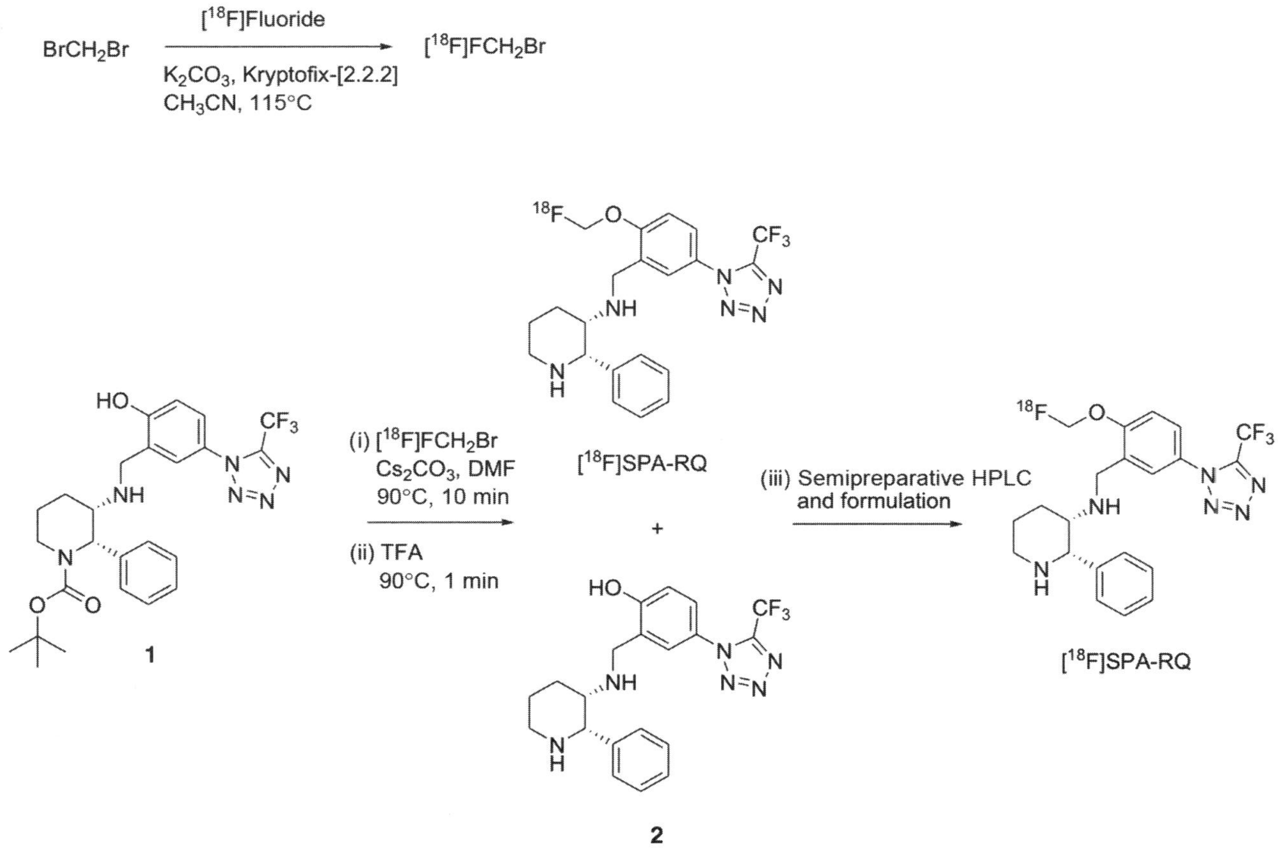

FIGURE 1 Synthesis of [^{18}F]fluoromethylbromide and [^{18}F]SPA-RQ.

was then displaced from the column with an aqueous solution of K_2CO_3 and Kryptofix-[2.2.2] (0.75 mL of a solution of 23 mg K_2CO_3 and 260 mg of Kryptofix-[2.2.2] in 0.35 mL milliQ water and 6.65 mL of CH_3CN) [16] into a conical reaction vessel fitted to a water-cooled condenser [17]. The solution was heated at 115°C under a flow of helium for 8 min to evaporate water and CH_3CN. The [^{18}F]fluoride was further dried by azeotropic distillation using anhydrous CH_3CN [18] (1 mL) at 115°C for 15 min until complete dryness.

2.3 Synthesis of [^{18}F]SPA-RQ

After cooling down to 80°C, a solution of dibromomethane [19] (50 μL) in anhydrous CH_3CN [18] (1 mL) was added to the dried [^{18}F]fluoride. The reaction mixture was heated to 115°C, and [^{18}F]FCH_2Br was distilled through a water-cooled condenser (for separation of CH_2Br_2 and CH_3CN) with a gentle flow of helium (5 mL/min) and bubbled into a second reaction vessel [20] containing the precursor **1** [21] (0.2–0.3 mg) and a small amount of Cs_2CO_3 [22] (1–2 mg) in anhydrous DMF [23] (0.2 mL). After distillation of a sufficient amount of radioactivity to the second vial (±10 min), the reaction mixture was heated at 90°C for 10 min. Subsequently, the DMF was removed using a high flow of helium while maintaining the temperature at 90°C. After removal of the DMF, hydrolysis of the *N*-Boc-protecting group was performed by adding TFA [24] (0.2 mL) and heating the mixture at 90°C for 1 min.

2.4 Purification and Formulation of [^{18}F]SPA-RQ

After deprotection, the crude radiolabeling mixture was cooled to room temperature and diluted with preparative high performance liquid chromatographic (HPLC) eluent (1.8 mL of a mixture of 0.01 M H_3PO_4/EtOH 77.5/22.5 v/v) [25]. The diluted reaction mixture was purified by semipreparative HPLC by injection on a Sunfire C18 column [26] eluted with a mixture [25] of 22.5% EtOH in 0.01 M H_3PO_4 at a flow rate of 1 mL/min. The corresponding HPLC chromatograms are shown in Fig. 2.

The fraction corresponding to [^{18}F]SPA-RQ eluted after about 16–17 min, with a yield of about 6–60 mCi. This HPLC-purified fraction was diluted (if volume <2.2 mL) with preparative HPLC eluent to a volume of 2.2 mL and passed through an apyrogenic sterile filter [27] under laminar air flow [28] into a sterile vial [29]. The final injection solution was obtained by diluting this 2.2 mL with 7.8 mL of buffer [30] containing 6.25% (w/v) D-glucose [31] and 2.8 mM Na_2HPO_4 [32], which was passed through the same membrane filter, and the vial was swirled to obtain a homogeneous solution. After purification, the semipreparative HPLC column was rinsed consecutively with 70% ethanol, 90% ethanol, and again 70% ethanol, on which the column was stored.

3 QUALITY CONTROL PROCEDURES

CAUTION: All radiopharmaceuticals produced for clinical use must have local regulatory approval before use in humans. Trained personnel must carry out

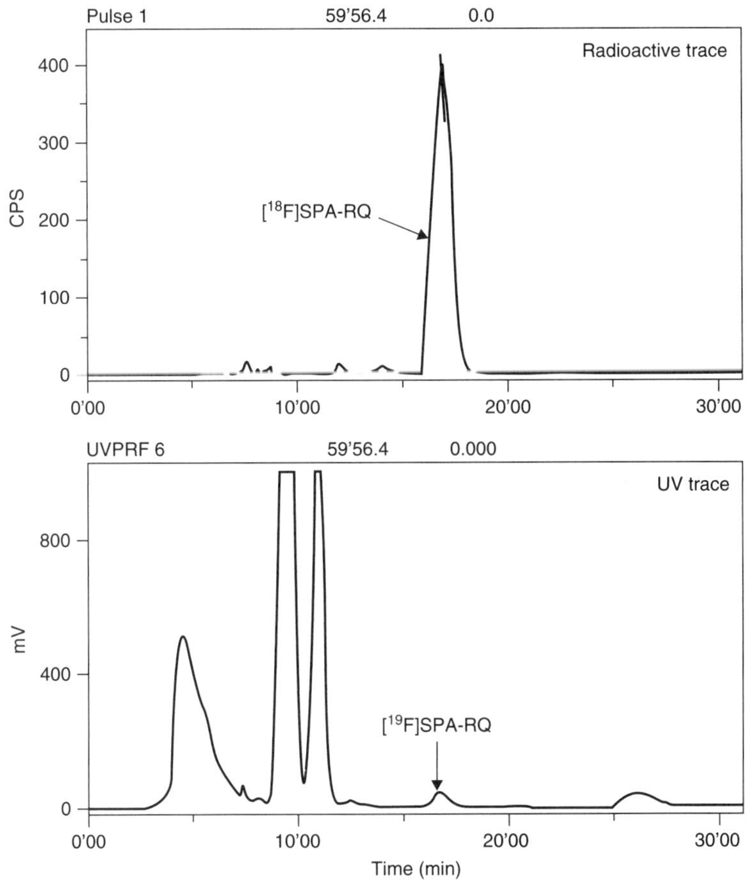

FIGURE 2 Semipreparative UV and radioactive HPLC traces for [^{18}F]SPA-RQ.

quality control (QC) procedures, and each dose must meet all established QC criteria before release to the clinic.

QC procedures for [^{18}F]SPA-RQ, based on the current requirements for radiopharmaceuticals laid out in the European Pharmacopeia (Ph. Eur.) [33], are summarized below. Complete QC data for three repeat batches of [^{18}F]SPA-RQ produced using the method described in this chapter are summarized in Table 1. Each of the three doses met all of the established QC criteria.

3.1 QC Control Performed before Batch Release

3.1.1 Visual Inspection

The [^{18}F]SPA-RQ dose is examined in a lead-shielded laminar airflow hood [28] and must be clear, colorless, and free of particulate matter.

TABLE 1 QC Data for Three Repeat Runs of [^{18}F]SPA-RQ

QC Test	Release Criteria	Run 1	Run 2	Run 3
Total radioactivity in batch at end production[a]	≤27 mCi (1000 MBq)[a]	23.9 (884)	21.7 (803)	6.2 (230)
Visual inspection	Clear, colorless	Conforms	Conforms	Conforms
Radiochemical identity	Retention time (RT) of [^{18}F]SPA-RQ (min)	6.56	7.42	7.05
	RT of the reference (min)	6.59	7.42	7.11
	Difference ≤10%	Conforms	Conforms	Conforms
Radiochemical purity (%)	Total radiolabeled side products ≤5	<1	<1	<1
Chemical purity	Maximum of unidentified chemical side products: 1 μg/injected dose[b]	Conforms	Conforms	Conforms
Amount of SPA-RQ in total volume (μg/injected dose)[c]	≤5	10	6	6
Amount of SPA-RQ deprotected precursor (**2**) in total volume (μg/injected dose)	≤1.5	<LOD[d]	<LOD[d]	<LOD[d]
Specific activity[e] (GBq/μmol)	No limit established	39	59	18
Residual solvent analysis	DMF <4.1 mg/injection	<LOD[f]	<LOD[f]	<LOD[f]
	Ethanol <10% v/v	4.9	4.4	4.5
Dose pH	pH of the finished product is 4–8	7	6	6
Sterile filter integrity test (bar)	Bubble point ≥3.45	3.7	3.8	3.8
Radionuclidic identity ($t_{1/2}$) (min)	105–115	110.9	108.0	109.1
Endotoxin analysis (IU/mL)	<7	Conforms	Conforms	Conforms
Sterility testing	No growth after 14 d incubation at 37°C	Conforms	Conforms	Conforms

[a]Batch volume of the final formulated [^{18}F]SPA-RQ is 10 mL. At a maximum concentration of 27 mCi (1000 MBq)/10 mL, stability of the drug product [^{18}F]SPA-RQ is warranted up to 6 h post production.
[b]Calculated using the UV response factor for SPA-RQ.
[c]If the mass of SPA-RQ in the total volume is ≥5 μg then the batch fraction that can be administered is limited so that the mass of SPA-RQ in the injected dose is ≤5 μg.
[d]LOD, limit of detection for SPA-RQ = 0.07 μg/mL; LOD for **2** = 0.10 μg/mL.
[e]At the end of synthesis.
[f]LOD, limit of detection for DMF = 0.0001% m/v.

3.1.2 *Radiochemical Identity*

The radiochemical identity of [^{18}F]SPA-RQ was checked using reverse phase HPLC and is based on using unlabeled SPA-RQ as an external reference material. The HPLC analysis was conducted on a Merck Hitachi L6200A pump [34] connected to a UV spectrometer [35] set at 254 nm. After passage through the UV detector, the HPLC eluate was led over a 3-inch NaI(Tl) scintillation detector [36] connected to a multichannel analyzer [37]. The output signal was recorded and analyzed using a GINA Star data acquisition system [38]. The analytical HPLC column that was used was a Waters XTerra RP18 column (5 μm, 4.6 mm × 250 mm) [39] eluted with a mixture of acetonitrile and 10 mM H_3PO_4 pH 3.9 [40] (50/50 v/v; 1 mL/min (RT = ~6–7 min). After injection and analysis of the reference solution [41], a blank injection [42] was performed to ensure a flat baseline before injection of the [^{18}F]SPA-RQ preparation. The retention time of [^{18}F]SPA-RQ should be the same (±10%) as the retention time observed for the [^{19}F]SPA-RQ reference standard. If a larger deviation is observed, a co-injection of [^{18}F]SPA-RQ and the reference solution should be performed. The retention time of the peak of [^{19}F]SPA-RQ observed in the UV channel should be identical (±0.3 min) to the retention time of the peak of [^{18}F]SPA-RQ observed in the radiometric channel.

3.1.3 *Radiochemical Purity*

Radiochemical purity was analyzed using the same HPLC system used to check the identity of [^{18}F]SPA-RQ [34–40]. The total of radiolabeled side products should be ≤5%. Representative analytical UV and radioactive HPLC traces are displayed in Fig. 3.

At a concentration of 2.79 mCi/mL (103 MBq/mL) of [^{18}F]SPA-RQ in a sterile solution of 5% (w/v) D-glucose solution containing 5% v/v ethanol and 4 mM phosphate buffer, no radiolysis was observed over a period of 6 h for all three validation runs. If longer storage periods or higher radioactive concentrations are used, the radiochemical purity of the batch needs to be reanalyzed before injection.

3.1.4 *Specific Radioactivity*

The specific radioactivity of [^{18}F]SPA-RQ was determined with the same HPLC system used for the radiochemical identity and radiochemical purity analysis [34–40]. Rather than setting a lower limit for specific activity, the maximum mass of SPA-RQ that is administered to a volunteer is limited to ≤5 μg. The mass of **2** should be ≤1.5 μg per administered dose. For the chemical purity of [^{18}F]SPA-RQ, the maximum mass of unidentified chemical impurities is 1 μg per administered dose [43] in analogy with the radiochemical purity of PET radiopharmaceutical monographs in the Ph. Eur [33].

3.1.5 *Residual Solvent Analysis*

Analysis of solvent levels (DMF and ethanol) in [^{18}F]SPA-RQ doses was conducted using a Shimadzu GC-2014 gas chromatograph [44] with a GC capillary

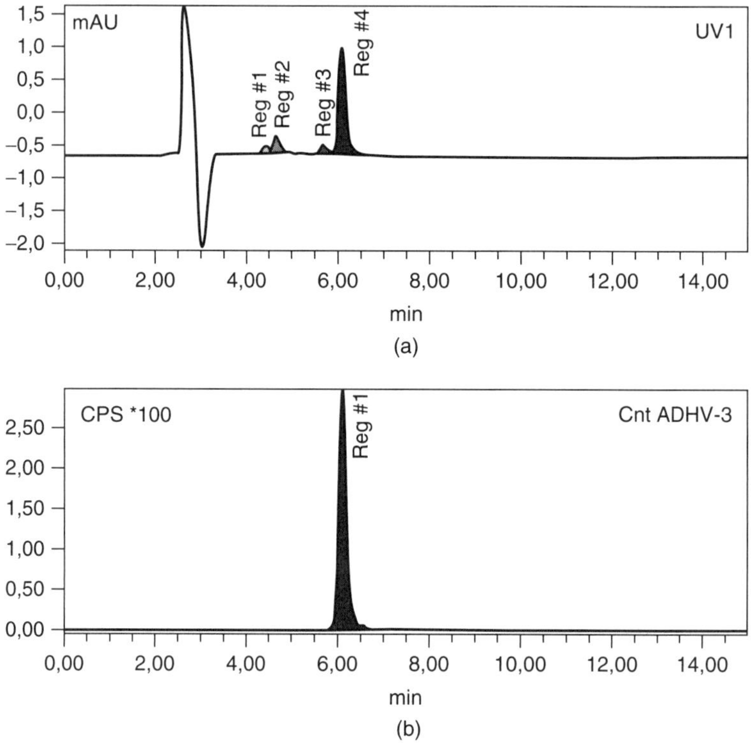

FIGURE 3 Analytical UV (a) and radioactive (b) HPLC traces for [^{18}F]SPA-RQ.

Supelco Equity™ column [45] and a Shimadzu AOC-20i autoinjector using the following column temperature gradient: 0–5 min, 40°C; 5–12 min, linear gradient from 40 to 200°C; 12–15 min, 200°C. H_2 is used as the carrier gas at a linear velocity of 40 cm/s. This method is specific for ethanol and DMF. For the residual class 2 solvent DMF, a limit of 4.1 mg per injection is set, as described in the Ph. Eur. and ICH guideline Q3C [46]. The amount of residual ethanol should be <10% v/v so that the product can be safely administered to the patient. Ethanol is needed to increase radiochemical stability [47] and to minimize the tracer being retained on the walls of the sterile filter, the vial, and the syringes used to administer the drug product to the patient [48].

3.1.6 Dose pH

The pH of the [^{18}F]SPA-RQ dose was determined by applying a small amount of the final formulated product to a pH 0–14 pH-indicator strip [49]. The drug product pH should be in the range 4–8.

3.1.7 Residual Kryptofix-[2.2.2] Analysis

This test is not performed in the QC of [^{18}F]SPA-RQ, as it was validated that Kryptofix does not codistil from the first reaction vial.

3.1.8 Sterile Filter Integrity Test

Testing of the integrity of the filter was done by bubble point determination. The sterile filter was connected to a nitrogen supply via a regulator and to a needle that was submerged in water. The nitrogen pressure was gradually increased until bubbles were observed, and the corresponding pressure was recorded as the bubble point pressure. The bubble point pressure for the particular filter used (Millex®-GV) [27] should be ≥3.45 bar.

3.2 QC Control Performed Post Batch Release

3.2.1 Radionuclide Identity

Determination of radionuclide identity and purity is based on a two-time point radioactivity measurement in a Capintec CRC®-15R dose calibrator [50]. The half-life $t_{1/2}$ was calculated using the equation

$$t_{1/2} = t\frac{\ln(2)}{\ln\left(\frac{A_1}{A_2}\right)}$$

where t = time interval between two radioactivity measurements, A_1 = radioactivity measured at first time point, A_2 = radioactivity measured at second time point.

The calculated half-life must be 105–115 min according to the Ph. Eur. guidelines [33].

3.2.2 Endotoxin Analysis

Bacterial endotoxin determination of the [^{18}F]SPA-RQ dose was done using the limulus amebocyte lysate (LAL) test [51] according to the Ph. Eur. guidelines [52]. The bacterial endotoxin limit for [^{18}F]SPA-RQ doses is <7 international units (IU)/mL. If the total batch volume (10 mL) were to be injected to one volunteer, it would correspond to 70 IU, which is well below the 175 IU per dose limit for radiopharmaceuticals specified in the Ph. Eur [52].

3.2.3 Sterility Testing

Samples of the final formulated [^{18}F]SPA-RQ (1 mL) were mixed with fluid tryptic soy broth (TSB) medium [53] and with fluid thioglycolate (in the presence of resazurin indicator) medium. TSB plates were used to test for aerobes and fungi, while thioglycolate plates were used to test for anaerobes. If oxygen is present in the thioglycolate plates, it will react with the resazurin, resulting in a pink cloudy color on top of the medium as an indication for false-positive results. For negative controls, TSB and thioglycolate media were incubated with 1 mL of sterile air. The [^{18}F]SPA-RQ plates were incubated together with positive and negative controls for 14 days at 37°C according to current Ph. Eur. guidelines [54]. Plates were visually inspected after 7 and 14 days of incubation and compared to the positive and negative standards. Positive standards must show growth (turbidity) on the plates. Growth at the top of the thioglycolate plate indicates presence of aerobic microorganisms; growth at the bottom indicates presence of anaerobic microorganisms. The negative controls and the [^{18}F]SPA-RQ plates must have no culture growth after 14 days of incubation to be indicative of sterility.

CHEMICAL ABSTRACTS NOMENCLATURE (REGISTRY NUMBER)

Acetonitrile (75-05-8)
Carbonic acid, cesium salt (1,2) (534-17-8)
Carbonic acid, potassium salt (1:2)(584-08-7)
Dibromomethane (74-95-3)
N,N-dimethylformamide (68-12-2)
Ethanol (64-17-5)
D-glucose monohydrate (5996-10-1)
1-Heptane sulphonic acid, sodium salt (1:1) (22767-50-6)
4,7,13,16,21,24-Hexaoxa-1,10-diazabicyclo[8.8.8]hexacosane (23978-09-8)
Phosphoric acid (7664-38-2)
Phosphoric acid, sodium salt (1:2) (7558-79-4)
Sodium hydroxide (1310-73-2)
Trifluoroacetic acid (76-05-1)

For detailed supplier information, see Appendix 1 (page 335).

REFERENCES AND NOTES

1. Otsuka M, Yoshioka K. Psychiol Rev 1993;73:229–308.
2. Hietala J, Nyman MJ, Eskola O, Laakso A, Grönroos T, Oikonen V, Bergman J, Haaparanta M, Forsback S, Marjamäki P, Lehikoinen P, Goldberg M, Burns D, Hamill T, Eng W-S, Coimbra A, Hargreaves R, Solin O. Mol Imaging Biol 2005;7:262–272.
3. Datar P, Srivastava S, Coutinho E, Govil G. Curr Top Med Chem 2004;4:75–103.
4. Huang SC, Korlipara VL. Expert Opin Ther Pat 2010;20:1019–1045.
5. Patel L, Lindley C. Expert Opin Pharmacother 2003;4:2279–2296.
6. Bergström M, Fasth K-J, Kilpatrick G, Ward P, Cable KM, Wipperman MD, Sutherland DR, Langström B. Neuropharmacology 2000;39:664–670.
7. Van der Mey M, Janssen CGM, Janssens FE, Jurzak M, Langlois X, Sommen FM, Verreet B, Windhorst AD, Leysen JE, Herscheid JDM. Bioorgan Med Chem 2005;13:1579–1586.
8. Wolfensberger SPA, van Berckel BNM, Airaksinen AJ, Maruyama K, Lubberink M, Boellaard R, Carey WDH, Reddingius W, Veltman DJ, Windhorst AD, Leysen JE, Lammertsma AA. Mol Imaging Biol 2009;11:241–245.
9. Hargreaves R. J Clin Psychiatry 2002;63:18–24.
10. Solin O, Eskola O, Hamill TG, Bergman J, Lehikoinen P, Grönroos T, Forsback S, Haaparanta M, Viljanen T, Ryan C, Gibson R, Kieczykowski G, Hietala J, Hargreaves R, Burns HD. Mol Imaging Biol 2004;6:373–384.
11. Bergström M, Hargreaves RJ, Burns HD, Goldberg MR, Sciberras D, Reines A, Petty KJ, Ögren M, Antoni G, Langström B, Eskola O, Scheinin M, Solin O, Majumdar AK, Constanzer ML, Battisti WP, Bradstreet TE, Gargano C, Hietala J. Biol Psychiatry 2004;55:1007–1012.

12. Rotem HYOX18 purchased from Rotem Industries, Beer Sheva, Israel, and used as received.
13. Niobium target purchased from IBA (Ion Beam Applications), Louvain La Neuve, Belgium.
14. 18/9 Cyclone cyclotron purchased from IBA, Louvain La Neuve, Belgium.
15. SepPak™ Light Accell plus QMA cartridges were purchased from Waters (Milford, MA, USA; part no. WAT023525). Conditioning was done by rinsing with 5 mL K_2CO_3 0.5 M and 2 × 5 mL of milliQ water followed by drying under a flow of nitrogen (1 bar) for 5 min.
16. Potassium carbonate (Ph. Eur. grade) was purchased from Certa (part no. FA075069) and used as received. Kryptofix-[2.2.2] 98%was purchased from ABX (product no. 800.1000) and used as received. Acetonitrile (H_2O <30 ppm) was purchased from VWR BDH Polabo(product no. 83639.320) and used as received.
17. Sovirelconical glass vial with side arm and cooler.
18. Dry acetonitrile 99.8% was purchased from Sigma-Aldrich (product no. 271004) and used as received.
19. Dibromomethane 99% was purchased from Acros Organics (product no. 112831000) and used as received.
20. Sovirelconical glass vial (no side arm, no cooler).
21. [^{18}F]SPA-RQ precursor ((2S,3S)-1-t-butoxycarbonyl-2-phenyl-3-[2-hydroxy-5-(5′-trifluoromethyltetrazo-1-yl)phenylmethylene-amino]-piperidine) was supplied by Merck Research Laboratories (New Jersey, USA) and used as received.
22. Cesium carbonate 99.5% was purchased from Acros Organics (product no. 192040250) and used as received.
23. Anhydrous DMF 99.8% was purchased from Sigma-Aldrich (product no. 227056) and used as received.
24. Trifluoroacetic acid 99.5% was purchased from Acros Organics (product no. 293810250) and used as received.
25. Semipreparative HPLC mobile phase: phosphoric acid 85.6% (Ph. Eur. grade) was purchased from Certa (product no. 10162). Ethanol 96% (Ph. Eur. grade) was purchased from Fagron (product no. FA073627) and used as received. Phosphoric acid 0.1 M was prepared under sterile conditions by diluting of 6.75 mL of concentrated phosphoric acid (85.6%) to 1000 mL with water p.i. (B.Braun, art no. 3620581). The solution was sterile filtered through a Pall VacuCap 60 filter (0.2 μm, 25 mm, Pall Corporation, USA, product no. TA4632). Phosphoric acid 0.01 M was prepared under sterile conditions by further diluting (10×) phosphoric acid 0.1 M with water p.i. (B.Braun). Preparation of the preparative mobile phase is done under sterile conditions. A volume of 775 mL of phosphoric acid 0.01 M and 225 mL of ethanol were stirred to mix and sterile filtered using a Pall VacuCap 60 filter (0.2 μm, 25 mm, Pall Corporation, USA, product no. TA4632). A total of 20 mL of this sterile mobile phase was additionally filtered through a sterile membrane filter (Millex®-GV, 0.22 μm, 13 mm, Millipore; part no. SLGV013SL) and used as the dilution buffer in the synthesis module.
26. Semipreparative HPLC column: Sunfire C18; 5 μm, 4.6 × 250 mm, purchased from Waters (part no. WAT186002560); flow rate: 1 mL/min; $\lambda = 254$ nm. The column was equilibrated with HPLC eluent for 25 min.
27. Millex-GV sterile filters (0.22 μm, 13 mm) were purchased from Millipore, Ireland (part no. SLGV013SL), and used as received.
28. Kojair Biowizard, TCPS, Belgium.
29. Sterile vials (25 mL) were purchased from Mallinckrodt (Covidien), Missouri, USA (part no. 281015). The stopper is swabbed with an Alkotip® tissue and vented with a sterile Millex-GV filter [27] with sterile needle (BD, Microlance™3, 0.6 × 25 mm).
30. The buffer containing 6.25% glucose and 2.8 mM Na_2HPO_4 was prepared under sterile conditions. To 70 mL of water p.i. (B.Braun) was added 5 mL of a solution of 40 mg Na_2HPO_4 in water p.i. (B.Braun) and 25 mL of D-glucose 25% solution. After homogenization, the solution was sterile filtered through a sterile membrane filter (Pall Acrodisc filter, 0.2 μm, 25 mm, Pall Corporation, product no. PN4192).
31. Glucose monohydrate (Ph. Eur. grade) was purchased from VWR BDH Prolabo (product no. 0269.1000). A 25% D-glucose solution was prepared under sterile conditions by dissolving

27.5 g of glucose monohydrate in 100 mL water p.i. (B.Braun). The solution was sterile filtered through a Pall Acrodisc filter (0.2 μm, 25 mm, Pall Corporation, product no. PN4192).

32. Disodium hydrogenphosphate (Ph. Eur. grade) was purchased from Merck (product no. FA074427).
33. European Pharmacopoeia 6.0—Radiopharmaceutical Preparations (0125).
34. Hitachi, Tokyo, Japan.
35. Waters 2487 Dual λ absorbance detector, Waters, Milford, USA.
36. Wallac 1480 Wizard, Wallac, Turku, Finland.
37. Medi-lab Select, Mechelen, Belgium.
38. Raytest, Straubenhardt, Germany.
39. Analytical HPLC column: XTerra C18; 5 μm, 4.6 mm × 250 mm, purchased from Waters (part no. 186000584/186000496); flow rate 1 mL/min; $\lambda = 254$ nm. The column was equilibrated with analytical HPLC eluent for 15 min.
40. Analytical mobile phase: 0.2% 1-heptanesulfonic acid (sodium salt) in CH_3CN/0.01 M H_3PO_4 50/50 v/v, pH adjusted to 3.9 with 1 N NaOH. Acetonitrile was purchased from VWR BDH Prolabo (product no. 83639.320). 1-Heptanesulfonic acid (sodium salt) was purchased from Acros Organics (product no. 411270250). NaOH (1 N) was purchased from Merck. The analytical HPLC eluent was prepared under nonsterile conditions; 500 mL of CH_3CN and 500 mL of 0.01 M phosphoric acid were mixed, and 2 g of 1-heptanesulfonic acid (sodium salt) was added. The pH of the mixture was adjusted to 3.9 with 1 N NaOH.
41. SPA-RQ authentic reference standard ([2-fluoromethoxy-5-(5-methyl-tetrazol-1-yl)-benzyl]-([2S,3S]-2-phenyl-piperidin-3-yl)-amine) was supplied by Merck Research Laboratories (New Jersey, USA) and used as received. A stock solution of SPA-RQ in anhydrous DMF at a concentration of 0.1 mg/mL is made. The reference solution is then prepared by further dilution of this stock solution in water p.i.to a final concentration of 2.5–5 μg/mL. A volume of 10 μL of this SPA-RQ reference solution is injected onto analytical HPLC.
42. The blank solution is prepared by mixing 7.8 mL of buffer containing 6.25% (w/v) D-glucose and 2.8 mM Na_2HPO_4 [17] with 2.2 mL of preparative HPLC mobile phase [13]. This solution is filtered through a sterile membrane filter [27]. A volume of 20 μL of this blank solution is injected onto analytical HPLC.
43. Calculated using the UV response factor for SPA-RQ.
44. Shimadzu Corporation, USA.
45. GC column: capillary Supelco Equity™ column; 5 μm film thickness, diameter 0.32 mm, length 30 m), purchased from Supelco (part no. 28062-U).
46. European Pharmacopeia 6.0—Chapter 5: Residual Solvents. ICH: International Conference on Harmonization of Technical Requirements for Registration on Pharmaceuticals for Human Use. Guidance for industry on Q3C impurities: residual solvents (CPMP/ICH/283/95). Available at: http://www.emea.europe.eu/pdfs/human/ich/028395/en.pdf
47. Kiselev MW, Tadino V. Eastern Isotopes, Inc. W.O. 2004/043497 A1. May 27. 2004.
48. Serdons K, Verbruggen A, Bormans G. J Nucl Med 2008;49:2071.
49. Merck, Darmstadt, Germany.
50. Capintec, Inc., USA.
51. Endosafe®-PTS™ 100 instrument, Charles River, Massachusetts, USA.
52. European Pharmacopoeia 6.0—Chapter 2.6.14: Bacterial Endotoxins.
53. Soya-bean casein digest medium.
54. European Pharmacopeia 6.0—Chapter 2.6.1: Sterility.

PART II

CARBON-11 LABELED RADIOPHARMACEUTICALS

CHAPTER 17

SYNTHESIS OF [^{11}C]CHOLINE CHLORIDE ([^{11}C]CHL)

BRIAN G. HOCKLEY, BRADFORD HENDERSON, AND XIA SHAO

Department of Radiology, University of Michigan School of Medicine, Ann Arbor, Michigan

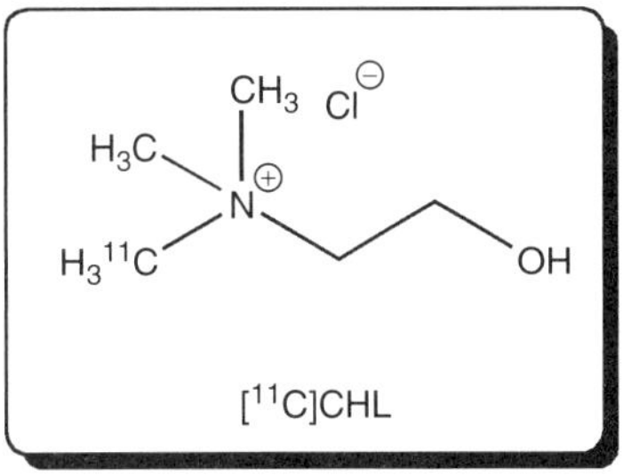

1 INTRODUCTION

[^{11}C]Choline is a particularly effective tool for imaging tumors localized in the brain, lung, esophagus, rectum, prostate, and urinary bladder [1]. In contrast to [^{18}F]fluorodeoxyglucose (FDG), [^{11}C]choline uptake in normal brain, heart, and urinary tract tissues is negligible, resulting in a greater sensitivity when imaging tumors located near those structures [2]. With the increasing demand for this tracer, several different synthetic approaches have been described in the literature, as well as different automated modules for production [3–7]. The most popular method is using a C18 Sep-Pak as a solid support for methylation and a CM Sep-Pak for purification. However, the back pressure generated by two adjacent Sep-Pak cartridges significantly slows down liquid transferring. The longer synthesis time results in lower yields and lower specific activity of the final dose because of the short half-life of carbon-11. A more effective method is available using only one CM Sep-Pak for both reaction and purification. The low back pressure of this system enabled efficient and reliable production of [^{11}C]choline using a TRACERlab $FX_{C\text{-}Pro}$ without the need to adjust incoming gas pressures. Thus the module can be utilized for solid-phase [^{11}C]methylation and easily configured for solution-phase [^{11}C]methylation. Furthermore, elimination of C18 Sep-Pak also

Radiochemical Syntheses, Volume 1: Radiopharmaceuticals for Positron Emission Tomography, First Edition. Edited by Peter J. H. Scott, Brian G. Hockley.

minimized the residual precursor, *N,N*-dimethylaminoethanol (DMAE), present in the final product, which may compete with the uptake of [^{11}C]choline into the cell.

2 SYNTHESIS PROCEDURES

CAUTION: All radiochemical syntheses must be carried out using appropriate equipment in a facility authorized for the use of radioactive materials. Personal protective equipment must be worn and all local radiation safety laws followed.

2.1 Production of [^{11}C]CO_2

[^{14}N]N_2 gas [8] was loaded into the [^{11}C]gas target [9] of a General Electric Medical Systems (GEMS) PETTrace cyclotron and pressurized to 250–255 psi. The target was bombarded with a proton beam (40 μA for 30 min) to generate ~3 Ci (111 GBq) of [^{11}C]CO_2.

2.2 Preparation of [^{11}C]Methyl Iodide

[^{11}C]CO_2 is delivered from the target via a 1/16″ Teflon delivery line by nitrogen pressure directly to a column packed with 0.3 g of molecular sieve [10] and 0.2 g of Shimalite–Nickle (Shimadzu Scientific pn: 22127719), where it is trapped at room temperature. The column is then sealed under hydrogen gas (atmospheric pressure) and heated to 350°C for 20 s to reduce the [^{11}C]CO_2 to [^{11}C]CH_4. The [^{11}C]CH_4 was passed through a column of phosphorus pentoxide desiccant (Fluka pn: 79610) and trapped on a column of carbosphere [11] at −78°C. Gaseous [^{11}C]CH_4 is released by heating the carbosphere column to 80°C. Once released, the methane enters a circulation loop that includes a gas pump [12], a column of iodine [13] at 100°C, the TRACERlab standard iodine reactor tube at 720°C, two adjacent columns of Ascarite II [14], and a column of Porapak Q [15] at room temperature. The gaseous mixture was circulated for 5 min, while [^{11}C]MeI accumulated on the Porapak column.

[^{11}C]MeI can be released from the Porapak column and delivered to the awaiting reactor by heating to 190°C.

2.3 Synthesis of [^{11}C]Choline

The reaction of gaseous [^{11}C]MeI with dimethylaminoethanol (Aldrich pn: 291263) takes place on a cation exchange Sep-Pak (Waters CM pn: WAT023531). The CM Sep-Pak is conditioned before use by flushing with 5 mL of 200 proof ethanol [16] followed by drying with nitrogen flow for 2–5 min. Dimethylaminoethanol (40 μL) is combined with 200 proof ethanol (20 μL) and added directly to the inlet side of the Sep-Pak immediately before synthesis without drying. [^{11}C]MeI is delivered from the methyl iodide synthesis unit to the CM Sep-Pak at a fixed flow rate of 20 mL/min. Reaction is immediate, and purification follows directly the completion of transfer of [^{11}C]MeI to the CM Sep-Pak. The flow path on the GE TRACERlab $FX_{C\text{-}Pro}$ was modified to accomplish this synthesis in an automated fashion [17] (Fig. 1).

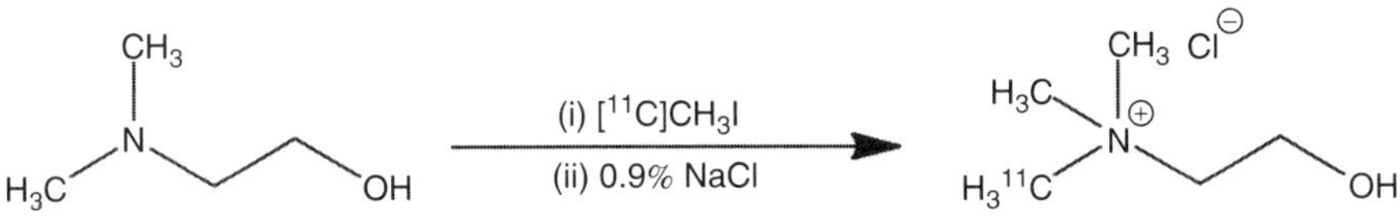

FIGURE 1 Synthesis of [^{11}C]choline.

2.4 Purification and Formulation of [^{11}C]Choline

Purification of [^{11}C]choline occurs exclusively on the CM Sep-Pak. Five milliliters of 200 proof ethanol is flushed through the Sep-Pak to remove the unreacted precursor, while the [^{11}C]choline is retained because of the affinity of the CM Sep-Pak for cations. The ethanol is rinsed from the Sep-Pak by flushing with 20 mL of United States Pharmacopeia (USP) grade sterile water for injection. The elution of [^{11}C]choline from the CM Sep-Pak is accomplished using 0.5 mL of USP grade 0.9% sodium chloride solution followed by 9.5 mL of USP grade sterile water for injection. This solution of [^{11}C]choline is then passed through a 0.22-μm sterilizing filter (Millipore Millex GS pn: SLGSV255F) into a 10-mL sterile, empty multidose vial (Hollister-Stier pn: 7515ZA).

3 QUALITY CONTROL PROCEDURES

CAUTION: All radiopharmaceuticals produced for clinical use must have local regulatory approval before use in humans. Trained personnel must carry out quality control (QC) procedures, and each dose must meet all established QC criteria before release to the clinic.

QC procedures for [^{11}C]choline are summarized below. Complete QC data for three repeat batches of [^{11}C]choline produced using the method disclosed herein are summarized in Table 1. Each of the three doses met all of the established QC criteria.

3.1 Visual Inspection

The [^{11}C]choline dose is examined in a clear glass vial and must be clear, colorless, and free of visible particulate matter.

3.2 Radiochemical Identity

Radiochemical identity is determined using a Shimadzu high performance liquid chromatography (HPLC) system with the following components: SCL-10Avp system controller, DGU-14A in-line degassing unit, LC-10ADvp pump, CDD-10Avp conductivity detector with temperature-controlled cell, CTO-20A oven, and Bioscan FC3300 flow count radioactivity detector. The system is not equipped with an ion suppressor. Chromatographic separation is performed on a Waters SCX column, IC-Pak™Cation M/D 3.9 × 150 mm, pn:

TABLE 1 QC Data for Three Repeat Runs of [^{11}C]Choline

QC Test	Release Criteria	Run 1	Run 2	Run 3
Yield/mCi (GBq)	N/A	164 (6.1)	206 (7.6)	287 (10.6)
Visual inspection	Clear, colorless	Clear, colorless	Clear, colorless	Clear, colorless
Radiochemical identity	RRT = 0.9–1.1	1.08	1.10	1.09
Radiochemical purity (%)	≥95	99.9	99.9	100
Specific activity	No limit established	NA	NA	NA
DMAE concentration (μg/mL)	<20	8.54	4.38	3.17
Dose pH	4.5–7.5	5.5	5.0	5.0
Sterile filter integrity test (psi)	>50	Pass	Pass	Pass
Radionuclidic identity ($t_{1/2}$) (min)	18.5–22.4	20.24	20.22	20.39
Endotoxin analysis (EU/mL)	≤17.5	≤2	≤2	≤2
Sterility testing	No colony growth out to 14 d	Pass	Pass	Pass

WAT036570, with a 5-mM HCl (Fisher Scientific) mobile phase and a flow rate of 1.0 mL/min. The retention time of [^{11}C]choline is compared to that of the [^{11}C]choline reference standard (choline chloride, Aldrich pn: 239941) and must be ±10% (relative retention time (RRT) must be 0.9–1.1).

3.3 Radiochemical Purity and Identity

The radiochemical purity of each batch is determined using the IC system outlined in Section 3.2. Radiochemical purity must be >95%. Representative analytical HPLC traces are displayed in Fig. 2. Radiochemical identity is determined by comparison of retention times of a known reference standard and the radioactive peak obtained. The RRT is calculated using the formula given below and must be between 0.9 and 1.1 to meet the release criteria.

$$\mathrm{RRT} = \mathrm{tr}([^{11}\mathrm{C}]\mathrm{choline})/\mathrm{tr}([^{12}\mathrm{C}]\mathrm{choline\ standard})$$

3.4 Specific Activity

The concentration of choline in each batch can be determined using the HPLC system described in Section 3.3. Choline can be detected at a minimum concentration of 1 μg/mL; however, quantitation can be readily achieved at 2 μg/mL with relative standard deviations of <10%. These values should be verified if this method is implemented, as instrument maintenance and reagent quality can dramatically affect sensitivity of electrochemical detectors. Although absolute specific activity cannot always be calculated using this method, a minimum specific activity can be determined based on the detection limit. There is no specific activity release limit for [^{11}C]choline indicated at our institution.

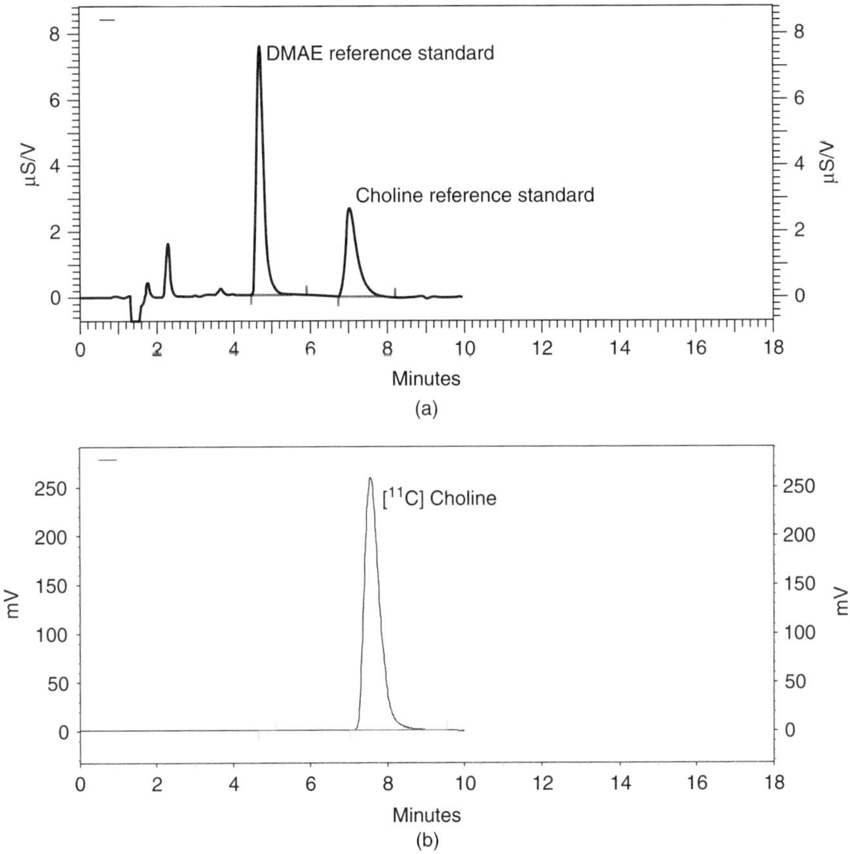

FIGURE 2 Analytical HPLC trace of (a) choline reference standard and precursor and (b) [^{11}C]choline.

3.5 Residual Dimethylaminoethanol

The concentration of DMAE in each batch can be determined using the HPLC system described in Section 3.3. DMAE can be detected at a minimum concentration of 0.2 μg/mL; however, quantitation can be readily achieved at 1 μg/mL with relative standard deviations of $<10\%$. These values should be verified if this method is implemented, as system maintenance column and reagent quality can dramatically affect sensitivity.

3.6 Dose pH

The pH of the [^{11}C]choline dose was analyzed by applying a small amount of the dose to ColorpHast® pH 2.0–9.0 nonbleeding pH-indicator strips [18] and determined by visual comparison to the scale provided. Dose pH must be 4.5–7.5.

3.7 Sterile Filter Integrity Test

The sterile filter from the [^{11}C]choline (with needle still attached) was connected to a nitrogen supply via a regulator. The needle was submerged in water, and the nitrogen pressure was gradually increased. If the pressure was raised above the filter acceptance pressure (50 psi) without seeing a stream of bubbles, the filter was considered intact. If a stream of bubble occurs at <50 psi, the test fails.

3.8 Radionuclidic Identity

Activities were measured using a Capintec CRC®-15R radioisotope dose calibrator, and half-life was calculated using Equation (1). Calculated half-life must be 18.4–22.4 min.

$$t_{1/2} = -\ln 2(\text{time difference}/(\ln(\text{ending activity}/\text{starting activity}))) \quad (1)$$

3.9 Endotoxin Analysis

Endotoxin content in doses of [^{11}C]choline was analyzed by a Charles River Laboratories EndoSafe® Portable Testing System and according to the US Pharmacopeia [19]. Doses must contain $\leq$175 endotoxin units (EU), which is equal to 17.5 EU/mL, as our formulation volume is 10 mL.

3.10 Sterility Testing

Samples of [^{11}C]choline were added to tubed fluid thioglycolate media (FTM) and tubed tryptic soy broth (TSB) media. FTM is used to test for anaerobes, aerobes, and microaerophiles, while TSB is used to test for nonfastidious and fastidious microorganisms. The media tubes were incubated along with positive and negative controls for 14 days. FTM is incubated at 30–35°C, and TSB is incubated at 20–25°C according to current USP guidelines [20]. Tubes were visually inspected on the third, eighth, and fourteenth days of the test period and compared to the positive and negative standards. The test article ([^{11}C]choline) and negative controls must have no growth after 14 days in order to pass the test for sterility. Positive controls can be established separately as a growth promotion test [20] according to USP guidelines.

CHEMICAL ABSTRACTS NOMENCLATURE (REGISTRY NUMBER)

Dimethylamino ethanol (108-01-0)

Ethanaminium, 2-hydroxy-*N,N,N*-trimethyl- (62-49-7)

Ethanol (64-17-5)

Hydrochloric acid (7647-01-0)

REFERENCES AND NOTES

1. Hara T. [^{11}C]Choline and 2-deoxy-2-[^{18}F]flour-D-glucose in tumor imaging with positron emission tomography. Mol Imaging Biol 2002;4:267–273.

2. Hara T, Kosaka N, Kondo T, Kishi H, Kobori O. Imaging of brain tumor, lung cancer, esophagus cancer, colon cancer, prostate cancer and bladder cancer with [C-11]choline. J Nucl Med 1997;38:250.
3. Hara T, Yuasa M. Automated synthesis of [11C]choline, a positron-emitting tracer for tumor imaging. Appl Radiat Isot 1999;50:531–533.
4. Pascali C, Bogni A, Iwata R, Cambie M, Bombardieri E. [11C]Methylation on a C18 Sep-Pak cartridge: a convenient way to produce [*N*-methyl-11C]choline. J Labelled Comp Radiopharm 2000;43:195–203.
5. Reischl G, Bieg C, Schmiedl O, Solbach C, Machulla H. Highly efficient automated synthesis of [11C]choline for multi dose utilization. Appl Radiat Isot 2004;60:835–838.
6. Quincoces G, Penuelas I, Valero M, Serra P, Collantes M, Marti-Climent J, Arbizu J, Garcia-Velloso M, Richter J. Simple automated system for simultaneous production of 11C-labeled tracers by solid supported methylation. Appl Radiat Isot 2006;64:808–811.
7. Cheung M, Ho C. A simple, versatile, low-cost and remotely operated apparatus for [11C]acetate, [11C]choline, [11C]methionine and [11C]PIB synthesis. Appl Radiat Isot 2009;67:581–589.
8. Ultra high purity nitrogen gas poisoned with 1% O_2 purchased from Purity Plus and used as received.
9. GEMS standard [^{11}C]carbon target with a 75-μm Havar target foil and GEHC standard vacuum foil.
10. Alltech is a division of Grace Davison Discovery Sciences (pn: 5624, 4A 80/100 mesh).
11. Carbosphere, 0.45 g, Alltech/Grace pn: 5680.
12. GE TRACERlab standard equipment.
13. Iodine crystals, 2.2 g, EMD pn: IX01203.
14. Three inches in TRACERlab standard column, Thomas Scientific pn: C049H40.
15. Porapak Q, 0.3 g, Alltech/Grace pn: 2700.
16. Decon Laboratories, Inc.
17. Shao X, Hockley BG, Hoareau R, Schnau P, Scott P. High efficiency, fully automated preparation and quality control of [11C]choline and [18F]fluormethylcholine for routine clinical application. Appl Radiat Isot 2010. DOI: 10.1016/j.apradiso.2010.09.022.
18. EMD Chemicals Inc., USA (part no. 9578-3).
19. U. S. Pharmacopeia <85> Bacterial Endotoxins Test. USP 32–NF 27. 2009.
20. U. S. Pharmacopeia <71> Sterility Tests. USP 32–NF 27. 2009.

CHAPTER 18

SYNTHESIS OF 2-(4-*N*-[^{11}C]METHYLAMINOPHENYL)-6-HYDROXYBENZOTHIAZOLE ([^{11}C]6-OH-BTA-1; [^{11}C]PIB)

CÉCILE PHILIPPE AND MARKUS MITTERHAUSER

Department of Nuclear Medicine, Medical University of Vienna, Vienna, Austria
Department of Pharmaceutical Technology and Biopharmaceutics, University of Vienna, Vienna, Austria

WOLFGANG WADSAK

Department of Nuclear Medicine, Medical University of Vienna, Vienna, Austria
Department of Inorganic Chemistry, University of Vienna, Vienna, Austria

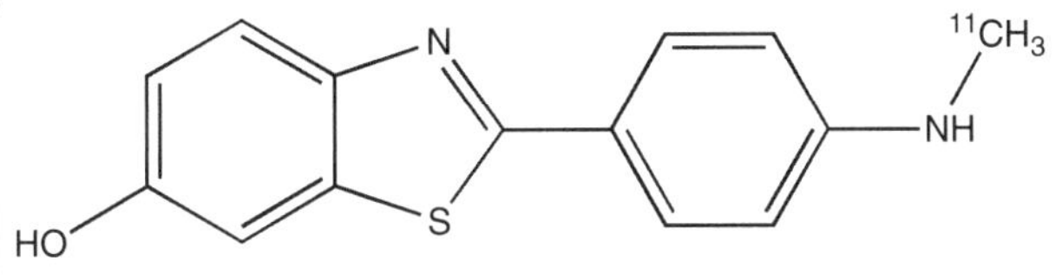

[^{11}C]6-OH-BTA-1 ([^{11}C]PIB)

1 INTRODUCTION

2-(4-*N*-[^{11}C]methylaminophenyl)-6-hydroxybenzothiazole, also known as [^{11}C] 6-OH-BTA-1 and [^{11}C] Pittsburgh compound B ([^{11}C]PIB; **2**) is a potent radiotracer for imaging amyloid plaques using positron emission tomography (PET) [1–4]. The formation of amyloid plaques in the brain is one of the main characteristics of Alzheimer's disease (AD). Therefore, amyloid plaques may be a useful target for selective biomarkers for differential diagnosis of AD and other dementias [5–7].

The original radiosynthesis of [^{11}C]PIB utilized the protected precursor 2-(4′-aminophenyl)-6-methoxymethoxybenzothiazole (6-MOMO-BTA-0) and consisted of labeling the aniline nitrogen with [^{11}C]CH_3I under forcing

Radiochemical Syntheses, Volume 1: Radiopharmaceuticals for Positron Emission Tomography, First Edition. Edited by Peter J. H. Scott, Brian G. Hockley.

conditions (125°C, KOH), followed by deprotection with hot acid [8]. Owing to the unsatisfactory low radiochemical yields, which were obtained through this method, further radiosyntheses with other methylation agents and precursors were tested. The application of [^{11}C]methyl trifluoromethanesulfonate ([^{11}C]methyl triflate, [^{11}C]CH_3OTf), a highly reactive methylation agent [9,10], offered an approach for simple and efficient labeling. The radiosyntheses described consist of direct radiolabeling of the aniline nitrogen of the unprotected precursor, 2-(4′-aminophenyl)-6-hydroxybenzothiazole (6-OH-BTA-0; **1**), with [^{11}C]CH_3OTf at 80°C in acetone [11] and at 20°C in 2-butanone [12], respectively. The following radiosynthesis describes a further optimized approach using 6-OH-BTA-0 (**1**) as the labeling precursor and [^{11}C]CH_3OTf as the methylation agent at 75°C in 2-butanone (see Fig. 1).

2 SYNTHESIS PROCEDURES

CAUTION: All radiochemical syntheses must be carried out using appropriate equipment in a facility authorized for the use of radioactive materials. Personal protective equipment must be worn and all local radiation safety laws followed.

2.1 Preparation of the Synthesis Module

All parts used before high performance liquid chromatography (HPLC) purification (reactor, HPLC injector, tubing) were rinsed with water [13], acetone [14], and 2-butanone [15] and then dried in a stream of helium. Parts used after preparative HPLC [16] (i.e., solid phase extraction (SPE) tubing, dilution flask, product collection vial, product outlet tubing) were cleaned using ethanol [17] and physiological saline (0.9%) [18]. A scheme of the synthesis module [19] is presented in Fig. 2.

Vessels were filled as follows:

Vial 2: 1 mL of HPLC eluent (i.e., acetonitrile [20] and 0.1 M ammonium acetate [21] (60/40; v/v) with 0.2% 2 M triethylammonium acetate [22]) for quenching of the reaction and dilution before HPLC injection;

Flask 3: 100 mL of sterile water [23] for dilution of the lipophilic HPLC fraction after peak cutting;

Vial 4: 5 mL of physiological saline (0.9%) [18] for final rinsing of the SPE cartridge [24] and dilution of the eluate;

Vial 5: 1.5 mL of ethanol [17] for elution of the SPE cartridge [24];

Vial 6: 10 mL of sterile water [13] for washing the SPE cartridge [24].

A C18 SPE cartridge [24] was preconditioned using 5 mL of ethanol [17] and 10 mL of sterile water [13], dried with air, and connected to its designated

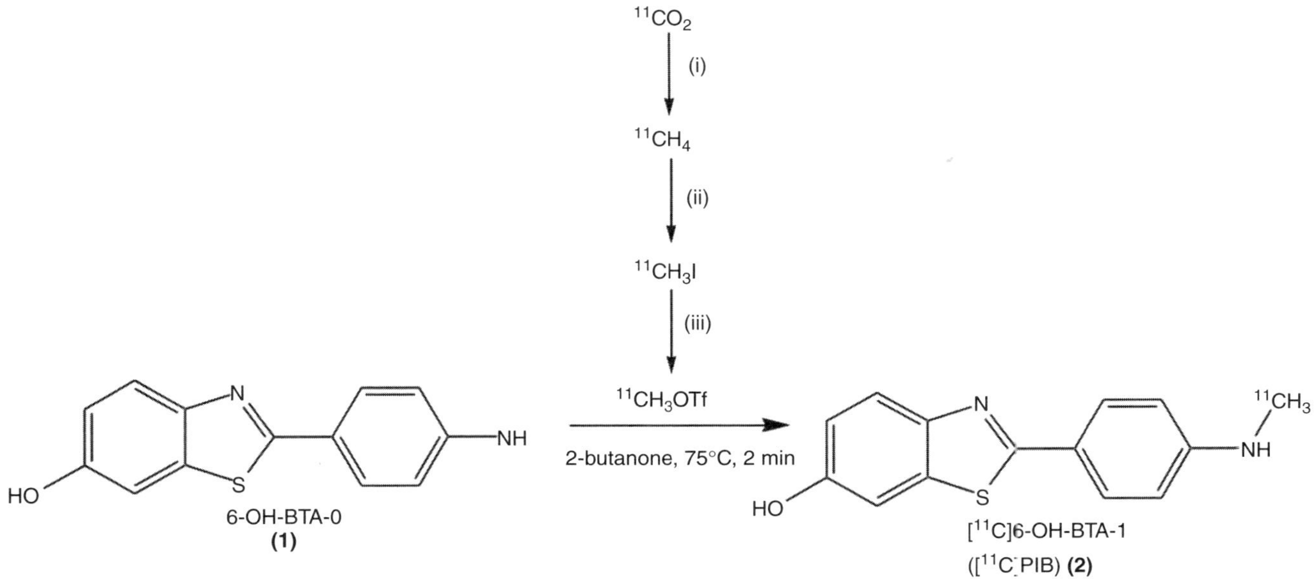

FIGURE 1 Radiosynthesis of [11C]PIB. (i) H_2/Ni, 360°C; (ii) I_2, 720°C, 5 min; (iii) AgOTf/C, 200°C, 2 min.

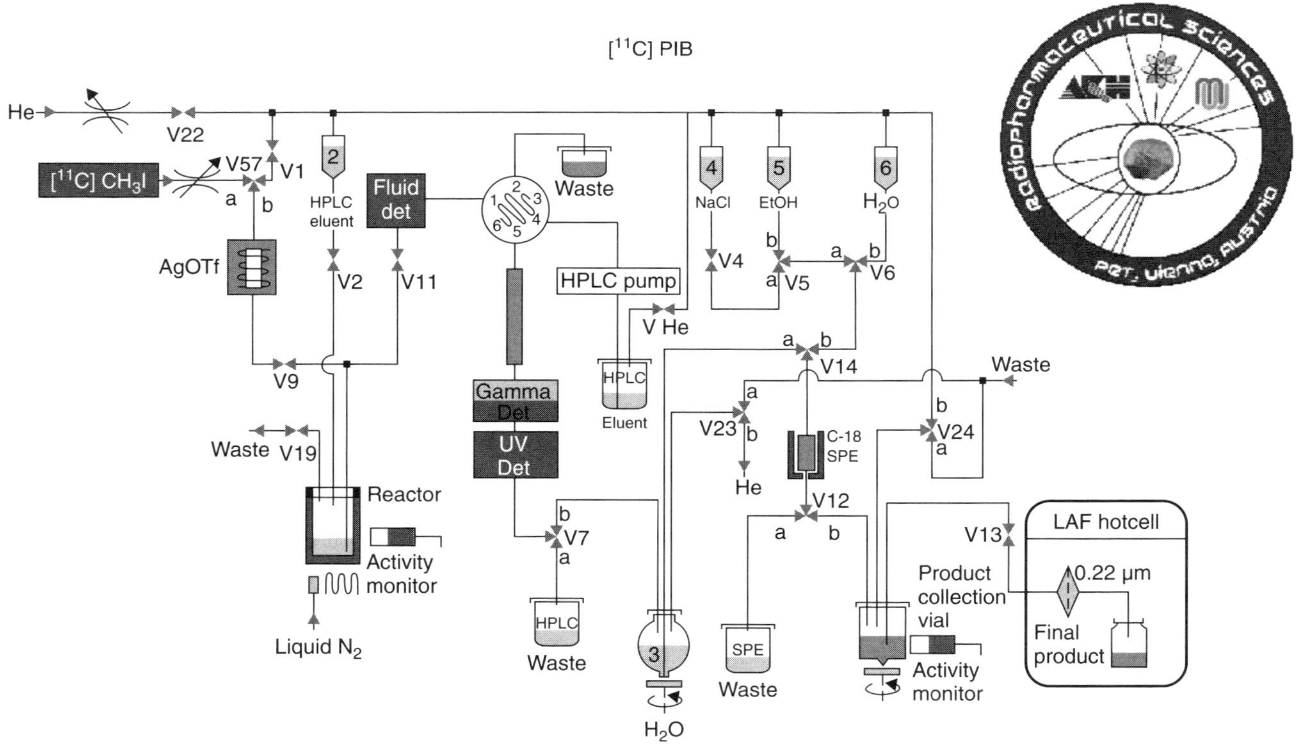

FIGURE 2 Illustration of the automated procedure for the radiosynthesis and purification of [^{11}C]PIB with a commercial C11-synthesizer.

position. The product collection vial was filled with 4 mL of physiological saline (0.9%) [18], 1 mL of 3% saline [25], and 1 mL of phosphate buffer (125 mM) [26]. The reactor was filled with 2 mg of 6-OH-BTA-0 [27] dissolved in 0.5 mL of 2-butanone [15] and subsequently placed into the heating block.

The AgOTf-oven, containing a tube filled with about 300 mg of silver-triflate-impregnated graphitized carbon [28], was preheated to 200°C.

2.2 Production of $[^{11}C]CO_2$

$[^{11}C]CO_2$ was produced in a GE PETtrace cyclotron [29] by irradiation of a gas target containing N_2 and 1% O_2 using the $^{14}N(p,\alpha)^{11}C$ nuclear reaction with 16.5 MeV protons. Typical beam currents were 48–58 μA, and the irradiation was stopped as soon as the desired activity level was reached (~1.75–2.15 Ci (65–80 GBq)).

2.3 Production of $[^{11}C]CH_3I$

$[^{11}C]CH_3I$ was produced using the gas phase conversion described by Larsen et al. in 1997 [30] in an automated synthesis module [31]. Thereby, $[^{11}C]CO_2$ was trapped on a molecular sieve (4 Å) and converted to $[^{11}C]CH_4$ within 120 s in the presence of Ni catalyst [32] and hydrogen at 360°C. Subsequently, $[^{11}C]CH_4$ was reacted with elemental iodine [33] at 720°C to get $[^{11}C]CH_3I$ and recirculated for 300 s. $[^{11}C]CH_3I$ was trapped online on Porapak N [34] in a small glass column and finally released by heating to 190°C in a stream of helium (45 mL/min) within 160 s.

2.4 Production of $[^{11}C]CH_3OTf$

$[^{11}C]CH_3OTf$ was produced in the previously prepared synthesizer system [19] and remotely controlled via GINAstar software [35]. $[^{11}C]CH_3I$ was converted into $[^{11}C]CH_3OTf$ by online passage through a heated tube (200°C) containing about 300 mg of silver-triflate-impregnated graphitized carbon [28].

2.5 Synthesis of $[^{11}C]$PIB

The fully automated preparation of $[^{11}C]$PIB was performed on the prepared radiosynthesis module [19] and remotely controlled via GINAstar software [35]. $[^{11}C]CH_3OTf$ was delivered through V9 and trapped online in the reactor containing 2 mg 6-OH-BTA-0 [27] in 500 μL 2-butanone [15] cooled to 13°C. The reactor was sealed and the reaction mixture heated to 75°C and kept at this temperature for 2 min. After cooling down to room temperature, the reaction mixture was quenched by the addition of 1 mL of the HPLC eluent (through V2) and subsequently transferred to the 5-mL injection loop of the HPLC system through V11 passing the fluid detector.

2.6 Purification and Formulation of $[^{11}C]$PIB

Injection of the crude reaction mixture to the semipreparative reverse phase high performance liquid chromatographic (RP-HPLC) column [36] was controlled by

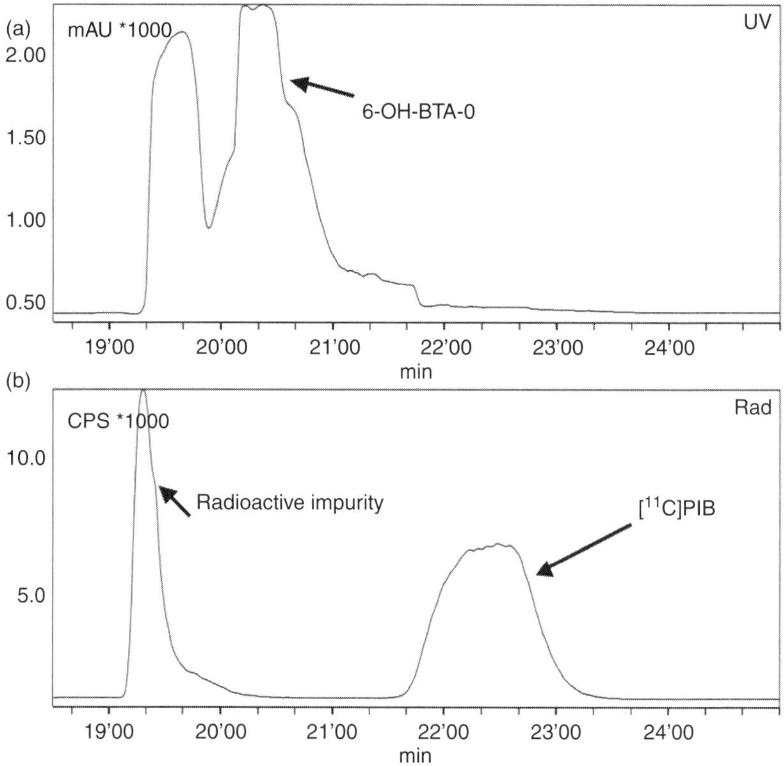

FIGURE 3 Semipreparative UV (a) and radioactive (b) HPLC traces for [^{11}C]PIB. X-axis shows overall synthesis time; retention time of [^{11}C]PIB is, therefore, ~4–5 min.

an automated fluid detector. Semipreparative HPLC [16] was performed with acetonitrile [20], 0.1 M ammonium acetate [21], and 2 M triethylammonium acetate [22] (60/40/0.2; v/v/v) as a mobile phase with a flow rate of 8 mL/min. Chromatograms were registered using a UV-detector (254 nm) and a NaI radioactivity detector in series. An example is given in Fig. 3.

The [^{11}C]PIB fraction (retention time, 4–5 min) was cut into flask 3 through V7 and diluted with 100 mL sterile water [23]. The resulting solution was then pushed over a C18 SPE cartridge [24]. After washing with 10 mL water [13] (Vial 6), the purified product was completely eluted with 1.5 mL ethanol [17] (Vial 5) and 5 mL 0.9% saline [18] (Vial 4). Formulation was done with a further 4 mL physiological saline (0.9%) [18], 1 mL saline solution (3%) [25], and 1 mL physiological phosphate buffer (125 mM) [26] in the product collection vial. This final solution was transferred to a shielded laminar air-flow hot cell and, there, sterile-filtered (0.22 μm) [37] into a sterile 25-mL vial [38] containing another 5 mL of saline solution 0.9% [18]. Hence, the final total volume was 17.5 mL. The total synthesis time was 32 min from end of bombardment (EOB) and 17 min from [^{11}C]CH_3OTf.

3 QUALITY CONTROL PROCEDURES

CAUTION: All radiopharmaceuticals produced for clinical use must have local regulatory approval before use in humans. Trained personnel must carry out quality control (QC) procedures, and each dose must meet all established QC criteria before release to the clinic.

QC procedures for [^{11}C]PIB, based on the current requirements for radiopharmaceuticals laid out both in the European Pharmacopoeia [39] and in the US Pharmacopoeia [40] are summarized below and can be performed within 10 min (except microbial and endotoxin testing). Complete QC data for three repeat batches of [^{11}C]PIB produced using the methods described herein are summarized in Table 1.

3.1 Visual Inspection

The [^{11}C]PIB dose is examined behind a 511-keV shielding lead glass window: the solution must be clear, colorless, and free of particles.

3.2 Radiochemical Identity and Radiochemical Purity

Chemical and radiochemical impurities were detected and quantified using analytical radio-HPLC [41]. Acetonitrile [20] and 0.1 M ammonium formiate [42] (50/50; v/v) was used as the mobile phase with a flow rate of 1.5 mL/min

TABLE 1 QC Data for Three Repeat Runs of [^{11}C]PIB

QC Test	Release Criteria	Run 1	Run 2	Run 3
Yield mCi (GBq)	N/A	197 (7.3)	141 (5.2)	95 (3.5)
Visual inspection	Clear and colorless	Pass	Pass	Pass
Radiochemical identity (min)	HPLC RT 4.7–5.3	4.87	4.92	4.88
Radiochemical purity (%)	≥95	100	100	99.5
Specific activity (GBq/μmol)	N/A	187.6	163.3	179.1
Precursor amount 6-OH-BTA-0 (μg/mL)	N/A	1.785	1.461	0.418
Residual solvent analysis	2-Butanone <5000 ppm	<1 ppm	<1 ppm	<1 ppm
	Acetonitrile <410 ppm	7.5 ppm	13 ppm	8.3 ppm
	Ethanol ≤10%	8.5%	8.5%	8.5%
Dose pH	5.0–8.5	7.3	7.3	7.2
Dose osmolality (mOsm/kg	190–370	307	289	311
Sterile filter integrity test (psi)	>50	Pass	Pass	Pass
Radionuclidic identity ($t_{1/2}$) (min)	20.3±2	19.9	20.2	19.7
Endotoxin analysis (IU/mL)	<17.5	<1	<1	<1
Sterility testing	No colony growth out to 14 d	Pass	Pass	Pass

N/A, not applicable; RT, retention time.

on a reversed phase column [43]. The whole HPLC analysis was completed within 8 min, the retention time of the precursor (6-OH-BTA-0) was 3.7–4.3 min ($k' = 1.2$–1.6), and the product [^{11}C]PIB was eluted with a retention time of 4.7–5.3 min ($k' = 1.8$–2.2). A potential radioactive by-product, 6-[^{11}C]methoxy-BTA-0, eluted at 7.5–8.4 min ($k' = 3.5$–4.1). The chemical identity of [^{11}C]PIB was determined by co-injection of the unlabeled reference compound, [^{12}C]PIB [44]. For calculation of radiochemical purity, the portion of [^{11}C]PIB in relation to total radioactivity is determined and must be $\geq$95%. Sample chromatograms are given in Fig. 4.

3.3 Specific Activity

The specific activity (a_S) was defined as the result of the division of measured radioactivity of [^{11}C]PIB (GBq) by amount of [^{12}C]PIB (μmol). The amount of [^{12}C]PIB was determined using HPLC. (A calibration curve was previously determined by measuring known concentrations (micrograms per milliliter) of an authentic PIB sample [44], recording the corresponding peak areas and fitting the data by linear regression). Thus, the specific activity is calculated by the following Equation (1):

$$a_S[\text{GBq}/\mu\text{mol}] = A[\text{GBq}]/(c_{\text{PIB}}[\mu\text{g/mL}] \times V[\text{mL}]/256.32\ [\text{g/mol}]) \quad (1)$$

where A is the measured radioactivity, c_{PIB} is the calculated concentration of unlabeled PIB through HPLC, and V is the volume of the product solution.

3.4 Residual Solvent Analysis

Residual solvents were analyzed by gas chromatography (GC) with flame ionization detector (FID) [45] (carrier gas: He, flow rate 2.7 mL/min, 45°C (2.5 min); 20°C/min to 110°C; 30°C/min to 200; 200°C (10 min); FID: 270°C).

The limits were <5000 ppm for 2-butanone, ≤10% for ethanol, and <410 ppm for acetonitrile [46].

3.5 Dose pH

The pH of the [^{11}C]PIB dose was analyzed by a WTW pH 537 Microprocessor pH meter [47]. The limits of the pH values for this parenteral preparation are set to be 5.0–8.5.

3.6 Dose Osmolality

The osmolality of the [^{11}C]PIB dose was analyzed by a Wescor osmometer 5520 [48]: a 10-μL aliquot of the dose was applied onto a paper filter inside the osmometer chamber. The limits of the osmolality for this parenteral preparation were set to be 190–370 mOsm/kg.

3.7 Sterile Filter Integrity Test

The sterile filter [37] from the aseptic filling of the final [^{11}C]PIB solution (with needle still attached) was connected to a nitrogen supply via a regulator. The needle was submerged in water, and the nitrogen pressure was gradually increased.

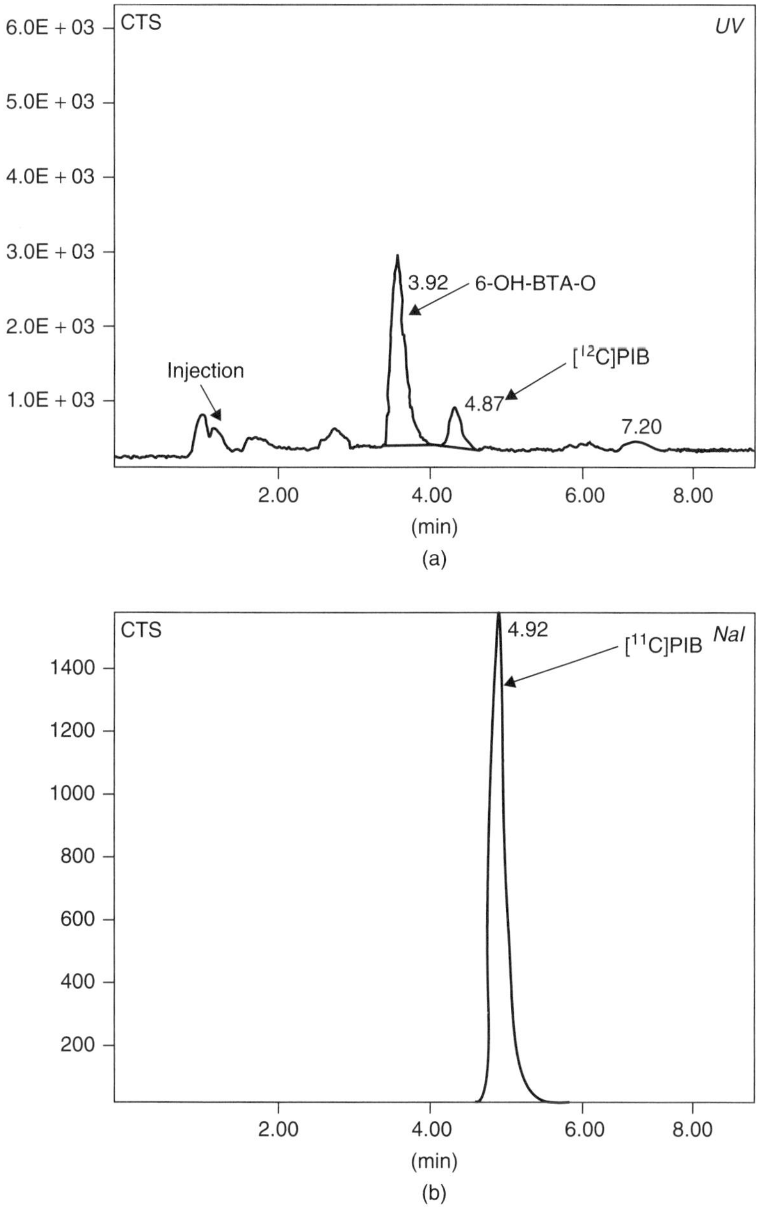

FIGURE 4 Analytical UV (a) and radioactive (b) HPLC traces for a final [^{11}C]PIB product sample.

If the pressure was raised above the filter acceptance pressure (50 psi) without seeing a stream of bubbles, the filter was considered intact.

3.8 Radionuclidic Identity

γ-Spectra were recorded within a Packard Cobra II Auto-Gamma counter [49] and peaks analyzed. Only at 511 and 1022 keV peaks were allowed. Since all PET radioisotopes yield the same γ trace, additionally, activities were measured at different time points using a Veenstra VDC404 radioisotope dose calibrator [50], and physical half-life was calculated using Equation (2). Calculated half-life must be 20.3±2 min.

$$t_{1/2} = -\ln 2 \cdot (t_2 - t_1)/\ln\left(\frac{A_2}{A_1}\right) \tag{2}$$

In Equation (2) A_1 is the starting activity at t_1 and A_2 is the end activity at t_2.

3.9 Endotoxin Analysis

Bacterial endotoxin content in doses of [^{11}C]PIB was analyzed (when the probes were physically decayed) by an external laboratory [51] where the analyses were performed according to the current European [52] and US Pharmacopeia [53] guidelines using the limulus amoebocyte lysate (LAL)-test [54]: For this, an aliquot of the decayed [^{11}C]PIB dose was added to the LAL and incubated at 37°C for 60 min; then the extent of clotting was compared against a negative control and a positive standard. The limit is <17.5 IU/mL. The determination of this parameter was always done after injection to the patient when radioactivity was decayed in accordance with the regulations laid down in the pharmacopeias [39,40].

3.10 Sterility Testing

Sterility testing was performed (when the probes were physically decayed) according to current European Pharmacopoeia [55] standard procedures by an external laboratory [51] whereby fluid thioglycolate media (FTM, sensitive for anaerobes and fungi) [55,56] and soybean casein digest media (SCDM, sensitive for aerobes) [55,56] were used and aliquots of the decayed [^{11}C]PIB dose were added. The FTM samples were incubated at 30–35°C for 14 days, and the SCDM samples were incubated at 20–25°C and were visually compared against a negative control and a positive standard after the 14-day incubation period. The determination of this parameter was always done after injection to the patient when radioactivity was decayed in accordance with the regulations laid down in the pharmacopoeias [39,40].

CHEMICAL ABSTRACTS NOMENCLATURE (REGISTRY NUMBER)

Acetic acid (64-19-7)
Acetone (67-64-1)
Acetonitrile (75-05-8)

Ammonium acetate (631-61-8)

Ammonium formate (540-69-2)

6-OH-BTA-0 (= 6-benzothiazolol, 2-(4-aminophenyl)-; = 2-(4′-aminophenyl)-6-hydroxybenzothiazole) (178804-18-7)

6-OH-BTA-1 (= 6-benzothiazolol, 2-[4-(methylamino)phenyl]-; = 2-[4′-(methylamino)phenyl]-6-hydroxybenzothiazole; = Pittsburgh Compound B; = PIB) (566169-93-5)

2-Butanone (78-93-3)

Disodiumhydrogenphosphate-dihydrate (10028-24-7)

Ethanol (64-17-5)

Iodine (7553-56-2)

6-Methoxy-BTA-0 (= benzenamine, 4-(6-methoxy-2-benzothiazolyl)-; = 2-(4′-aminophenyl)-6-methoxybenzothiazole) (43036-17-5)

Ni (Shimalite 80/100 reduced) (7440-02-0)

Porapak N, Porapak™ Porous Polymer Adsorbent matrix, 80-100 mesh (no CAS no. available)

Silver trifluoromethanesulfonate (2923-28-6)

Sodium chloride (7647-14-5)

Sodium dihydrogenphosphate-monohydrate (10049-21-5)

Triethylamine (121-44-8))

REFERENCES AND NOTES

1. Nordberg A. Lancet Neurol 2004;3:519–527.
2. Nordberg A. Eur J Nucl Med Mol Imaging 2008;35 Suppl 1:S46–S50.
3. Herholz K, Carter SF, Jones M. Br J Radiol 2007;80 Spec No 2:S160–S167.
4. Rabinovici GD, Jagust WJ. Behav Neurol 2009;21:117–128.
5. Tolboom N, van der Flier WM, Boverhoff J, Yaqub M, Wattjes MP, Raijmakers PG, Barkhof F, Scheltens P, Herholz K, Lammertsma AA, van Berckel BNM. J Neurol Neurosurg Psychiatry 2010;81:882–884.
6. Roe CM, Mintun MA, Ghoshal N, Williams MM, Grant EA, Marcus DS, Morris JC. Neurology 2010;75:42–48.
7. Grimmer T, Henriksen G, Wester HJ, Förstl H, Klunk WE, Mathis CA, Kurz A, Drzezga A. Neurobiol Aging 2009;30:1902–1909.
8. Mathis CA, Wang Y, Holt DP, Huang G-F, Debnath ML, Klunk WE. J Med Chem 2003;46: 2740–2754.
9. Jewett DM. Int J Rad Appl Instrum A 1992;43:1383–1385.
10. Holschbach M, Schüller M. Appl Radiat Isot 1993;44:897–898.
11. Solbach C, Uebele M, Reischl G, Machulla H-J. Appl Radiat Isot 2005;62:591–595.
12. Wilson AA, Garcia A, Chestakova A, Kung H, Houle S. J Labelled Comp Radiopharm 2004;47:679–682.
13. Aqua bidest. "Fresenius" was purchased from Fresenius Kabi (Graz, Austria).
14. Acetone (HPLC grade) was purchased from Merck (Darmstadt, Germany).
15. 2-Butanone, CHROMASOLV® (for HPLC, $\geq$99.7%) was purchased from Sigma Aldrich (Vienna, Austria).

16. The preparative HPLC system consisted of a Rheodyne injector, a Sykam S1021 pump, a UV detector (254 nm), and a radioactivity detector in series.
17. Ethanol (part. no. 1.00986.1000) was purchased from Merck (Darmstadt, Germany) and used as received.
18. Saline solution 0.9% was purchased from B.Braun (Melsungen, Germany).
19. The fully automated synthesizer was a C11-methylation module originally purchased from Nuclear Interface® (now: General Electric Medical Systems, Uppsala, Sweden).
20. Acetonitrile (part. no. 1.00029.1000) in HPLC grade was purchased from Merck (Darmstadt, Germany) and used as received.
21. Ammonium acetate solution 0.1 M was prepared by dissolving 0.1 mol of ammonium acetate (Fluka, Buchs, Switzerland) in 100 mL of bidistilled water.
22. Triethylammonium acetate 2 M was prepared by adding 2.78 mL triethylamine (Fluka, Buchs, Switzerland) and 1.14 mL acetic acid (Fluka, Buchs, Switzerland). The solution was diluted with water to 10 mL.
23. "Meditrade" Aqua ad injectabilia; Meditrade Medicare Medizinprodukte (Kufstein, Austria).
24. C18plus SepPak® cartridges were purchased from Waters® (Milford, MA, USA).
25. Saline solution 3% was obtained from Landesapotheke Salzburg, Austria.
26. Phosphate buffer 125 nM was prepared by dissolving 0.224 g sodium dihydrogenphosphate monohydrate and 1.935 g disodiumhydrogenphosphate dihydrate in 100 mL bidistilled water.
27. The labeling precursor, 6-OH-BTA-0, was produced under full GMP and was purchased from ABX-Advanced Biochemical Compounds (Radeberg, Germany; product no. 5101).
28. Silver triflate impregnated carbon was prepared by dissolving 1 g of silver trifluoromethanesulfonate (Sigma Aldrich, Vienna, Austria) in 20 mL acetonitrile (for DNA synthesis, $\leq$10 ppm H_2O, Merck, Darmstadt, Germany). To this solution, 3 g of Graphpac-GC (80/100 mesh, Alltech, Deerfield, Illinois, USA) were added, and the suspension was stirred in the dark for about 30 min. The solvent was removed under reduced pressure, and the obtained powder was further dried in the dark for 2 h (rotary evaporator).
29. The GE PETtrace cyclotron was from General Electric Medical Systems (Uppsala, Sweden).
30. Larsen P, Ulin J, Dahlstrom K, Jensen M. *Appl Radiat Isot* 1997;48:153–157.
31. MeI-Microlab, General Electric Medical Systems (Uppsala, Sweden).
32. Ni catalyst (Shimalite Ni reduced, 80/100 mesh) was purchased from Shimadzu (Kyoto, Japan).
33. Iodine (sublimated GR for analysis ACS, ISO, Reag. Ph Eur, product no. 1.04761.0100) was purchased from Merck (Darmstadt, Germany).
34. Poropak® N (80/100 mesh) was obtained from Alltech Associates, Inc. (Grace GmbH & Co. KG, Vienna, Austria).
35. Raytest Isotopenmessgeräte GmbH (Straubenhardt, Germany).
36. Supelcosil™ LC-ABZ$^+$; 5 μm, 250 mm × 10 mm; (No. 59179) from Supelco (Bellfonte, PA, USA).
37. Millex-GV® PVDF 0.22-μm filters (part. no. SLGV 033RS) were purchased from Millipore (Bedford, MA, USA) and used as received.
38. DRN 4370 evacuated vials 25 mL were purchased from Mallinckrodt Medical B.V. (Petten, Holland).
39. Pharmacopoeia Europea <6.0/0125>. Radioactive pharmaceuticals. Ph.Eur. 6, 2008.
40. U.S. Pharmacopeia <823>. Radiopharmaceuticals for Positron Emission Tomography Compounding. USP 32-NF 27, 2009.
41. Merck Hitachi (Darmstadt, Germany).
42. 0.1 M ammonium formiate solution was prepared by dissolving 0.1 mol of ammonium formiate (Merck, Darmstadt, Germany) in 100 mL of bidistilled water.
43. Prodigy 5 μm Phenyl-3 (PH-3); 250 × 4.6 mm was obtained from Phenomenex (Aschaffenburg, Germany).
44. [^{12}C]PIB, 6-OH-BTA-1, free base (product no. 5140) was purchased from ABX-Advanced Biochemical Compounds (Radeberg, Germany).

45. Gas chromatography was performed using an HP 6890 series system (Agilent, Santa Clara, CA, USA) with flame ionization detector (FID) on a HP-Innowax capillary column (30 m × 0.32 mm × 0.25 μm (no. 19091N-113)).
46. European Pharmacopoeia 6th edition (2.4.24), Identification and Control of Residual Solvents. Ph.Eur. 6, Strasbourg, France, 2008.
47. The WTW pH 537 pH Meter was purchased from WTW (Weilheim, Germany).
48. The Wescor osmometer 5520 was purchased from Sanova Medical Systems (Vienna, Austria).
49. The Cobra II Auto-Gamma counter was purchased from Packard (Meriden, CT, USA).
50. The Veenstra VDC404 activimeter was purchased from Veenstra Instruments (Joure, Netherlands).
51. The external laboratory that performed both endotoxin analyses and sterility testings was the research laboratory of the Department for Infectious Diseases and Tropical Medicine of the Medical University of Vienna, Vienna, Austria.
52. European Pharmacopoeia 6th edition (2.06.14), Determination of Bacterial Endotoxins. Ph.Eur. 6, Strasbourg, France, 2008.
53. U.S. Pharmacopeia <85>. Bacterial Endotoxins Test. USP 32-NF 27, 2009.
54. The Pyrochrome® Limulus Amoebocyte Lysate (LAL) Test was purchased from Cape Cod (East Falmouth, MA, USA).
55. European Pharmacopoeia 6th edition (2.06.01), Determination of Sterility. Ph.Eur. 6, Strasbourg, France, 2008.
56. U.S. Pharmacopeia <71>. Sterility Tests. USP 32-NF 27, 2009.

CHAPTER 19

SYNTHESIS OF [^{11}C]-meta-HYDROXYEPHEDRINE ([^{11}C]MHED)

FILIPPO LODI

PET Radiopharmacy, Nuclear Medicine Unit, Policlinico S.Orsola-Malpighi, Bologna, Italy

ASSUNTA CARPINELLI

IBFM-CNR, Nuclear Medicine Department, University of Milan Bicocca, San Raffaele Scientific Institute, Milano, Italy

CLAUDIO MALIZIA AND STEFANO BOSCHI

PET Radiopharmacy, Nuclear Medicine Unit, Policlinico S.Orsola-Malpighi, Bologna, Italy

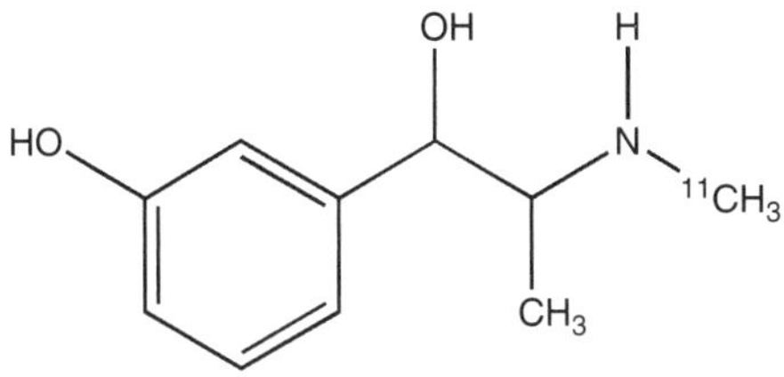

[^{11}C]MHED

1 INTRODUCTION

Positron emission tomographic (PET) imaging of sympathetic innervation of the heart can involve the radiolabeling of true adrenergic neurotransmitters or catecholamine analogs (false neurotransmitters) with radiopharmaceuticals that allow the study of heart diseases [1]. [^{11}C]-*meta*-Hydroxyephedrine ([^{11}C]MHED) (**2**) is a norepinephrine analog that accumulates in nerve terminals but is not metabolized by catechol-*O*-methyl transferase (COMT) and is also resistant

Radiochemical Syntheses, Volume 1: Radiopharmaceuticals for Positron Emission Tomography, First Edition. Edited by Peter J. H. Scott, Brian G. Hockley.

to oxidative deamination by monoamine oxidase (MAO) [2]. Owing to its charateristics, [^{11}C]MHED is a putative radiopharmaceutical for noninvasive assessment of sympathetic neuronal integrity of the heart in clinical PET [3].

2 SYNTHESIS PROCEDURES

2.1 Production of [^{11}C]CO_2

[^{11}C]CO_2 was produced by nuclear reaction $^{14}N(p,\alpha)^{11}C$ on a mixture of nitrogen (N60 purity grade) and 1% oxygen using a high performance ^{11}C target [4] in a PETtrace cyclotron (16.5 MeV protons) [4]. The target was irradiated (70 μA current for 25 min) to produce ~5.6 Ci (207 GBq) of [^{11}C]CO_2.

2.2 [^{11}C]CO_2 Trapping and Reduction to [^{11}C]CH_4

[^{11}C]CO_2 was delivered into TRACERlab $FX_{C\text{-}Pro}$ module [4] modified for the [^{11}C]MHED synthesis. [^{11}C]CO_2 was trapped at room temperature in Shimalite-nickel [5] catalyst/molecular sieves [6] (0.2 g/0.3 g) column, previously conditioned at 350°C for 10 min with hydrogen flow, and flushed with helium. The column was filled with hydrogen (6.0 purity grade) and warmed at 350°C. [^{11}C]CH_4 produced was trapped at −90°C in a Carbosphere [7] column, and residual [^{11}C]CO_2 and water were trapped in a P_2O_5 (Sicapent) [8] /NaOH (Ascarite) [9] column.

2.3 Synthesis of Methylating Agent [^{11}C]CH_3I from [^{11}C]CH_4

Carbosphere was warmed at 90°C, and the released [^{11}C]CH_4 was pulled under helium flow and reacted with molecular iodine [10], sublimated at 100°C, in a furnace at 734°C [11,12]. [^{11}C]CH_3I produced by radical iodination was trapped in Porapak Q column [13].

2.4 Synthesis of [^{11}C]MHED

Porapak Q column was warmed at 190°C, and [^{11}C]CH_3I was delivered under helium flow (10 mL/min) in a reactor containing metaraminol free base (**1**) [14,15] as precursor dissolved in *N,N*-dimethylformamide (DMF) (≥99.5%) [16] (1 mg in 200 μL of DMF). The reactor was warmed at 100°C for 5 min (Fig. 1).

2.5 Purification of [^{11}C]MHED

The reactor was cooled at 25°C, and the reaction mixture was diluted with 1.3 mL of semipreparative HPLC mobile phase: CH_3COONH_4 0.1 M (Ph. Eur.): ethanol (Ph. Eur.) 95:5 [17]. The reactor solution was delivered in a semipreparative HPLC for the purification of [^{11}C]MHED as shown in Fig. 2 (Synergi Polar-RP column, 10 μm, 250 × 10 mm [18], flow 5 mL/min, $\lambda = 280$ nm). The fraction corresponding to [^{11}C]MHED (typically eluting between 8.5 and 10 min) was collected for 1 min into a collecting vial and was sterilized with a 0.22-μm filter [19] in a sterile vial [20]. The sterile vial, filter, and needle assembly was prepared in a class A isolator. After synthesis was complete, the semipreparative

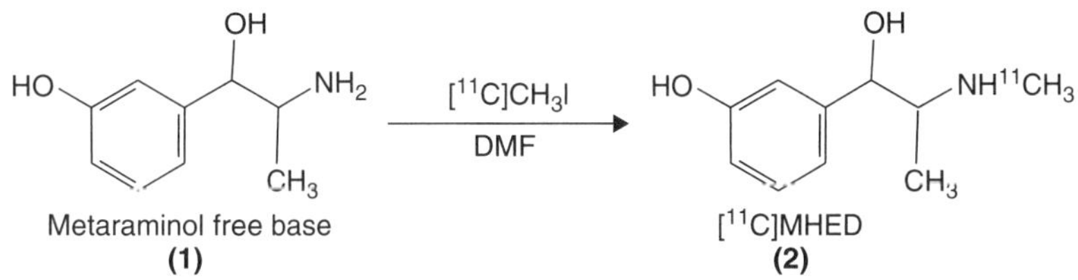

FIGURE 1 Synthesis of [^{11}C]MHED.

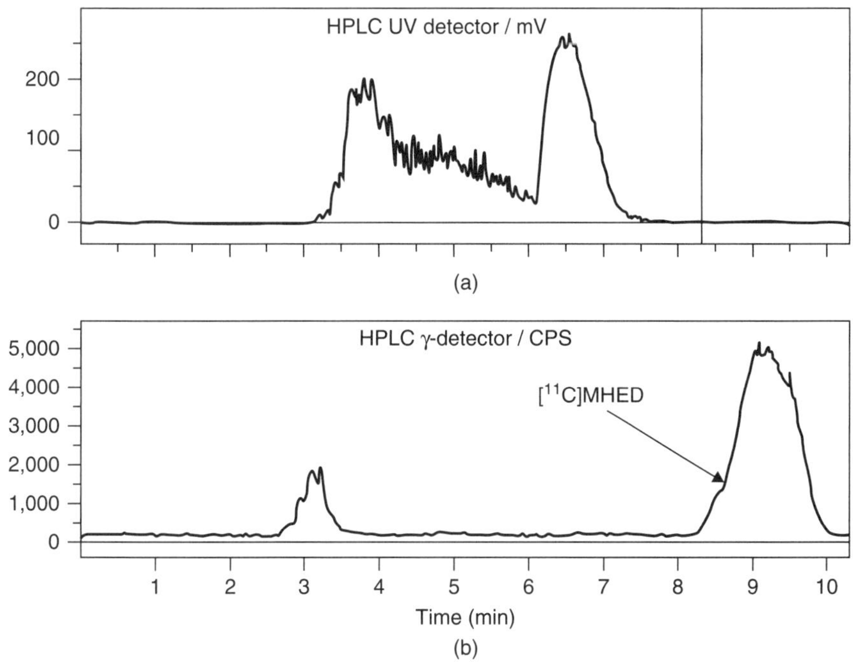

FIGURE 2 Semipreparative UV (a) and radioactive HPLC traces (b) for [^{11}C]MHED.

HPLC column was flushed with 70% ethanol [21]. The overall synthesis time was 35 min. An automated cleaning procedure of the synthesis module, using ethanol 70% and acetone, was performed after complete radiation decay.

2.6 Synthesis of [^{11}C]MHED Using $LiAlH_4$/HI Method

Cyclotron-produced [^{11}C]CO_2 (1.2 Ci, 44.4 GBq) was delivered into FX ^{11}C methylation module [4] and trapped in a coil cooled at −150°C with liquid nitrogen for purification. After coil warming, [^{11}C]CO_2 was bubbled under helium flow in a solution of $LiAlH_4$ (0.1 mL, 0.1 M in THF) [22] in the first reactor. After elimination of the solvent, 0.2 mL of HI 57% [23] was added, and the reactor was warmed at 140°C. Produced [^{11}C]CH_3I was distilled under a stream of helium (20 mL/min), purified by Ascarite/Sicapent column, and transferred into a second reactor containing precursor metaraminol free base [15] (**1**) dissolved in *N,N*-dimethylformamide [16] (1 mg in 100 μL of DMF). The reactor was warmed at 100°C for 5 min. The reaction mixture was diluted with 1 mL of semipreparative HPLC mobile phase: CH_3COONH_4 0.1 M : ethanol 95:5 and injected in a semipreparative HPLC column for the purification of [^{11}C]MHED as described above. The overall synthesis time was 35–40 min.

3 QUALITY CONTROL PROCEDURES

CAUTION: All radiopharmaceuticals produced for clinical use must have local regulatory approval before use in humans. Trained personnel must carry out quality control (QC) procedures, and each dose must meet all established QC criteria before release to the clinic.

QC of [^{11}C]MHED was based on Radiopharmaceutical Preparations monograph reported in the European Pharmacopoeia (Ph. Eu.) [24]. Complete QC results for three batches of [^{11}C]MHED produced with the methods previously described are reported in Table 1. Each of the three batches met the QC criteria.

3.1 Visual Inspection

[^{11}C]MHED injectable solution must be clear and colorless.

TABLE 1 QC Data for Three Repeat Runs of [^{11}C]MHED

QC Test	Release Criteria	Run 1	Run 2	Run 3
Yield mCi (GBq) (DC, EOB)	N/A	558 (20.6) 467 (17.3)[a]	594 (22) 470 (17.4)[a]	578 (21.4) 468 (17.3)[a]
Visual inspection	Clear, colorless	Clear, colorless	Clear, colorless	Clear, colorless
Radiochemical identity (min)	RT = 7.9	RT = 8	RT = 7.8	RT = 8
Specific activity (Ci/μmole)	No limit established	0.4–1.5[a]	0.4–1.5[a]	0.4-1.5[a]
Radiochemical purity (%)	≥95	98 >97%[a]	99.3 >97%[a]	98 >97%[a]
pH	4.5–8.5	6.0	6.0	6.0
Residual solvents analysis	Ethanol <50 mg/*V* Acetone <50 mg/*V*	<50 mg/mL Absent	<50 mg/mL Absent	<50 mg/mL Absent
Radionuclidic purity (%)	≥99	>99.99	>99.99	>99.99
Radionuclidic identity ($t_{1/2}$) (min)	19.9–20.9	20.5	20.4	20.4
Endotoxin analysis	<175/V IU/mL	Pass	Pass	Pass
Sterile Filter Integrity Test (psi)	>56	>56	>56	>56
Sterility testing	According to Ph. Eu.	Pass	Pass	Pass

[a]Synthesis and QC data from San Raffaele Scientific Institute starting from 1200 mCi (44.4 GBq) of [^{11}C]CO_2 produced by using the $LiAlH_4$/HI method.
N/A, not applicable; DC, decay corrected; EOB, end of bombardment; RT, retention time.

3.2 Radiochemical Identity

Radiochemical identity was determined with the HPLC method: Synergi Polar-RP column [25] with 0.1 M CH_3COONH_4: ethanol 90:10 as mobile phase at the flow of 1 mL/min. HPLC was equipped with radio and UV detectors ($\lambda = 280$ nm). The retention time of [^{11}C]MHED was compared to that of the MHED reference standard [26] and must be ±10%.

3.3 Radiochemical Purity

HPLC analysis of radiochemical purity was conducted using Synergi Polar-RP column (4 μm, 250 × 4.6 mm) [25] with 0.1 M CH_3COONH_4: ethanol 90:10 as mobile phase at a flow rate of 1 mL/min. HPLC was equipped with UV ($\lambda = 280$ nm) and radio detectors. Radiochemical purity must be minimum 95%. Representative analytical HPLC traces are displayed in Fig. 3.

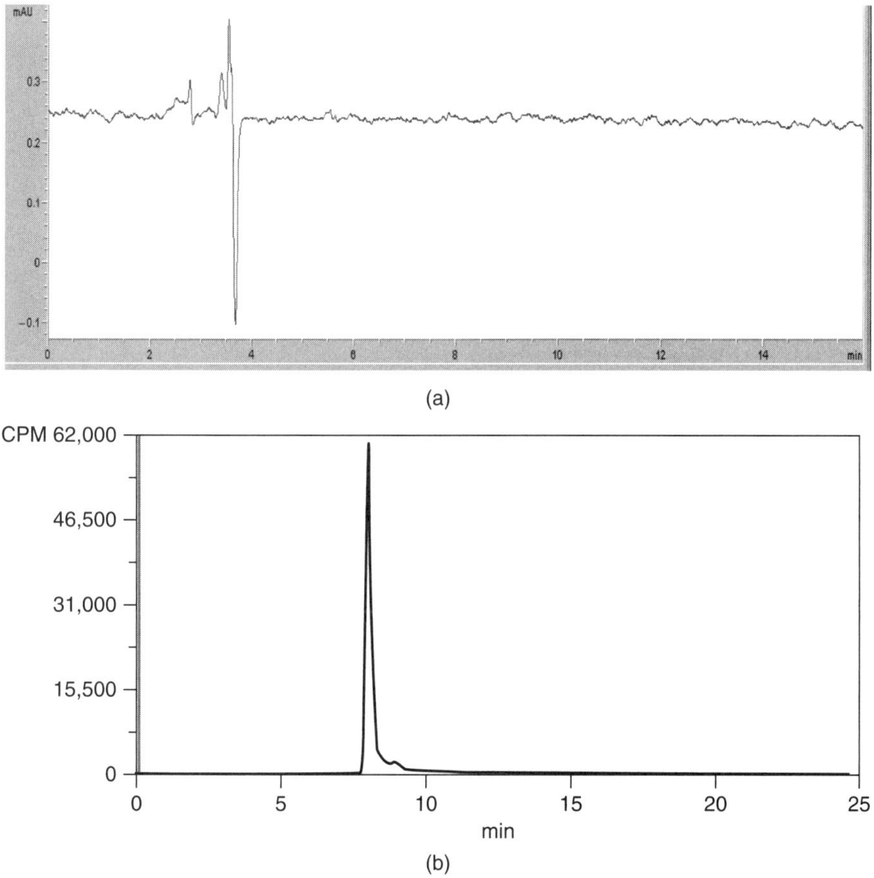

FIGURE 3 UV (a) and radioactive (b) analytical HPLC traces for [^{11}C]MHED.

3.4 Residual Solvents Analysis

Residual organic solvents were measured by a gas chromatograph [27] on a capillary column RTX-200 (30 m $\times$ 0.3 mm) [28]. The limits for the residual solvents reported in the residual solvents monograph of the European Pharmacopoeia are 50 mg/V for acetone (from TRACERlab $FX_{C\text{-}Pro}$ module cleaning procedure) and ethanol (class 3). V is the maximum injected volume.

3.5 pH

A small amount of [^{11}C]MHED solution was applied to pH 0–14 non-bleeding pH-indicator strips [29], and the pH was determined by visual comparison to the scale provided. The pH of the injectable solution must be between 4.5 and 8.5.

3.6 Radionuclidic Identity

Half-life (19.9–20.9 min) was calculated by measuring activities using Capintec CRC® -15R radioisotope dose calibrator [30] for about three half-life intervals and using the equation

$$t_{1/2} = -\ln 2(\text{time difference}/(\ln(\text{ending activity}/\text{starting activity})))$$

3.7 Radionuclidic Purity

Radionuclidic purity was determined by high resolution γ-ray spectrometry [31]. The HPGe detector has a 30% relative efficiency and a resolution of 1.8 keV at 1332 keV. The spectrometry system was calibrated in the 59–1836 keV range, using a multiradionuclide certified reference solution, obtained from an accredited standardization laboratory [32]. γ-Ray spectra were acquired for 600 s. Each sample was prepared and counted in an adequate geometry, allowing to give a total counting frequency of <2000 counts per second, in order to avoid excessive pile up.

The spectrum obtained with [^{11}C]MHED solution does not differ significantly from that obtained with a standardized fluorine-18 solution. Radiochemical purity must be minimum 99% of the total radioactivity.

3.8 Sterile Filter Integrity Test

Sterile filter, with needle attached, from the [^{11}C]MHED solution was connected to a nitrogen supply via regulator. The needle was submerged in water, and the nitrogen pressure was gradually increased. If the pressure was raised above the filter acceptance pressure (56 psi) without seeing a stream of bubbles, the filter was considered intact. If a stream of bubbles occurs <56 psi, the test fails.

3.9 Endotoxin Analysis

Endotoxin contents were measured by the conventional limulus amebocyte lysate (LAL) test, using a gel clotting assay [33]. The limit is less than 175/V IU (international units)/mL.

3.10 Sterility Testing

Tests for sterility were done according to the European Pharmacopoeia [34].

Samples of [^{11}C]MHED were placed onto fluid thioglycolate media (FTM) plates and soybean casein digest media (SCDM) plates. [^{11}C]MHED plates were incubated along with positive and negative controls for 14 days. FTM plates were incubated at 32°C, and SCDM plates were incubated at 22°C according to current Ph. Eu. guidelines.

CHEMICAL ABSTRACTS NOMENCLATURE (REGISTRY NUMBER)

Acetic acid, ammonium salt (1:1) (631-61-8)

Benzenemethanol, α-[(1*S*)-1-aminoethyl]-3-hydroxy-, (α*R*)- (54-49-9)

Benzenemethanol, 3-hydroxy-α-[(1*S*)-1-(methyl-[^{11}C]amino)ethyl]-, (α*R*)- (130440-31-2)

Ethanol (64-17-5)

Formamide, *N,N*-dimethyl- (68-12-2)

REFERENCES AND NOTES

1. Bengel FM, Schwaiger M. Assessment of cardiac sympathetic neuronal function using PET imaging. J Nucl Cardiol 2004;11:603–616.
2. DeGrado TR, Hutchins GD, Toorongian SA, Wieland DM, Schwaiger M. J Nucl Med 1993;34:1287–1293.
3. Schwaiger M, Kalff V, Rosenspire K, Haka MS, Molina E, Hutchins GD, Deeb M, Wolfe E Jr, Wieland DM. Circulation 1990;82:457–464.
4. G.E. Healthcare, USA. High performance [^{11}C]CO_2 target has the same interface/dimension as the standard G.E. target made of different aluminum alloy, with superior surface properties in target chamber. The target uses thicker foil. Improved helium and water cooling as well as higher fill and operating pressure are used. No preconditioning of the target. The gas target was freshly loaded before irradiation.
5. Shimalite-nickel was obtained from Shimadzu (product no. 221–27719).
6. Molecular sieves 80/100 were purchased by Alltech (product no. 5624).
7. Carbosphere 60/80 was purchased from Alltech (product no. 5680).
8. Sicapent was purchased from Merck (product no. 1.00543).
9. Ascarite II 20–30 mesh was purchased from Sigma Aldrich (product no. 223921).
10. Iodine resublimed was obtained from Merck (product no. 1.04761).
11. Link JM, Clark JC, Larsen P, Krohn KA. J Labelled Comp Radiopharm 1995;37:76–78.
12. Larsen P, Ulin J, Dahlstrom K, Jensen M. Appl Radiat Isot 1997;48:153–157.
13. Porapak Q 50–80 mesh was purchased from Alltech.
14. Rosenspire KC, Haka MS, Van Dort ME, Jewett DM, Gildersleeve DL, Schwaiger M, Wieland DM. J Nucl Med 1990;31:1328–1334.
15. Metaraminol free base (1 mg) was purchased from ABX, (Germany) (product no. 3380).
16. *N,N*-dimethylformamide ($\geq$99.5%) was purchased from Sigma-Aldrich (product no. 40228) and ABX (Germany).
17. Semipreparative HPLC mobile phase: 5% ethanol : 95% 0.1 M CH_3COONH_4. Absolute ethanol was purchased from Carlo Erba Reagenti (Ph. Eur. grade, product no. 4146052) and CH_3COONH_4 (Ph. Eur. grade) was obtained from Riedel-De Haën (product no. 32301). Preparation: CH_3COONH_4 (7.71 g) was dissolved in water (1000 mL). Absolute ethanol

(50 mL) was added to 950 mL of CH_3COONH_4 solution, and the mobile phase was stirred to mix and then passed through a sterile 0.22-μm filter before use.

18. Semipreparative HPLC column: Phenomenex Synergi Polar-RP, 10 μm, 250 × 10 mm, purchased from Phenomenex (part no. 00G-4351-N0). To obtain reproducible RT, it is mandatory to equilibrate the semipreparative column with at least 200–300 mL of mobile phase before synthesis.
19. Millex-GP sterile 0.22-μm filters (PES membrane) were purchased from Millipore (part no. SLGP033RS).
20. DRN4370 sterile and evacuated vials were purchased from Mallinckrodt.
21. Absolute ethanol (Ph. Eur. grade) was purchased from Carlo Erba Reagenti and diluted with water to 70% (product no. 4146052).
22. $LiAlH_4$ 0.1 M in THF was obtained from ABX (Germany) (product no. 802.0001).
23. Hydroiodic acid redistilled was obtained from ABX (Germany) (product no. 830).
24. European Pharmacopoeia 6th edition.
25. Analytical HPLC column: Phenomenex Synergi Polar-RP, 4 μm, 250 × 4.6 mm, purchased from Phenomenex (part no. 00G-4336-E0); analytical column was equilibrated with mobile phase for 30–60 min before use.
26. MHED hydrochloride reference standard was purchased from ABX, Germany (product no. 3400).
27. Varian, CP3380.
28. GC capillary column RTX-200 was purchased from Restek (product no. 15054).
29. pH 0–14 non-bleeding pH-indicator strips were obtained from Merck (product no. 1.09535.0001).
30. Capintec, Inc., USA.
31. Camberra, distributed in Italy by Technology Nuclear Electronics, Milan, Italy.
32. Areva CERCA LEA, Pierrelatte cedex, France.
33. Lonza, Verviers, Belgium.
34. European Pharmacopoeia: Sterility chapter (2.6.1).

CHAPTER 20

SYNTHESIS OF L-[methyl-^{11}C]METHIONINE ([^{11}C]MET)

AMY L. VĀVERE AND SCOTT E. SNYDER

Department of Radiological Sciences, St. Jude Children's Research Hospital, Memphis, Tennessee

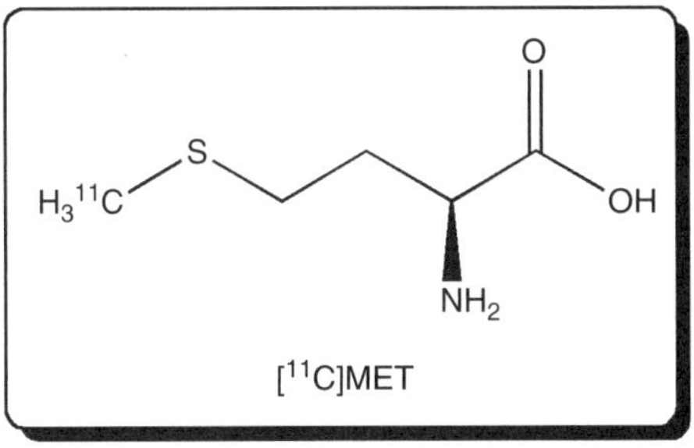

1 INTRODUCTION

Methionine is a naturally occurring amino acid necessary for protein synthesis. As tumors are rapidly growing, the formation of new proteins and cellular growth is amplified. Positron emission tomographic (PET) imaging with [^{11}C]methionine ([^{11}C]MET) measures amino acid uptake and protein synthesis following the route of its carbon-12 counterpart. In the brain, this includes specifically targeting the L-type amino acid transporter system, which allows it to cross the blood–brain barrier. A variety of solid tumors show accumulation of this tracer through increased transport [1]; however, the majority of its utilization in the clinical setting has been for brain tumors [2]. In slowly proliferating tissue, including the brain, the accumulation of [^{11}C]MET is minimal [3], which allows easier delineation of areas of increased uptake. Differentiation of benign versus malignant tumors is observed with high specificity and sensitivity, leading this tracer to be employed for the management of gliomas and other tumors of the cranial cavity, when available [4].

While not as common, other amino acid tracers are also utilized for PET, including labeled versions of tyrosine such as [^{18}F]fluoroethyltyrosine (Chapter 12) [5].

Radiochemical Syntheses, Volume 1: Radiopharmaceuticals for Positron Emission Tomography, First Edition. Edited by Peter J. H. Scott, Brian G. Hockley.

2 SYNTHESIS PROCEDURES

CAUTION: All radiochemical syntheses must be carried out using appropriate equipment in a facility authorized for the use of radioactive materials. Personal protective equipment must be worn and all local radiation safety laws followed.

The following is an outline of the traditional method for production of L-[*methyl*-^{11}C]methionine ([^{11}C]MET) utilizing the standard "wet" chemistry method for the production of [^{11}C]methyl iodide and a chemistry module with transfer of reagents from vials to a main reactor for the production of methionine. [^{11}C]Methyl iodide can also be efficiently produced in the gas phase from [^{11}C]CH_4 with higher specific activity but lower overall yields. Chemical synthesis of [^{11}C]MET is also possible on a solid phase support [6,7] or by using a closed-loop system [8].

2.1 Production of [^{11}C]CO_2

A target gas of N_2 + 0.5% O_2 [9] is loaded into an IBA Molecular Nirta® Gas target [10] and filled to a pressure of 25 bar. The target is bombarded with protons at 21 μA to induce a $^{14}N(p,\alpha)^{11}C$ nuclear reaction using an IBA Cyclone® 18/9 cyclotron [11]. A small amount of ^{13}N is also produced by the $^{14}N(p,pn)^{13}N$ reaction. The [^{11}C]CO_2 is remotely transferred to an IBA switching box, where it is diverted to be concentrated and purified in a liquid nitrogen cold trap (400 mL/min). Unloading of the target takes ~3 min. Before initiation of synthesis, helium [12] is passed over the frozen [^{11}C]CO_2 to remove small amounts of ^{13}N.

2.2 Synthesis of [^{11}C]Methyl Iodide

[^{11}C]CO_2 is converted to [^{11}C]methyl iodide ([^{11}C]CH_3I) using the standard wet chemistry method [13]. The trapped [^{11}C]CO_2 is transferred to a PETChem Solutions methyl iodide synthesis module [14]. The initial step involves bubbling the [^{11}C]CO_2 through a vessel (at 8–16 mL/min) containing 1–2 mg of lithium aluminum hydride (LAH) [15] and 200 μL of anhydrous tetrahydrofuran (THF) [16] chilled to −10°C [17]. The trapping of the [^{11}C]CO_2 (as the aluminum methoxide complex) takes ~3 min with an average trapping of 80% of the theoretical amount of radioactivity unloaded from the cyclotron at end of bombardment.

Once the trapping is complete, the THF is evaporated with a steady stream of helium and heating to 130°C for 1 min. The vessel is then cooled to below 0°C with a stream of compressed air, after which 12% phosphoric acid [18] is added drop-wise (~150 μL) to the reaction vessel via a peristaltic pump. The phosphoric acid and [^{11}C]CH_3OH mixture is mixed, and then the temperature is raised to 130°C, and a valve opens to a vessel containing 800 μL hydroiodic acid (57%) [19]. The [^{11}C]CH_3OH generated from the reaction with the phosphoric acid is bubbled through the hydroiodic acid (140°C), where it reacts to

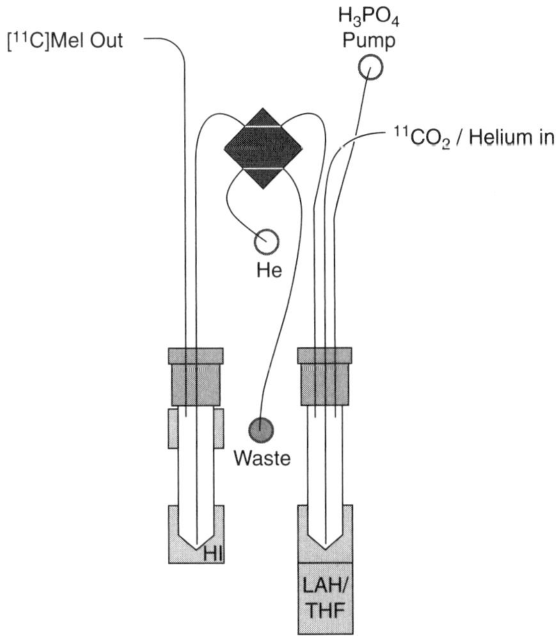

FIGURE 1 Diagram of [^{11}C]CH_3I production module.

form [^{11}C]CH_3I and continues, without stopping, to the chemistry module for methionine production (see Fig. 1).

A soda lime trap [20] is placed in-line between the HI outlet and the methionine reaction vessel to remove any unreacted [^{11}C]CH_3OH or water vapor produced from the boiling hydroiodic acid. Average corrected yields for [^{11}C]CH_3I production are 60–80%, and time of synthesis is 4–5 min from transfer of trapped [^{11}C]CO_2.

3 SYNTHESIS OF L-[methyl-^{11}C]METHIONINE

Synthesis of [^{11}C]MET is performed by the method of Ishiwata et al. [21], starting from [^{11}C]CH_3I and L-homocysteine thiolactone hydrochloride (Fig. 2), and adapted for preparation on a PETChem Solutions automated synthesis module [14] (Fig. 3).

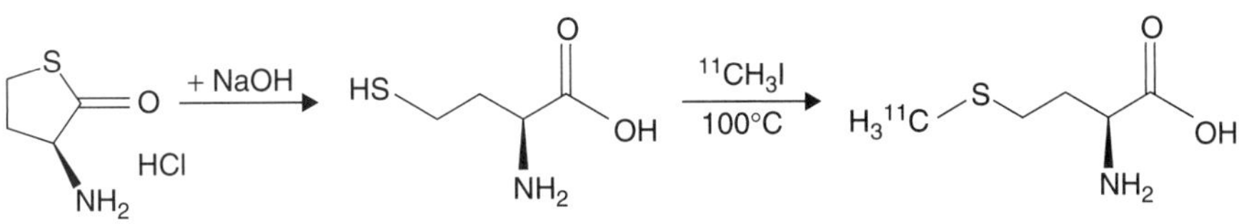

FIGURE 2 Synthesis of L-[*methyl*-^{11}C]methionine.

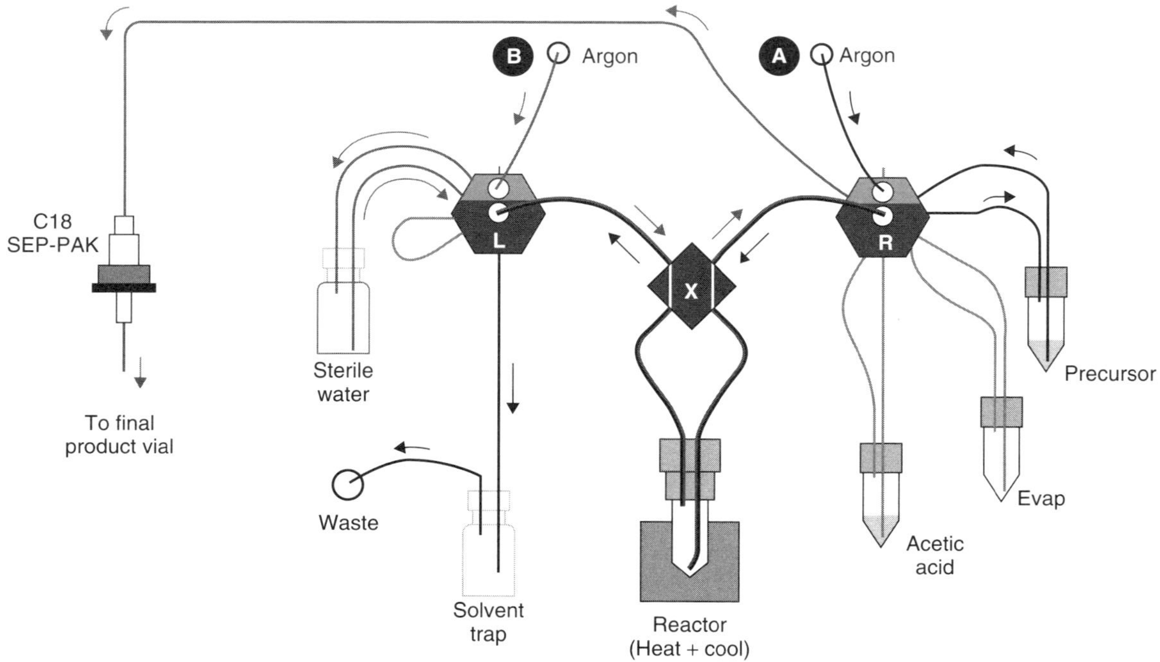

FIGURE 3 Diagram of the methionine synthesis module setup. Reagents are added to the reactor by argon pressure "A." Following the black pathway, the argon is sent through valve R and pushes the liquid toward the reactor (or the vial is used as a pass-through for evaporation). The reactor is vented through valve L and out to waste. Isolation valve X in the middle can be switched to seal the reactor during heating. Once synthesis is complete, pressure from argon "B" is sent via valve L (following the gray pathway) to flush the sterile water through the reactor, through valve R, and out for Sep-Pak purification, mixing, and final filtration for patient preparation.

The synthesis of [^{11}C]MET begins with trapping of previously generated [^{11}C]CH_3I bubbled through 250 μL of chilled [22] acetone [23] (<0°C). The distillation and trapping takes 3 min, and once this is complete, 200 μg of L-homocysteine thiolactone hydrochloride [24] dissolved in 100 μL of 32 mM sodium hydroxide (NaOH) [25] is added to the reactor vial. The reactor vial is sealed and heated to 100°C for 2 min. Ishiwata and colleagues [21] have shown that by limiting the concentration of NaOH to <250 mM, <4% D-methionine is formed. We have confirmed this finding as outlined in Section 3.3. At this point in the synthesis, the chemical reaction is complete and the acetone needs to be removed for patient formulation.

4 PREPARATION OF L-[methyl-^{11}C]METHIONINE FOR INJECTION

After brief cooling, the reaction vial is opened to argon flow through the reaction mixture with heating to 40°C for 3 min to evaporate the acetone. While evaporation of acetone can be achieved in less time at higher temperatures, we observed the formation of a high boiling impurity during our volatile organic analysis (Section 3.6), which was identified as diacetone alcohol (DAA). DAA results from a self-aldol condensation of acetone under basic conditions (NaOH solution of the precursor—see Fig. 4), but by keeping the temperature below 50°C during acetone evaporation, DAA levels remain minimal (typically <10 ppm) and acetone levels do not exceed release criteria.

To neutralize the solution, 320 μL of 10 mM acetic acid [26] is added to the solution, and the reaction mixture is mixed with bubbling and passed through a C18 Sep-Pak® Plus [27] and collected in an intermediate vial. Sterile water [28] (10 mL) is added to this vial with thorough mixing before the entire solution is passed through a 0.22-μm filter [29] to the final product vial in a lead-shielded collection drawer outside of the hot cell. Using appropriate radiation safety measures, the sterilizing filter is removed for bubble-point testing, and the final product vial is transported to a dose-dispensing station where a sample is withdrawn for quality control (QC) testing. Overall synthesis time from target unloading to delivery of product is 18–20 min. Average synthesis yield (calculated over ~100 syntheses) is 19±3% uncorrected from theoretical target production to final product at

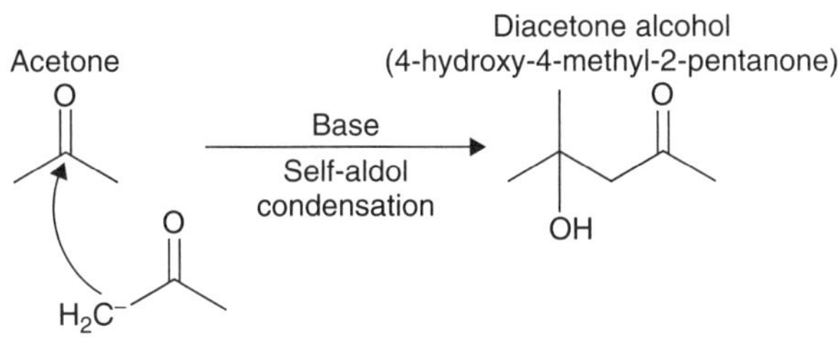

FIGURE 4 Self-aldol condensation of acetone to form diacetone alcohol.

end of synthesis (EOS). For patient production, 1.1 Ci [^{11}C]CO_2 unloaded results in 200–250 mCi of [^{11}C]MET for patient use.

5 QUALITY CONTROL PROCEDURES

CAUTION: All radiopharmaceuticals produced for clinical use must have local regulatory approval before use in humans. Trained personnel must carry out QC procedures, and each dose must meet all established QC criteria before release to the clinic.

QC procedures for [^{11}C]MET are performed in compliance with United States Pharmacopeia (USP) General Chapter <823> [30] and in accordance with acceptance criteria specified in our approved Investigational New Drug (IND) Application (#104987). These acceptance criteria are based on, but not limited to, the current USP Monograph entitled "Methionine C 11 Injection" and are summarized below. Complete QC data for three representative batches of [^{11}C]MET produced using the method described herein are summarized in Table 1. Each of the three doses met all of the established QC criteria.

5.1 Visual Inspection

The [^{11}C]MET dose is examined remotely through the lead-glass window of an Amercare A^2 PET Dispensing Isolator [31] and must be clear, colorless, and free of any particulate matter. This is doubly confirmed by the QC technician by visual inspection of the QC sample syringe.

5.2 Radiochemical Purity and Identity

According to our approved IND, the patient dose of [^{11}C]MET must be $\geq$90% radiochemically pure and have a retention time within 15% of the nonradioactive L-methionine standard. High performance liquid chromatographic (HPLC) analysis is conducted using an Agilent 1200 Series LC System with a UV detector [32] and a Bioscan Flow Count 3300 radioactivity detector [33]. Two reference standard injections (L-methionine solution [34], 1 μL) are performed before analysis of the patient sample on a Phenomenex Luna 10 μm SCX 100 Å column (4.6 $\times$ 250 mm plus SCX guard cartridge) [35], using a mobile phase of 0.1 M potassium phosphate, pH 2.6, [36] at a flow rate of 2.0 mL/min, with UV absorbance monitored at 210 nm. Peak integration limits are set to the following: slope sensitivity = 5, peak width = 0.05, area reject = 5, and height reject = 1. System suitability is demonstrated by a difference of 10% or less between the retention times of the two standard injections. Two patient samples (1 μL injection each), a co-injection sample [37] (Fig. 5), and a third reference standard to confirm system suitability are analyzed. Radiochemical purity is determined by the percentage of the methionine peak in the patient sample, while radiochemical identity is confirmed by comparison of the radioactive peak to the UV peak in the co-injection sample.

TABLE 1 Representative Quality Control Data

Test	Acceptance Criteria	Batch 1	Batch 2	Batch 3
Yield, uncorrected (corrected)[a](%)	NA	20.2 (38.5)	15.5 (31.7)	16.4 (33.5)
Appearance and color	Clear and colorless	Clear and colorless	Clear and colorless	Clear and colorless
Radiochemical purity and identity (%)	≥90 [^{11}C]MET RT within 15% of standard	96.34	99.11	98.03
Chemical purity (annually) (%)	≤4 D-methionine	1.30	1.39	1.08
Radionuclidic purity (annually) (%)	>95	99.75	99.78	99.73
Specific activity in Ci/mmol (GBq/mmol)	NA	1649 (61,013)	1629 (60,273)	1576 (58,312)
Volatile organic impurities %	Acetone <0.5 THF <0.072 Ethanol <0.5 DAA <0.04	0.085 <0.001 0.002 0.003	0.084 <0.001 0.003 0.005	0.100 <0.001 0.003 0.004
pH	4.5–7.5	6.0	6.0	6.0
Filter integrity (psi)	≥50 breaking pressure	56	55	59
Radionuclidic identity (min)	18.36–22.44 (20.4±10%)	20.12	20.21	20.02
Bacterial endotoxin (LAL) (EU/mL)	≤5	<2.50	<2.50	<2.50
Sterility	No colony growth out to 14 d	Pass	Pass	Pass

[a]Uncorrected overall yield at EOS from theoretical target production of [^{11}C]CO_2. Parenthetical yield is corrected to SOS.

5.3 Chemical Purity

The acceptance criteria require that the final product be enantiomerically pure L-methionine with ≤4% D-methionine. Since our synthetic method for [^{11}C]MET has been shown to produce <4% D-[^{11}C]methionine [21], confirmation of chiral purity is limited to annual validation. Enantiomeric purity is assessed by HPLC using a Phenomenex Chirex-33126 D-penicillamine column (4.6 × 50 mm) with a mobile phase of 1 mM copper sulfate:2-propanol, 95:5 [38]. The flow rate is 0.5 mL/min, and the UV absorption of the reference standards is monitored at 254 nm. Retention times of the L- and D-methionine standards are ~6.9 and 8.6 min, respectively, which provides baseline separation and easy quantification of isomers. Patient samples are analyzed, and the radioactive retention times of the isomers present are compared to standards of L-methionine alone, D-methionine alone, and a mixture of D- and L-methionine [39] (Fig. 6).

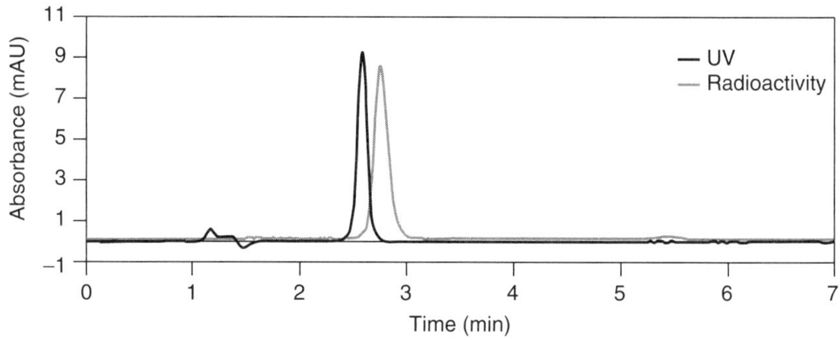

FIGURE 5 HPLC analysis of a co-injection of L-methionine reference standard (100 μg/mL) with [^{11}C]MET. Displayed is an overlay of the UV absorbance at 210 nm (black) and the radioactivity signal (gray).

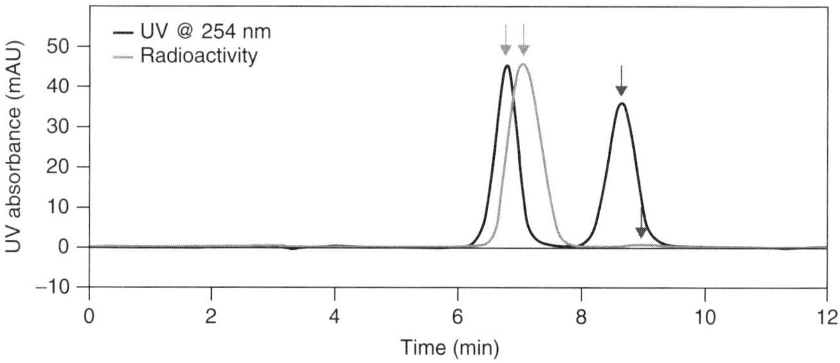

FIGURE 6 Chiral HPLC analysis of a standard reference mixture of L-(gray arrow) and D-methionine (black arrow) (retention times of 6.86 and 8.62 min, respectively) overlayed with the radioactive trace of a patient sample run showing <1% D-methionine.

5.4 Radionuclidic Purity

Acceptance criteria requires a radionuclidic purity of ≥95% ^{11}C for the patient sample of [^{11}C]MET, and this is tested annually using a Raytest Mucha* multichannel analyzer (MCA) [40] and its accompanying Gina Star 4 software. A small amount of patient sample is analyzed in the MCA, counting all decays between 0 and 2048 keV for four 30 sec collection periods. While the USP monograph for [^{11}C]methionine does not specify how this percentage should be determined, other monographs (i.e., [^{18}F]FDG) list that all Compton, 511-keV, and 1022-keV decays are integrated as positron emissions.

5.5 Specific Activity

According to our approved IND, there is no specific activity requirement for [^{11}C]MET; however, we perform annual validation testing to verify that the mass concentrations of homocysteine and methionine in the final dose product remain below 20 μg/mL. Since uptake of [^{11}C]MET is not a saturable process, we monitor these levels only to ensure that our chemistry is consistent. By integrating the mass peak of the UV trace of the patient sample and using the standards as a reference, the mass of [^{11}C]MET injected on the HPLC can be determined. The radioactivity of a known volume of sample is then measured in a dose calibrator to determine the amount of radioactivity injected. The specific activity at EOS is reported in curies per millimole [^{11}C]MET.

5.6 Volatile Organic Impurities Analysis

Possible residual solvents in this preparation are acetone, ethanol, THF, and diacctonc alcohol. Rcmaining acctonc could be present from the reaction solution, ethanol is used to sterilize the final product line, and THF is used in the production of methyl iodide. Diacetone alcohol results from a self-aldol condensation of acetone under basic conditions (vide supra). Acetone and ethanol have been established as class 3 volatile solvents and are limited to 0.5%, whereas THF is a class 2 solvent and is limited to 0.072%. Diacetone alcohol (4-hydroxy-4-methyl-2-pentanone) has no dose limit listed by the USP; however, it has a Hazardous Materials Identification System (HMIS) rating of 2 and is a Workplace Hazardous Materials Identification System (WHMIS) class D-2B, similar to acetone, acetonitrile, and THF. On the basis of the available toxicity data, diacetone alcohol (rat oral LD_{50} = 2520 mg/kg) seems closest in hazard to acetonitrile (rat oral LD_{50} = 2460 mg/kg), but less toxic than THF (rat oral LD_{50} = 1650 mg/kg). To be conservative, we have set the release criterion for diacetone alcohol at 0.04%, equal to the USP limit for acetonitrile.

Two analyses of a standard solution containing 0.1% acetone, 0.01% THF, 0.1% ethanol, and 0.01% diacetone alcohol are run before analysis of the patient sample to ensure system suitability. Testing for organic volatile impurities in the patient dose is performed on two injections of the patient sample (0.5 μL each) and compared to prerun and an additional postrun reference standard using an Agilent 7890A GC System with a 7683B injector [32] and a J & W Durabond-Wax column [41] (Fig. 7). Integration limits are set to a peak area equivalent to 0.001% or 2.5% of the lowest release limit.

5.7 Dose pH

The pH of the final dose must be between 4.5 and 7.5 for release. The pH of MET is analyzed by a depositing a small volume (~100 μL) of patient sample on a pHydrion™ Dip Stik [42] and determined by visual comparison to the scale provided by the manufacturer.

5.8 Sterile Filter Integrity Test

The integrity of the sterilizing filter is confirmed by a bubble-point test. The filter is considered intact if a flow of nitrogen through the filter reaches greater than 50 psi before the filter fails. The sterile filter (and needle) from the production of [^{11}C]MET is attached to tubing connected to house compressed air with a

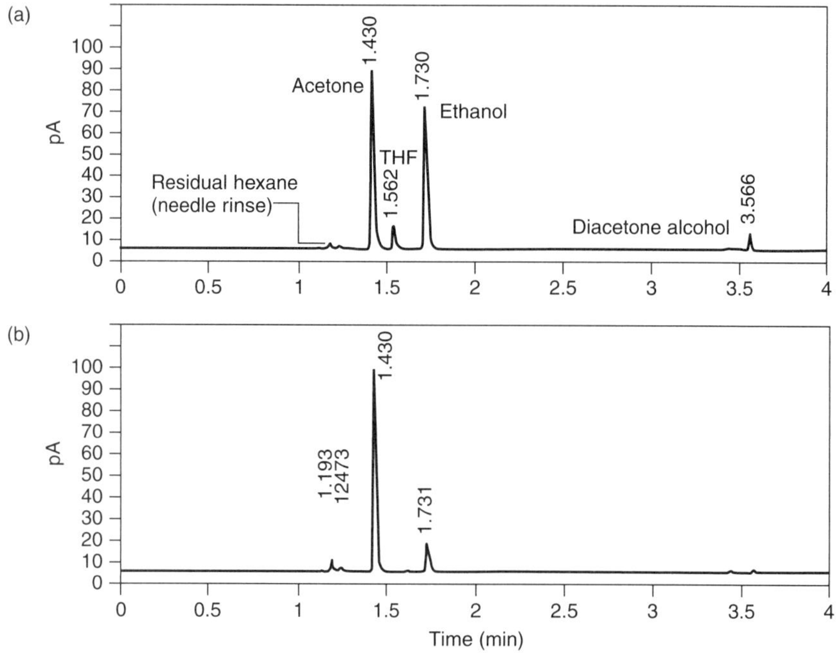

FIGURE 7 Volatile organic impurities analysis by GC of a standard containing 0.1% acetone, 0.01% THF, 0.1% ethanol, and 0.01% diacetone alcohol (a). A representative patient sample is also shown (b).

regulator to control flow. After submerging the needle in water, the compressed air flow is slowly increased until bubbles are observed in the water. The pressure at which this occurs is noted, and if this is >50 psi, the filter is considered intact.

5.9 Radionuclidic Identity

Confirmation of the identity of the radionuclide used in the production of [^{11}C]MET is made by measurement of the radioactive half-life. The acceptable range for carbon-11 is 18.36–22.44 min, which is the known half-life (20.4 min) ±10%. This measurement is made using a Capintec CRC®-15PET Dose Calibrator [43] by noting the millicurie (megabecquerel) of a small sample of radioactivity at the start and then every 5 min for a total of four measurements. The half-life from 0 to 5, 0 to 10, and 0 to 15 min is calculated (Eq. 1), and the average of the three half-life measurements is determined as the official value.

$$t_{1/2} = \frac{(\Delta t \times \ln 2)}{\ln\left(\frac{N_0}{N_t}\right)} \tag{1}$$

where $t_{1/2}$ = half-life, Δt = elapsed time, N_0 = initial radioactivity, and N_t = radioactivity at time t

5.10 Endotoxin Analysis

Bacterial endotoxin testing is performed to ensure that the final product contains ≤5 EU (endotoxin units)/mL, which, assuming injection of the total 10 mL of the final product, would remain well within the USP bacterial endotoxin limit for radiopharmaceuticals of <175 EU/V, where V is the maximum volume of injection. Testing is performed using the Charles River Endosafe—portable test system (PTS) [44]. This Food and Drug Administration (FDA)-approved system utilizes disposable cartridges [45] that examine the patient sample and a positive control in duplicate. The lower detection limit under our conditions is 2.50 EU/mL.

5.11 Sterility

Within 24 h of production, a sample of [^{11}C]MET is injected into a tube containing fluid thioglycolate medium (FTM) [46] and also into a tube containing trypticase soy broth (TSB) [47] in a laminar flow hood using aseptic technique. FTM supports the growth of a variety of organisms including anaerobes and aerobes, whereas TSB will cultivate fastidious and nonfastidious microorganisms. After thorough mixing by inversion, the FTM tube is incubated at 30–35°C [48] and the TSB tube is incubated at room temperature (20–25°C) according to current USP guidelines [49]. Tubes are visually inspected on the third, seventh, and fourteenth days of incubation for turbidity and color change by comparison to a lighted white and lighted black background. Samples that show no culture growth at the end of the 14-day test period are considered to be sterile.

CHEMICAL ABSTRACTS NOMENCLATURE (REGISTRY NUMBER)

Acetic acid (64-19-7)
Acetone (67-64-1)
Argon (7440-37-1)
Carbon dioxide (124-38-9)
Copper (II) sulfate pentahydrate (7758-99-8)
Ethanol (64-17-5)
Helium (7440-59-7)
L-Homocysteine thiolactone hydrochloride (31828-68-9)
Hydroiodic acid (10034-85-2)
4-Hydroxy-4-methyl-2-pentanone (diacetone alcohol) (123-42-2)
Lithium aluminum hydride (16853-85-3)
D-Methionine (348-67-4)
L-Methionine (63-68-3)
Nitrogen (7727-37-9)
Phosphoric acid (7664-38-2)
Potassium phosphate monobasic (7778-77-0)

2-Propanol (67-63-0)
Soda lime (8006-28-8)
Sodium hydroxide (1310-73-2)
Tetrahydrofuran (109-99-9)

REFERENCES AND NOTES

1. Miyazawa H, Arai T, Iio M, Hara T. J Nucl Med 1993;34:1886–1891.
2. Galldiks N, Kracht LW, Berthold F, Miletic H, Klein JC, Herholz K, Jacobs AH, Heiss WD. J Neuro Oncol 2010;96:231–239.
3. O'Tuama LA, Phillips PC, Smith QR, Uno Y, Dannals RF, Wlison AA, Ravert HT, Loats S, Loats HA, Wagner HNJ. J Nucl Med 1991;32:16–22.
4. Jacobs AH, Dittmar C, Winkeler A, Garlip G, Heiss WD. Mol Imaging 2002;1:309–335.
5. Jager PL, Vaalburg W, Prium J, de Vries EGE, Langen K-J, Piers DA. J Nucl Med 2001;42:432–445.
6. Mitterhauser M, Wadsak W, Krcal A, Schmaljohann J, Eidherr H, Schmid A, Viernstein H, Dudczak R, Kletter K. Appl Radiat Isot 2005;62:441–445.
7. Lodi F, Rizzello A, Trespidi S, Di Perro D, Marengo M, Farsad M, Fanti S, Al-Nahhas A, Rubello D, Boschi S. Nucl Med Commun 2008;8:736–740.
8. Gomez V, Gispert JD, Amador V, Llop J. J Labelled Comp Radiopharm 2008;51:83–86.
9. $N_2 + 0.5\%\ O_2$ was purchased as a custom gas mixture from Nexair (Memphis, TN) and used as received.
10. Aluminum target body and window with target dimensions of 110 mm length, 16-mm diameter, and a volume of 25 cm^3. Target is water cooled, and windows are helium cooled.
11. IBA Molecular, Louvain-la-Neuve, Belgium.
12. Helium grade 6.0 (99.9999% pure) was purchased from Nexair (Memphis, TN) and used as received.
13. Comar D, Cartron J-C, Maziere M, Marazano C. Eur J Nucl Med 1976;1:11–14.
14. PETChem Solutions,Chelsea, MI.
15. Lithium aluminum hydride was purchased from Sigma Chemical Company (St. Louis, MO) and used as received. Individual tubes were prepared under argon atmosphere with 1–2 mg LAH and closed with a rubber stopper. These were stored in batches in a -20°C freezer (sealed in a container with drierite) until the day of use. Tubes were fully thawed and dried before use with the synthesis module.
16. Anhydrous tetrahyrofuran was prepared by elution of THF (Sigma, St. Louis, MO) through a dual column solvent drying system (Glass Contour Solvent Systems by SG Water USA, Nashua, NH).
17. Cooling by a stream of compressed air (house) directed past the reaction vessel.
18. Phosphoric acid (ACS reagent, $\geq$85 wt% in H_2O) was purchased from Sigma-Aldrich (cat. no. 438081) and diluted to 12% with Milli-Q water.
19. Hydroiodic acid (57 wt% in H_2O, distilled, 99.95%) from Sigma-Aldrich (part no. 210021) and used as received.
20. Soda lime (granular, +100 mesh, A.C.S. reagent) was purchased from Sigma Aldrich (cat. no. 266434) and crushed to 60 mesh or smaller, with large pieces and soda lime dust sifted out. A small amount of glass wool was packed into a 1-mL chemical resistant syringe (Henke Sass Wolf, Norm-ject®, cat. no. 4010.200V0) with the plunger removed, followed by filling the syringe up to the 9-mL mark with the prepared 60-mesh soda lime. The prepared soda lime traps were then stored in a desiccator until use.
21. Ishiwata K, Ido T, Vaalburg W. Int J Radiat Appl Instrum Part A 1988;39:311–314.

22. Chilling in the PETChem Solutions Chemistry Module is achieved by passing a stream of liquid CO_2 past the reactor vial. The liquid CO_2 was purchased from Nexair (Memphis, TN) and used as received.
23. Acetone (CHROMASOLV®, for HPLC, ≥99.9%) was purchased from Sigma-Aldrich (part no. 270725) and used as received.
24. L-Homocysteine thiolactone hydrochloride (≥99 was purchased from Fluka (Sigma cat. no. 53527) and used as received.
25. Sodium hydroxide (10 N solution) was purchased from Fisher Scientific (cat. no. SS267) and diluted to 32 mM with Milli-Q water.
26. Acetic acid (glacial, HPLC) was purchased from Fisher Scientific (part no. A35-500) and diluted to 10 mM using Milli-Q water.
27. C18 Sep-Pak® Plus cartridges were purchased from Waters (part no. WAT020515) and conditioned with 10 mL ethanol and 10 mL water before use.
28. Sterile Water for Injection, USP, was purchased from APP Pharmaceuticals, LLC (cat. no. 401753E) and used as received.
29. Millex® -GS Sterilizing Filter Unit from Millipore (part no. SLGS 025 OS).
30. U.S. Pharmacopoeia <823>. Radiopharmaceuticals for Position Emission Tomography Compounding. USP 32-NF 27, 2009.
31. A^2 PET Isolator with a Class 100 air environment and a remote Automated Dose Dispenser from Amercare, Ltd., Oxfordshire, UK.
32. Agilent Technologies, Atlanta, GA.
33. Bioscan, Inc., Washington, DC.
34. L-Methionine (reagent grade, ≥98%) was purchased from Sigma Chemical (cat. no. M9625) and dissolved in mobile phase (0.1 M potassium phosphate, pH 2.6) at 100 μg/mL. Individual vials containing 1 mL were aliquoted, sealed, and stored at −80°C until the day of use.
35. Analytical HPLC column: Luna 10-μm SCX 100 Å column (4.6 × 250 mm) purchased from Phenomenex (cat. no. 00G-4401-E0) with an SCX guard cartridge (cat. no. AJ0-4308) and a Security Guard Cartridge System (cat. no. KJ0-4282) also from Phenomenex. The column was equilibrated for at least 20 min with mobile phase at 2 mL/min.
36. Analytical mobile phase: potassium phosphate monobasic (Sigma, cat. no. 60218) was diluted in water to make a 0.10 M solution. Phosphoric acid was used to titrate to a pH of 2.6.
37. A co-injection sample of [^{11}C]MET was prepared by adding 50 μL of the patient sample to a vial of the L-methionine standard solution (1 mL). One microliter of this mixture was then injected on the HPLC for analysis.
38. Copper sulfate pentahydrate (crystalline/technical) purchased from Fisher Chemical (cat. no. C489) was weighed out to make a 1.0-mM solution in Milli-Q water, then filtered (0.22 μm) and degased. This was mixed in-line at 95:5 by the HPLC pump with 2-propanol (CHROMASOLV®, for HPLC, 99.9%) from Sigma-Aldrich (cat. no. 650447).
39. L-Methionine (reagent grade, ≥98%—Sigma cat. no. M9625) and D-Methionine (puriss., ≥99% Fluka–Sigma cat. no. 64330) were separately dissolved in the chiral HPLC eluent solution (95:5, 1 mM copper sulfate : 2-propanol) at a concentration of 2.0 mg/mL. Equal volumes of these standard solutions were combined to form the D-,L-methionine standard solution.
40. Mucha* multi-channel analyzer was purchased from Raytest USA, Inc. (Wilmington, NC).
41. Oven temperature was 60°C for 1 min, then ramped to 140°C at a rate of 40°C/min with a total run time of 4 min. Injection volume was 0.5 μL, and the front inlet heater was at 250°C. The split ratio was 15:1, hydrogen flow was 40 mL/min, and the air flow was 400 mL/min. The column used was a J & W (part no. 122–7032) at 30 m (length) × 250 μm (i.d.) × 0.25 μm (film).
42. pHydrion™ Brilliant Dip Stik 5–9 was purchased from Cole-Parmer (cat. no. 59200-14) and used as received.
43. Capintec CRC® -15PET Dose Calibrator was purchased from Capintec, Inc. (Ramsey, NJ).
44. Charles River Laboratories International, Inc., Wilmington, MA.

45. Limulus amebocyte lysate (LAL) test cartridges (cat. no. PTS20F) were purchased from Charles River Endosafe, a division of Charles River Laboratories (Charleston, SC) and used as received.
46. BBL™ thioglycolate medium without Indicator-135C, 10 mL (cat. no. 221797).
47. BBL™ trypticase soy broth, 8 mL (cat. no. 221092).
48. Incubation took place in a Thelco Laboratory Incubator (Thermo).
49. U S Pharmacopeia <71>. Sterility Tests. USP 32–NF 27, 2009.

CHAPTER 21

SYNTHESIS OF (+)-α-[^{11}C]DIHYDROTETRABENAZINE ([^{11}C]DTBZ)

MICHAEL R. KILBOURN

Department of Radiology, University of Michigan Medical School, Ann Arbor, Michigan

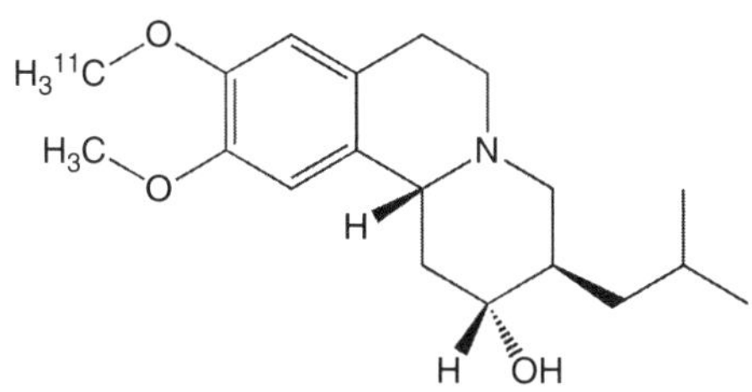

1 INTRODUCTION

In neurodegenerative diseases there are typically characteristic losses of particular types of neurons in the human brain: for example, progressive losses of dopaminergic neurons in the basal ganglia are the hallmark of Parkinson's disease. Among a number of approaches to in vivo imaging of such neuronal losses, the targeting of the vesicular monoamine transporter type 2 (VMAT2) has been particularly successful using radioligands such as (+)-α-[^{11}C]dihydroterabenazine ([^{11}C]DTBZ: (2*R*,3*R*,11b*R*)-(1,3,4,6,7,11b-hexahydro-9-[^{11}C]methoxy-10-methoxy)-3-(2-methylpropyl)-2-hydroxy-2*H*-benzo[a]quinolizine). DTBZ is the active metabolite of the clinically used drug tetrabenazine (TBZ) developed and first used in patients in the 1950s [1] as an antipsychotic, but used today for control of involuntary movements in diseases such as Huntington's disease.

The VMAT2 is an integral protein of the presynaptic vesicles and serves a function to transport monoamines from the cell cytosol into the storage vesicle, from where they can be subsequently released into the synapse [2]. The

Radiochemical Syntheses, Volume 1: Radiopharmaceuticals for Positron Emission Tomography, First Edition. Edited by Peter J. H. Scott, Brian G. Hockley.

VMAT2 is not specific for any monoamine but is a common protein handling the transport of dopamine, norepinephrine, serotonin, and histamine. The utility of VMAT2 imaging in neurodegenerative disease is made possible by the compartmentalization of neuronal types in the human brain: for example, dopaminergic nerve terminals predominate in the basal ganglia and thus provide the biochemical specificity for examining losses of such terminals in Parkinson's disease [3].

The long history of TBZ use and the availability of information on its pharmacokinetics and metabolism [1] provided a rich background for development of radioligands for the VMAT2.

2 SYNTHESIS PROCEDURES

CAUTION: All radiochemical syntheses must be carried out using appropriate equipment in a facility authorized for the use of radioactive materials. Personal protective equipment must be worn and all local radiation safety laws followed.

2.1 Production of [^{11}C]CO_2

[^{14}N]N_2 gas [4] was loaded into the [^{11}C]gas target [5]of a General Electric Medical Systems (GEMS) PETTrace cyclotron and pressurized to 250–255 psi. The target was bombarded with a proton beam (40 μA for 30 min) to generate ~3Ci (111 GBq) of [^{11}C]CO_2.

2.2 Preparation of [^{11}C]Methyl Iodide

All procedures were done using a GEMS TRACERlab $FX_{C\text{-}Pro}$ synthesis system. [^{11}C]CO_2 is delivered from the target via a 1/16″ Teflon delivery line by nitrogen pressure directly to a column packed with 0.3 g of molecular sieve [6] and 0.2 g of Shimalite–Nickel [7] where it is trapped at room temperature. The column is then sealed under hydrogen gas (atmospheric pressure) and heated to 350°C for 20 s to reduce the [^{11}C]CO_2 to [^{11}C]CH_4. The [^{11}C]CH_4 was passed through a column of phosphorus pentoxide desiccant [8] and trapped on a column of carbosphere [9] at −78°C. Gaseous [^{11}C]CH_4 is released by heating the carbosphere column to 80°C. Once released, the methane enters a circulation loop that includes a gas pump, a column of iodine [10] at 100°C, the TRACERlab standard iodine reactor tube at 720°C, two adjacent columns of Ascarite II [11], and a column of Porapak Q [12] at room temperature. The gaseous mixture was circulated for 5 min while [^{11}C]MeI accumulated on the Porapak column. [^{11}C]MeI can be released from the Porapak column and delivered to the awaiting reactor by heating to 190°C.

2.3 Synthesis of [^{11}C]DTBZ

All steps (Fig. 1) through the sterile filtration of the final product solution are performed using the integrated functions of the GE TRACERlab $FX_{C\text{-}Pro}$ synthesis system. The reaction of gaseous [^{11}C]MeI with (2*R*,3*R*,11b*R*)-9-*O*-desmethyldihydrotetrabenazine [12] takes place in the FXcPro reaction vessel. The vessel is charged with 1.0 mg of precursor, 10 μL of 10 M NaOH, and 100 μL

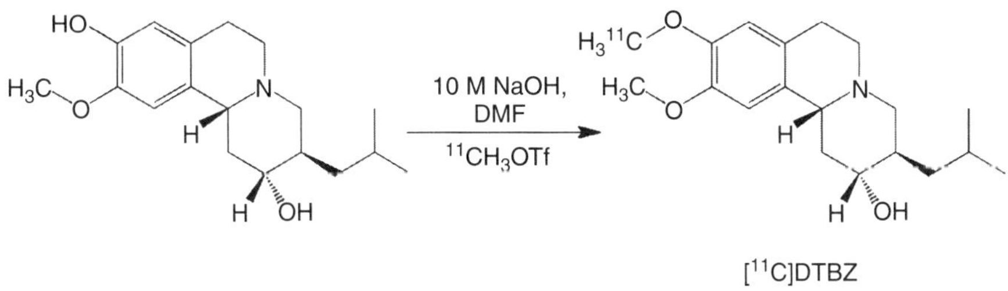

FIGURE 1 Synthesis of [^{11}C]DTBZ.

of dimethylformamide [13]. [^{11}C]MeI is delivered from the methyl iodide synthesis unit and bubbled into the reaction vessel at a fixed flow rate of 15 mL/min for 3 min; then the flow is stopped, and the solution is stirred for an additional 2 min. The reaction mixture is diluted with 1 mL of 85/25 20 mM NaH_2PO_4/ethanol and injected into a preparatory HPLC column (Luna C8(2) [14], 10 × 100 mm, eluted with 85/25 20 mM NaH_2PO_4/ethanol at 3 mL/min). The product fraction (1 mL) is collected at 10–11 min, diluted with isotonic saline (9 mL), and passed through a 0.22-mm sterilizing filter [15] into a sterile 10-mL multidose vial.

3 QUALITY CONTROL PROCEDURES

CAUTION: All radiopharmaceuticals produced for clinical use must have local regulatory approval before use in humans. Trained personnel must carry out quality control (QC) procedures, and each dose must meet all established QC criteria before release to the clinic.

QC procedures for [^{11}C]DTBZ are summarized below. Complete QC data for three repeat batches of [^{11}C]DTBZ produced using the method described herein are summarized in Table 1. Each of the three doses met all of the established QC criteria.

TABLE 1 Quality Control Analysis Data for Three Repeat Runs of [^{11}C]DTBZ

QC Test	Release Criteria	Run 1	Run 2	Run 3
Yield mCi (GBq)	N/A	76 (2.81)	79 (2.92)	80 (2.96)
Visual inspection	Clear	Pass	Pass	Pass
Radiochemical identity	0.9–1.1	1.01	1.01	1.02
Radiochemical purity (%)	>93	100	100	100
Specific activity (GBq/μmol)	N/A	515	277	463
Dose pH	4.5–7.5	4.5	5.0	5.0
Radionuclidic identity ($t_{1/2}$) (min)	18.5–22.4	Pass	Pass	Pass
Residual solvent analysis (ppm)	Acetone (<5000)	35	19	25
	DMF (<880)	160	3	Not detected
Endotoxin analysis (EU/mL)	<17.5	<2.00	<2.00	<2.00
Sterility testing	No growth at 14 d	Pass	Pass	Pass

N/A, not applicable.

3.1 Visual Inspection

The [^{11}C]DTBZ dose is examined in a clear glass vial and must be clear, colorless, and free of visible particulate matter.

3.2 Radiochemical Identity

Radiochemical identity is determined using a Shimadzu HPLC system with the following components: SCL-10Avp system controller, DGU-14A in-line degassing unit, LC-10ADvp pump, CTO-20A oven, and Bioscan FC3300 flow count radioactivity detector. Chromatographic separation is performed on a Luna 5-µm C-8 column [14] (2 × 100 mm) with a mobile phase of 75/25 20 mM ammonium acetate/methanol (pH 4.5) and a flow rate of 0.70 mL/min. The retention time of [^{11}C]DTBZ was 1.7–2.0 min. Radiochemical identity is determined by comparison of retention times of a known DTBZ reference standard [12] and the radioactive peak obtained. The relative retention time is calculated using the formula below and must be 0.9–1.1 to meet the release criteria.

$$\text{RRT} = \text{tr}([^{11}\text{C}]\text{DTBZ})/\text{tr}([^{12}\text{C}]\text{DTBZ standard})$$

3.3 Radiochemical Purity and Identity

The radiochemical purity of each batch is determined using the HPLC system outlined in Section 3.2. Radiochemical purity must be >93%. Representative analytical HPLC traces are displayed in Fig. 2.

3.4 Specific Activity

The specific activity of each batch of [^{11}C]DTBZ is determined using the HPLC system outlined in Section 3.2. Before each radiochemical synthesis, system suitability (standard peak identity by retention time and UV detector response function) is established by injections of three samples of a fixed volume and standard concentration of DTBZ. Specific activity analysis is then done by injection of an identical volume of [^{11}C]DTBZ product solution. Mass of the product is calculated using the ratio of the integrated detector response for the product relative to the average integrated detector response for the standard injections.

3.5 Radionuclidic Identity

Radioactivity was measured using a Capintec CRC®-15R radioisotope dose calibrator, and half-life was calculated using Equation 1. Calculated half-life must be 18.4–22.4 min.

$$t_{1/2} = -\ln 2(\text{time difference}/(\ln(\text{ending activity}/\text{starting activity}))) \quad (1)$$

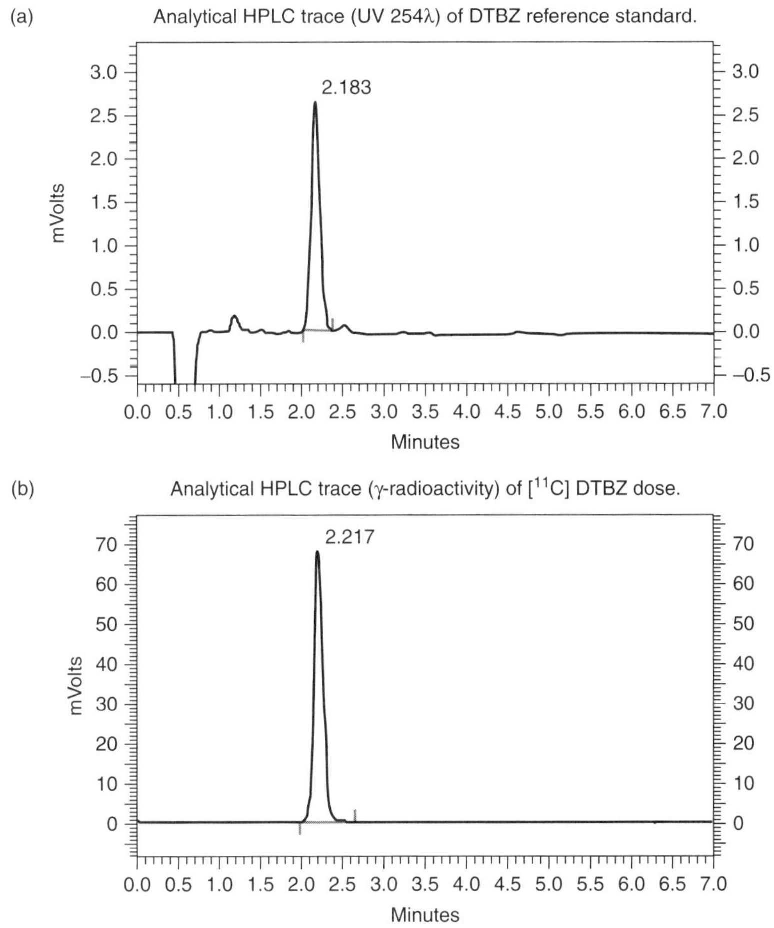

FIGURE 2 Analytical HPLC trace of DTBZ reference standard (a) and [^{11}C]DTBZ (b).

3.6 Residual Solvent Analysis

Gas chromatography (GC) is used to determine the residual solvents. The GC system used was a Shimadzu GC-2010 with an AOC-20i autoinjector, split/splitless inlet, a flame ionization detector (FID) and a Restek column (Stabilwax 30 m × 0.25 mm, 0.25 μm G16 stationary phase). Chromatographic conditions are outlined below:

Injection volume: 0.25 μL
Split ratio: 1:15
Injector temperature: 180°C
Linear velocity: 11.7 cm/s

Detector temperature: 250°C
Oven temp:
Initial temperature: 55°C
Hold at 55°C for 6.5 min
Heat to 180°C at 80°C/min
Hold at 180°C for 8.5 min

3.7 Endotoxin Analysis

Endotoxin content in doses of [^{11}C]DTBZ was analyzed by a Charles River Laboratories EndoSafe® Portable Testing System and according to the United States Pharmacopeia (USP) [16]. Doses must contain ≤175 endotoxin units (EU), which is equal to 17.5 EU/mL in a formulation volume of 10 mL.

3.8 Sterility Testing

Samples of [^{11}C]DTBZ were added to tubed fluid thioglycolate media (FTM) and tubed tryptic soy broth (TSB) media. FTM is used to test for anaerobes, aerobes, and microaerophiles, while TSB is used to test for nonfastidious and fastidious microorganisms. The media tubes were incubated along with positive and negative controls for 14 days. FTM is incubated at 30–35°C, and TSB is incubated at 20–25°C according to current USP guidelines. Tubes were visually inspected on the third, eighth, and fourteenth days of the test period and compared to the positive and negative standards. The test article ([^{11}C]DTBZ) and negative controls must have no growth after 14 days in order to pass the test for sterility. Positive controls can be established separately as growth promotion tests according to USP guidelines [17].

4 DISCUSSION

The procedure described above is a method for the synthesis of [^{11}C]DTBZ in a closed reaction vessel followed by HPLC purification; alternatives for reaction solvent and isolation using solid phase extraction have been reported [18]. This radiochemical can also be synthesized using solid-supported radiochemical methods [19] including performing the [^{11}C]methylation reaction in a loop system.

CHEMICAL ABSTRACTS NOMENCLATURE (REGISTRY NUMBER)

(2*R*,3*R*,11b*R*)-9-*O*-desmethyldihydrotetrabenazine (956903-21-2)
N,N-Dimethylformamide (68-12-2)
Ethanol (64-17-5)
Iodomethane (74-88-4)

Sodium dihydrogenphosphate (7558-80-7)
Sodium hydroxide (1310-73-2)

REFERENCES AND NOTES

1. Guay DRP. Tetrabenazine, a monoamine-depleting drug used in the treatment of hyperkinetic movement disorders. Am J Geriatr Pharmacother 2010;8:331–373.
2. Wimalasena K. Vesicular monoamine transporters: structure-function, pharmacology, and medicinal chemistry. Med Res Rev 2010. In press. DOI: 10.1002/med.20187.
3. Frey KA, Koeppe RA, Kilbourn MR. Imaging the vesicular monoamine transporter. Adv Neurol 2001;86:237–247.
4. Ultra high purity nitrogen gas with 1% O_2 purchased from Purity Plus and used as received.
5. General Electric Medical Systems standard [^{11}C]carbon target with a 75-μm Havar target foil and standard vacuum foil.
6. Alltech is a division of Grace Davison Discovery Sciences (pn: 5624, 4A 80/100 mesh).
7. Shimadzu Scientific pn: 22127719.
8. Fluka/Sigma Aldrich pn: 79610.
9. Alltech/Grace pn: 5680.
10. Iodine crystals, 2.2 g, EMD Chemicals pn: IX01203.
11. Three inches in TRACERlab standard column, Thomas Scientific pn: C049H40.
12. MonomerChem pn MC7042006 or ABX Biochemicals GmbH pn: 4710.
13. Aldrich pn: 227056.
14. Phenomenex.
15. Millipore Millex-GV, pn: SLGV33RS.
16. U. S. Pharmacopeia <85>. Bacterial Endotoxins Test. USP 32–NF 27, 2009.
17. U. S. Pharmacopeia <71>. Sterility Tests. USP 32–NF 27, 2009.
18. Quinoceres G, Collantes M, Catalan R, Ecay M, Prieto E, Martino E, Blesa FJ, Alvarez-Erviti L, Areses P, Arbizu J, Obeso JA, Marti-Climant JM, Richter JA, Penuelas I. Quick and simple synthesis of (11C)-(+)-alpha-dihydrotetrabenazine to be used as a PET radioligand of vesicular monoamine transporters. Rev Esp Med Nucl 2008;27:13–21.
19. Jewett DM, Kilbourn MR, Lee LC. A simple synthesis of [^{11}C]dihydrotetrabenazine. Appl Radiat Isot 1997;24:197–199.

CHAPTER 22

SYNTHESIS OF [^{11}C]FLUMAZENIL([^{11}C]FMZ)

ROBERTO CANALES-CANDELA, PATRICK J. RISS, AND FRANKLIN I. AIGBIRHIO

Wolfson Brain Imaging Centre, Department of Clinical Neurosciences, University of Cambridge, Addenbrooke's Hospital, Cambridge, UK

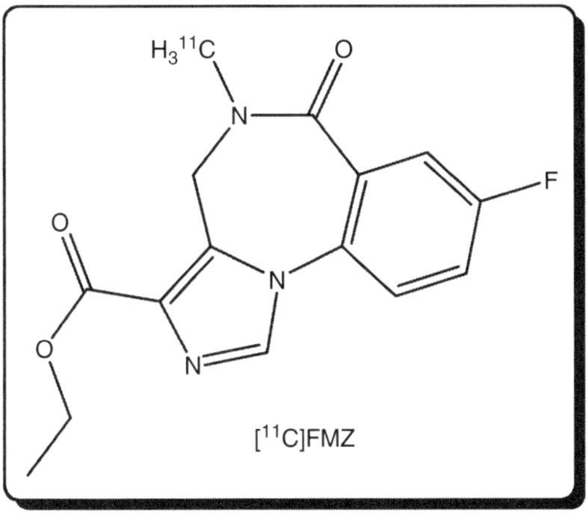

1 INTRODUCTION

Flumazenil is a benzodiazepine antagonist introduced in the late 1980s by Hoffmann-La Roche under the trade name Anexate. It acts as a competitive inhibitor at the benzodiazepine binding site on the $GABA_A$ receptor and has therefore been used as an antidote in the treatment of benzodiazepine overdoses [1]. Early work by Ehrin and others highlighted the potential use of flumazenil as positron emission tomographic (PET) imaging agent [2–4]. Several methods have been published in the literature describing the radiosynthesis of [^{11}C]FMZ ([^{11}C]$t_{1/2}$ = 20.9 min) [5–8], and it is now a well established procedure that is carried out routinely in numerous centers around the world. [^{11}C]FMZ is one of

Radiochemical Syntheses, Volume 1: Radiopharmaceuticals for Positron Emission Tomography, First Edition. Edited by Peter J. H. Scott, Brian G. Hockley.

the few radiotracers for PET imaging with a pharmacopeial monograph, which should be consulted when planning to use the tracer in human subjects. The radiosynthesis and quality control (QC) methods described in the current monograph have been used at our center for over five years during which time it has proved to be a reliable method for benchmarking the efficiency of methylation reactions.

2 SYNTHESIS PROCEDURE

CAUTION: All radiochemical syntheses must be carried out using appropriate equipment in a facility authorized for the use of radioactive materials. Personal protective equipment must be worn and local radiation safety rules followed.

The central benzodiazepine ligand N^5-[^{11}C]methylnorflumazenil ([^{11}C]FMZ) is produced by methylation with [^{11}C]iodomethane of desmethylflumazenil in the presence of a base (NaOH), using the captive solvent method [9] in a dedicated "loop" system installed in a modified General Electrics (GE) TRACERLab™ FX$_c$ synthesizer (Fig. 1).

The preparation consists of passing a flow of [^{11}C]iodomethane in a stream of helium gas for 3–6 min through a high performance liquid chromatography (HPLC)-injection loop with a column prefilter assembly that contains a thin film of a solution of the precursor and KOH(aq) in *N,N*-dimethyl formamide (DMF). The precursor/KOH solution is loaded through a manual HPLC injection port with a prefilter (Alltech, cat. 28640) and switched to the inject position to be connected in series to the preparative HPLC injection loop on the GE TRACER-Lab FX$_c$ system (Fig. 2). After completion of trapping of the [^{11}C]iodomethane, the FX-C HPLC valve is switched to the inject position, and the reaction mixture is separated by preparative HPLC on a reversed phase column. The mobile phase is 15% ethanol(aq), and the [^{11}C]flumazenil fraction is collected into a vial containing a physiological saline solution and subjected to sterilizing filtration.

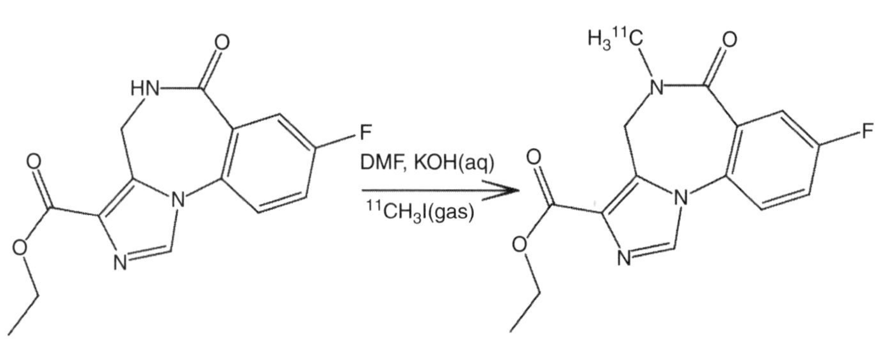

FIGURE 1 Radiosynthesis of [^{11}C]FMZ.

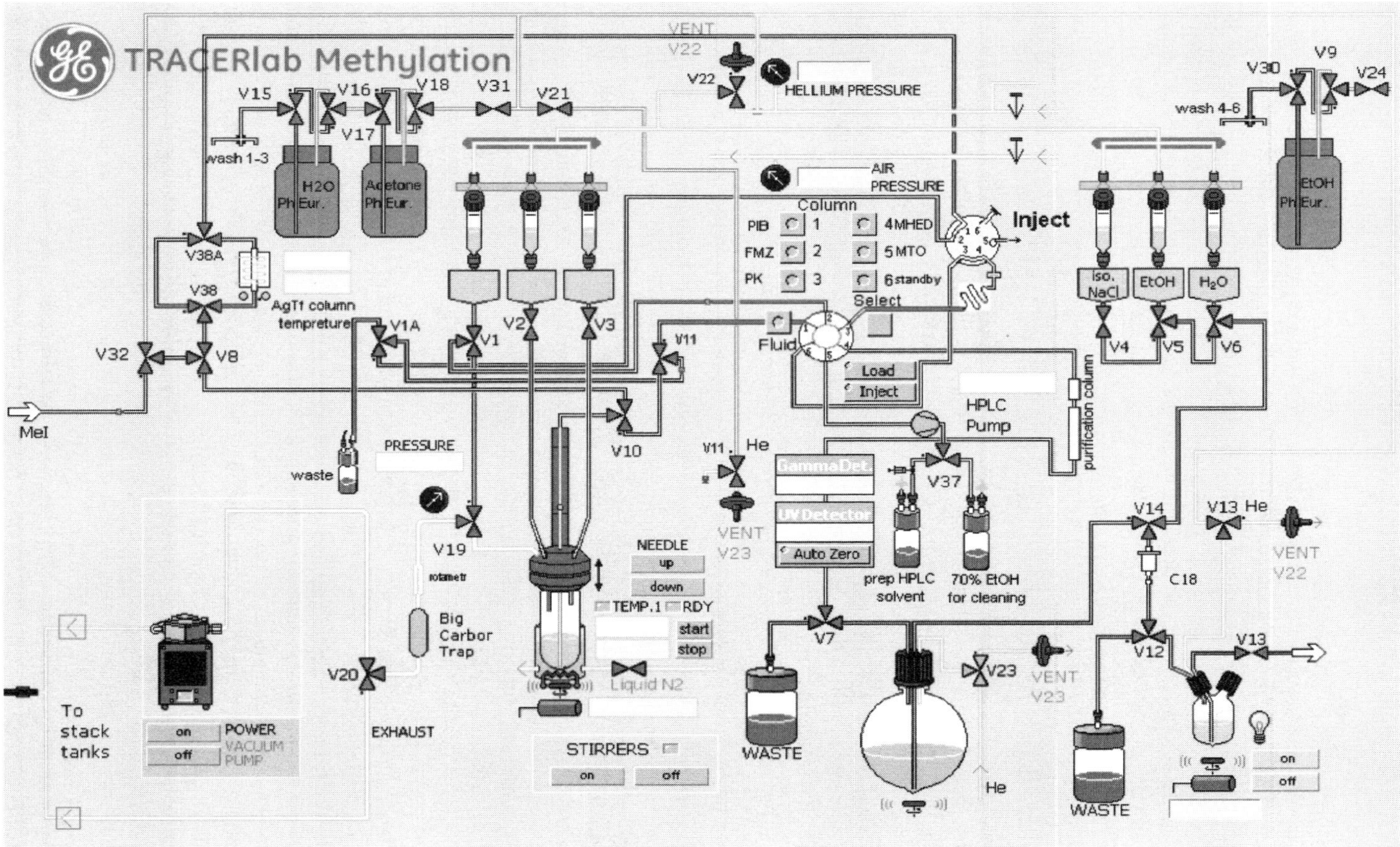

FIGURE 2 TRACERlab FX_c display of the customized synthesizer used in the automated ^{11}C radiotracer productions.

2.1 Radiosynthesis Preparation

2.1.1 Cyclotron

The GE PETtrace cyclotron used in our center is a negative ion particle accelerator with an acceleration energy of 16.5 MeV for protons and a maximal extracted beam on target of 45 μA. [^{11}C]Carbon dioxide is produced by the $^{11}N(p,\alpha)^{11}C$ nuclear reaction in a gas target filled with $^{nat}N_2$ containing 0.5% O_2. The standard parameters for [^{11}C]CO_2 production are 40 μA for 40 min, with an estimated yield at end of bombardment (EOB) of ~2.3 Ci (85 GBq).

2.1.2 [^{11}C]Iodomethane Module

[^{11}C]MeI is produced by the dry chemistry method in a GE MeI Microlab system. Production of [^{11}C]MeI is achieved in the gas phase by free radical iodination with recirculated iodine vapor at elevated temperatures (700°C). Before starting each synthesis, a leak-test is performed by applying helium pressure into the system. The preparation for each run also includes a conditioning sequence, which is run automatically from the MeI module control box during or before the cyclotron run. The run sequence in the Microlab MeI system must be set to optimal transfer values between the MeI Microlab module and the FX-C synthesizer. The radiolabeling reaction with [^{11}C]methyl iodide is given below.

$$^{11}N(p,\alpha)^{11}C \rightarrow [^{11}C]CO_2 \rightarrow [^{11}C]CH_3I \rightarrow R\text{-}[^{11}C]CH_3$$

For productions for human use, the iodine column is normally changed after five or six runs; for research or testing applications, up to 10 syntheses are acceptable.

2.1.3 Preparative HPLC System

The reaction mixture is directly injected into the TRACERlab FX_c module HPLC system for purification of the radiotracer. The HPLC system must be switched on at least 30 min before production to allow equilibration conditions to be achieved; the mobile phase is ethanol/water 15/85% (v/v), with a flow rate of 2 mL/min, and uses an analytical column with a stationary phase made of a spherical octadecylsilyl silica gel. Two detectors, a UV detector set at 254 nm and a NaI(Tl) radiation detector, are connected in series and send the output signals directly to the TRACERlab software, which helps the operator to manually select the product peak for collection.

The product is collected in the formulation vial, so it is important to maintain aseptic conditions in the vial and transfer lines. This can be attained by keeping valves, column, and lines in 70% ethanol. At the time of synthesis, a small preprogrammed routine can condition the product lines by passing through the production mobile phase to lower the concentration of ethanol in the final formulation. Aseptic manipulation is of paramount importance in order to reduce the bioburden in the system. All reagents, mobile phase, and cleaning solutions should be of the highest quality possible, preferably of pharmacopeia grade (Table 2 and Appendix). To further reduce the possibility of introducing contaminating microorganisms into the system, all mobile phases are filtered just before use in a sterilizing filter unit (0.22-μm, sterile polyethersulfone (PES) membrane) and transferred aseptically to the controlled area. Before sterilization, the mobile phase is prepared in a clean and depyrogenized 1-L glass bottle and degassed in an ultrasound bath by applying vacuum for at least 15 min.

2.1.4 *FX-C Module*

The standard configuration of the module was modified to introduce the reaction loop, and the flow path of the [^{11}C]MeI was rerouted, allowing for "captive solvent" methylations [9–11]. Modifications include the inclusion of new valves both in the low and high pressure sides of the system, allowing the use of multiple HPLC columns for the purification of several radiotracers. A bypass of the triflate column was also installed using a capillary tubing of 10 cm length (0.02 mm internal diameter) to reduce the flow rate of the incoming gas from the MeI Microlab system [12]. The TRACERlab control software of the module is an integral part of the system that communicates with the synthesis module through a PLC interface and can be used to control the customized hardware configuration.

For each radiochemical synthesis, a routine program can be stored controlling the sequence of commands that will operate the valves in the system. The first step to start the synthesis is to load this sequence into the operation mode, after which the software will present a list of preparation steps to be performed and checked before it will proceed to the next synthesis phase. Each step should be confirmed manually in the reference field of the production preparation window before proceeding to the next step. Following this stepwise procedure, the operator checks that the inlet and outlet of gases are properly connected, the HPLC system is ready, the formulation solution is added, product lines are well fitted, the environmental monitoring plate and radiomonitoring system are in place, and receiving vials are installed. The routine also instructs the operator to dissolve 0.3 mg of desmethylflumazenil precursor in 100 μL of DMF adding 5 μL of 15 M KOH(aq) in a cone-shaped vial. Vortex the mix, and leave the dissolved precursor on the bench for around 15 min, while any slurry falls onto the bottom, and inject the top layer. When all the set up steps have been checked, the program will place the TRACERlab FX_c system in standby mode waiting for the [^{11}C]CO_2 to be released from the cyclotron (EOB) and start the synthesis.

2.2 Radiochemical Synthesis

On completion of irradiation, a production report is printed out and the system must be checked visually before delivering radioactivity to the [^{11}C]MeI Microlab synthesizer module. The target is then emptied, and the [^{11}C]MeI production is started manually using the Microlab control box. At this point, no intervention into the process is possible or needed until [^{11}C]MeI is delivered into the methylation loop of the TRACERlab FX_c. The production and delivery of [^{11}C]MeI can be monitored graphically on a PC screen using an in-house program, with data acquired installed radiodetectors. The radiodetector placed near the ascarite trap in the [^{11}C]MeI MicroLab will show a decreasing sinusoidal wave at the recirculation step, and the radiodetector close to the product line will provide a Gaussian peak shape when the activity is transferred to the FX-C synthesizer.

When the release of [^{11}C]MeI is finished, the system is closed for 1.5 min to complete the methylation reaction. As soon as the methylation reaction is completed, the HPLC system is automatically switched to the inject mode. All the radioactivity is now transferred to the HPLC column, and the chromatograph traces from both detectors can be followed using the TRACERlab FX_c software (Fig. 3). The [^{11}C]flumazenil fraction is collected by pressing the "start peak" tab on top of the HPLC trace display (alternatively, V7 can be operated manually from the screen to deviate the HPLC outlet into the product vial).

Generally there are three radioactive products formed. After a few seconds, [^{11}C]methanol is eluted; subsequently unreacted [^{11}C]MeI is eluted (peak at 2.7 min) just before the precursor is eluted (3.8 min; UV signal peak up to ~1000 mV). Flumazenil is eluted after the precursor (peak maximum ~6 min), and this peak must be collected manually into the FX-C formulation vial. Usually it takes about 60 s to collect the peak (peak volume is ~2 mL). A typical preparative HPLC chromatogram for the purification of [^{11}C]FMZ is shown in Fig. 3. The final reaction product consists of [^{11}C]flumazenil in a mixture of 10 mL of saline and ~2–4 mL of 15% ethanol (the final solution containing <5 vol% ethanol).

After collection of the [^{11}C]flumazenil fraction in the FX-C formulation vial, the program will automatically start product transfer to the final sterilizing filtration step. When the product transfer is successfully finished and the product has passed through the sterile filter into sterile product vials, the process comes to an end. Transfer is usually complete after 3–4 min. Using the appropriate engineering controls and working practice to minimize contamination and dose, the product and QC vials are transferred to the dose calibrator for measurement of the product radiochemical yield. At this point, the synthesis routine is finished and the program will switch off all the valves and devices in the FX-C synthesizer. The radioactivity of the vials is measured and noted on the production record sheet. A 500-µL QC sample for immediate use should be withdrawn from the QC vial using aseptic technique and left in the dose calibrator area for further analysis.

After the adequate decay time (1–2 h depending on the initial radioactivity and the radiochemical yield), the system can be automatically cleaned. For this, an automatic cleaning routine program is used, which conveniently prepares the system for consecutive radiotracer productions. The program will perform automatic cleaning with ethanol and acetone, dry the system, and turn the column select switch to the standby position. After the cleaning routine is finished, the system is ready for the next production.

3 QUALITY CONTROL (QC) PROCEDURES

CAUTION: All radiopharmaceuticals produced for clinical use must have local regulatory approval before use in humans. Trained personnel must carry out QC procedures, and each dose must meet all established QC criteria before release to the clinic.

Owing to the short half-life of carbon-11 (20.4 min) and high levels of radioactivity, the QC method is divided into prerelease tests and postrelease

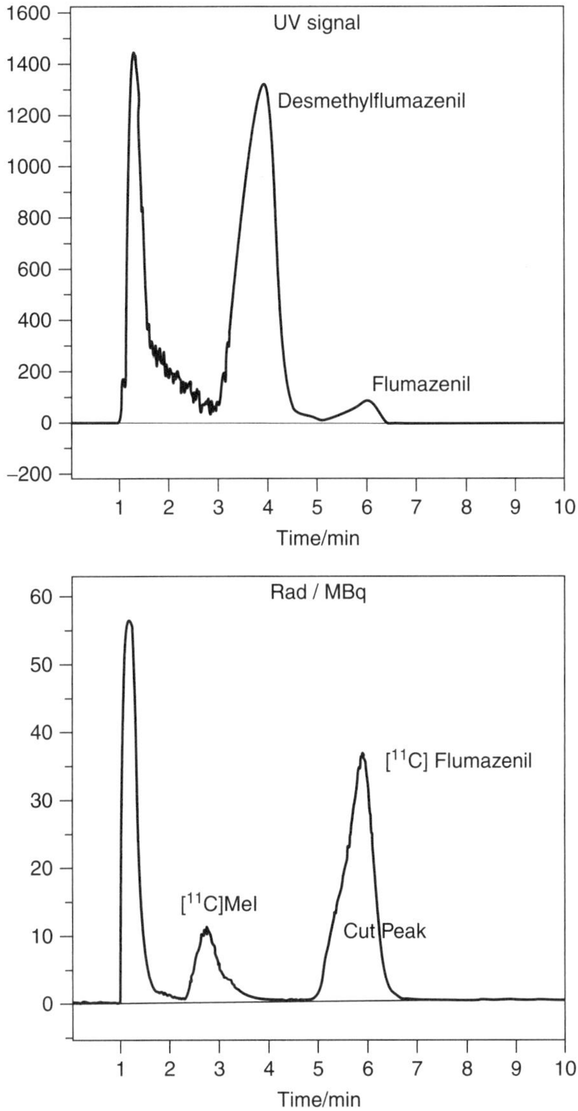

FIGURE 3 Chromatograms of the HPLC purification of [^{11}C]FMZ.

tests. Prerelease tests have to be performed before release, and all tests have to be passed for release of the production batch to be allowed. These postrelease tests are performed after the release of the product; however, if any of these tests fails, the production of [^{11}C]FMZ has to stop immediately and the reason for the failure has to be investigated. The production procedure has to be revalidated before production of [^{11}C]FMZ for human use is allowed again. The QC results for three typical [^{11}C]FMZ productions are shown in Table 1.

TABLE 1 Quality Control Data for Three Routine Productions of [^{11}C]FMZ

QC Test	Release Criteria	Run 1	Run 2	Run 3
Yield mCi (GBq)	≥13.5(0.5) in 10 mL	150 (5.56)	160 (5.93)	70 (2.60)
Visual inspection	Clear colorless	Pass	Pass	Pass
Radiochemical identity (s)	$\Delta T = t_{R11C} - t_{Rstd}$ $3 < \Delta T < 9$	6	6	5
Radiochemical purity	≥95%	99.53	100	96.13
Chemical amount (μg/dose)	Desmethylflumazenil ≤5	0.7	1.4	2.2
	FMZ ≤50	6.5	7.4	5.0
Specific activity	NA	132	146	88
Residual solvent analysis (mg/mL)	DMF ≤0.88	ND	ND	ND
Formulation solvent (%)	Ethanol< 10	4.1	4.4	3.7
Dose pH	5–8	7.0	7.0	6.5
Sterile filter integrity test (bar)	≥3.4	5.0	5.1	4.9
Radionuclidic identity ($t_{1/2}$) (validation test) (min)	20.4±0.5	nd	nd	nd
Endotoxin analysis (EU/mL)	LAL< 17.5	1.3	1.0	0.5
Sterility testing	Sterile	Sterile	Sterile	Sterile

NA, not applicable; ND, not detected; nd, not determined; t_R, retention time.

3.1 Prerelease Tests

3.1.1 Visual Inspection

A visual check of the product is performed behind a lead window; the product must be clear, colorless, and free of particulate matter.

3.1.2 pH Measurement

The pH of the product is determined using pH indicator sticks and should be within the range 6–8 pH units.

3.1.3 Radiochemical Identity and Chemical Purity

The radiochemical purity should be >95% and is determined using the following HPLC method:

Column: Luna C-18 (2) mercury cartridge, 20 × 2 mm 3 μm
Injection loop: 20 μL
Injection volume: 25 μL
Mobile phase: Ethanol/0.001 M H_3PO_4(aq) (10:90 v/v)
Flow rate: 1.0 mL/min
Radioactivity detector amplifier: 0.70 kV
UV detector: λ = 245 nm; AUFS = 0.05.

The suitability of the system is checked before each production by injecting a standard mix solution of 5 μg/mL of FMZ and 2 μg/mL desmethylflumazenil. The chemical amounts of FMZ and desmethylflumazenil are determined by quantification of the UV absorbance peaks of an injection of the QC sample. A spiked sample is prepared with FMZ standard solution (20 μg/mL) and QC sample (50:50) for radiochemical identity determination. The [^{11}C]FMZ and the

TABLE 2 Reagents Used in the Synthesis of [^{11}C]FMZ

Reagent	Amount Required Per Production
Desmethylflumazenil (precursor)	~0.3 mg
Anhydrous DMF	100 ⊥ 10 μL
15% Ethanol(aq)	~150 mL
15 M KOH solution	5 μL
0.9% w/v NaCl solution	10 mL

reference compound peaks should differ by a certain acceptance criteria, which corresponds to the dead volume between the UV and γ-detectors. All reference standards are purchased from a GMP-complaint manufacturer. In the final QC report, the maximum allowable injection volume and specific activity (SA) are calculated using an Excel spreadsheet based on the maximum amount of FMZ and desmethylflumazenil acceptance criteria. By default, the maximum allowable injection volume of Wolfson Brain Imaging Center (WBIC) products is limited by the product vial size to 10 mL.

The SA is calculated using a validated QC Excel sheet.

The activity of the sample is measured in a dose calibrator using a fixed volume.

3.2 Postrelease Tests

3.2.1 Filter Integrity Test

The integrity of the filter is tested using the bubble point test; to pass the test, the bubble point has to be ≥3.86 bar (56 psi). (Council of Europe. European Pharmacopoeia. 7th ed. Strasbourg; 5.1.1. Methods of preparation of sterile products; 01/2008:50101.).

3.2.2 Sterility Test

The test for sterility is performed by an external contractor upon decay of the radioactivity in the product. The sterility of the final product is tested by the membrane filtration method (Council of Europe. European Pharmacopoeia. 7th ed. Strasbourg; 2.6.1 Sterility; 07/2010:20601.).

3.2.3 Bacterial Endotoxins

Endotoxin testing is performed using a chromogenic technique with Endosafe limulus amebocyte lysate (LAL) system in a portable test system (PTS) Kinetic Reader (Charles Rivers Laboratories, Charleston, SC). The LAL test is performed immediately after release of the product, and the acceptance criteria for release is that the endotoxin concentration is below 17.5 EU/mL (Council of Europe. European Pharmacopoeia. 7th ed. Strasbourg; 2.6.14. Bacterial endotoxins; 01/2010:20614.).

3.2.4 Analysis of Formulation and Residual Solvents

Residual solvents in the final product are analyzed using a gas chromatography instrument using the following conditions:

Column: Alltech Gas Chrom 254 80/100, 6′ × 1/8″ × 0.85″
Detector: flame ionization detector (FID)
Gas flows:
- Helium (carrier gas): ~25 mL/min
- Hydrogen (FID): ~45 mL/min
- Air (FID): ~350 mL/min
- Specification limits: residual solvent DMF <0.88 mg/mL and formulation solvent ethanol <10%.

CHEMICAL ABSTRACTS NOMENCLATURE (REGISTRY NUMBER)

Desmethylflumazenil: ethyl 8-fluoro-6-oxo-5,6-dihydro-4*H*-imidazo[1,5-*a*][1,4]benzodiazepine-3-carboxylate (79089-72-8)

[^{11}C]Flumazenil: ethyl 8-fluoro-5-[^{11}C]methyl-6-oxo-5,6-dihydro-4*H*-imidazo[1,5-*a*][1,4]benzodiazepine-3-carboxylate (78755-81-4)

Appendix Reagents and Consumables for Production of [^{11}C]FMZ

Item	Specification	Supplier/Catalogue No./Size	Storage
Ethyl 8-fluoro-6-oxo-5,6-dihydro-4*H*-imidazo[1,5-*a*][1,4]benzodiazepine-3-carboxylate (precursor)	>95% white solid	ABX Biochemicals/1700.0001/ 1 mg per vial	−20°C, protected from light
N,N-dimethylformamide (DMF)	>99.5% anhydrous	Sigma-Aldrich/40228/ 250 mL	Store at RT over dried molecular sieves
Water for injections	For i.v. injection	Hameln Pharmaceuticals PL 01502-0003R	Store at RT
Sodium hydroxide pellets	>99.0%, Ph.Eur.	Sigma, Riedel de Haën/30620/500 g	Keep tightly closed at room temperature
Ethanol 70%	Ph.Helv., sterile	FLUKA/02877/1 L	Store at RT
Prep HPLC column: Luna 3 μm C18(2) 100A	Meets analytical specification	Phenomenex/00A-4251-E0/30 × 4.6 mm	Store at RT in 70% EtOH
NaCl solution for injection 0.9% w/w	BP, sterile	Brownes/PL24598/ 0002/10 mL	Store at RT
Acetone	Puriss. P.A., Ph.Eur.	Fluka, Riedel de Haën/00570-32201/2.5 L	Store at RT
Product vial 11 mL TechneVial	Sterile	Mallinckrodt/ DRN4357	Store at RT Store at RT
Sterilization filter Millex GP	Sterile, 0.22 μm, MF-Millipore membrane	Millipore/ SLGVSV255F	Store at RT
Mobile phase filter	Sterile, 0.22 μm, PES membrane	Corning/431097	Store at RT

i.v., intravenous; RT, room temperature.

REFERENCES

1. Beer MH, Porter RS, Jones TV. The Merck manual of diagnosis and therapy. 18th ed. Philadelphia: Merck Sharp & Dohme Corp.; 2007.
2. Ehrin E, et al. Preparation of carbon-11 labelled Ro 15-1788, a selective benzodiazepine receptor antagonist, and preliminary positron emission tomography studies. In: Medicinal chemistry. Dahlboom R. and Nilsson JLG, Ed. Uppsala; Sweden: Swedish Pharmaceutical Press; 1984.
3. Wolff J. The effect of benzodiazepine antagonist Ro 15-1788 on cerebral blood flow. In: 9th World Congress of Anaesthesiologists. Washington (DC) [n.p.]; 1988.
4. Lauven PM. The efficacy of the benzodiazepine antagonist flumazenil (RO 15-1788) on unconscious intubated intensive care patients. In: 9th World Congress of Anaesthesiologists. Washington (DC) [n.p.]; 1988.
5. Maziere M, et al. Synthesis of ethyl 8-fluoro-5,6-dihydro-5-[11C]methyl-6-oxo-4H-imidazo [1,5-a] [1,4]benzodiazepine-3-carboxylate (RO 15.1788-11C): a specific radioligand for the in vivo study of central benzodiazepine receptors by positron emission tomography. Int J Appl Radiat Isot 1984;35(10):973–976.
6. Ehrin E, et al. Preparation and preliminary positron emission tomography studies of 11C-Ro 15-1788, a selective benzodiazepine receptor antagonist. Acta Pharm Suec 1984;21(3):183–188.
7. Suzuki K, et al. Computer-controlled large scale production of high specific activity [11C]RO 15-1788 for PET studies of benzodiazepine receptors. Int J Appl Radiat Isot 1985;36(12):971–976.
8. Naagren K, Halldin C. Methylation of amide and thiol functions with [11C]methyl triflate, as exemplified by [11C]NMSP, [11C]flumazenil and [11C]methionine. J Labelled Comp Radiopharm 1998;41(9):831–842.
9. Wilson AA, et al. Radiotracer synthesis from [11C]-iodomethane: a remarkably simple captive solvent method. Nucl Med Biol 2000;27(6):529–532.
10. Wilson AA, et al. Further progress on a remarkably simple captive solvent method for [11c]-methylations. J Labelled Comp Radiopharm 2001;44:S999–S1001.
11. Cleij MC. Rapid preparation of [11C]flumazenil: captive solvent synthesis combined with purification by analytical sized columns. J Labelled Comp Radiopharm 2007;50(1):19–24.
12. Kuznetsov, et al. Captive solvent methylations in the injector loop of the Tracerlab carbon-11 module. J Labelled Comp Radiopharm 2009;52(S1):S313.

CHAPTER 23

SYNTHESIS OF [^{11}C]HEXADECANOIC ACID ([^{11}C]PALMITIC ACID)

Adam C. Runkle, Xia Shao, and Peter J. H. Scott

Department of Radiology, University of Michigan School of Medicine, Ann Arbor, Michigan

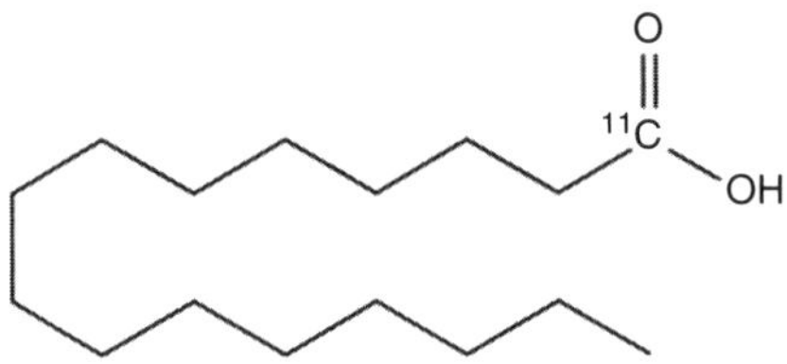

[^{11}C]Palmitic acid

1 INTRODUCTION

The increased utilization and metabolic necessity of fatty acids in the myocardium was described in the 1970s after investigators began to study myocardial metabolism [1–3]. Fatty acids were known to be utilized more frequently by the heart and, in fact, were shown to comprise 60–70% of the primary energy source for myocardial ATP production [4]. Therefore, [^{11}C]palmitic acid (2) was identified as a potentially valuable radiopharmaceutical for the assessment of myocardial metabolism and function [5–7]. In 1981, a brief description of the synthesis of [^{11}C]palmitic acid was offered by Lerch et al. in a study to determine the clearance of activity from canine cardiac tissues

Radiochemical Syntheses, Volume 1: Radiopharmaceuticals for Positron Emission Tomography, First Edition. Edited by Peter J. H. Scott, Brian G. Hockley.

[8]. Shortly thereafter, detailed descriptions of the synthesis were published [9–11], making it possible to reliably synthesize and produce [^{11}C]palmitic acid. However, although the production protocols for radiopharmaceuticals such as [^{11}C]acetic acid [12] and [^{11}C]carfentanil [13] have been described recently, and the use of [^{11}C]palmitic acid as a valuable research radiopharmaceutical remains high, the most recent description of its production appeared more than 15 years ago [14]. While the synthetic steps and preferred methods of production have not changed significantly since the original synthesis [6], various new technologies to improve purity, yield, and production time have become readily available.

The synthesis described herein utilizes some of the most recent technologies available to production chemists and affords several advantages including high radiochemical purity and dose activity while minimizing residual solvents and other impurities. It is the intention of the authors to offer a detailed overview of the synthesis for developing laboratory protocols for the production of [^{11}C]palmitic acid.

2 SYNTHESIS PROCEDURES

2.1 Production Apparatus Construction

While many radiopharmaceuticals can be synthesized to completion on automated, premanufactured radiopharmaceutical modules, [^{11}C]palmitic acid required the construction of a separate valve box (Fig. 1). The General Electric Medical Systems (GEMS) TRACERlab $FX_{C\text{-}Pro}$ was used to receive the [^{11}C]CO_2 from the cyclotron, [15] record the amount of activity received at the module [16] and modified to redirect the activity to the custom valve box. Modifications made to the $FX_{C\text{-}Pro}$ included the introduction of two three-way valves between **V27** and the **MeI Trap** (**3W-1** and **3W-2**, Fig. 1), the replacement of the **MeI Trap** column with a second column (**MSC2**) filled with molecular sieve [17], as well as the conversion of the previously dedicated HPLC eluent valve (**V30**) to a valve capable of directing activity. For the purposes of this synthesis, **V30** directed the activity released from **MSC2** to the adjacent hot cell, which contained the custom valve box.

The [^{11}C]CO_2 was first received on the custom valve box by **VC1** and directed to the reaction vessel. The reaction vessel (**Rxn vessel**) was prepared by inserting two sterile needles through the septum of a standard 10-mL sterile vial [18] and attaching a nitrogen source to one of the needles. Nitrogen was flushed through the vial for ~90–120 s, at which point all atmospheric gases were evacuated. This step was critical to the success of the synthesis, as the Grignard reagent [19] used for the reaction was easily oxidized by atmospheric gases and moisture. In order to prevent atmospheric contamination, the vent needle was removed, or capped, before removing the nitrogen source. Sterile 10-mL vials were also used for the evaporation vessel as well as the sterile product vial, and the sterile product vial was prepared according to the laboratory standard operating procedure [20].

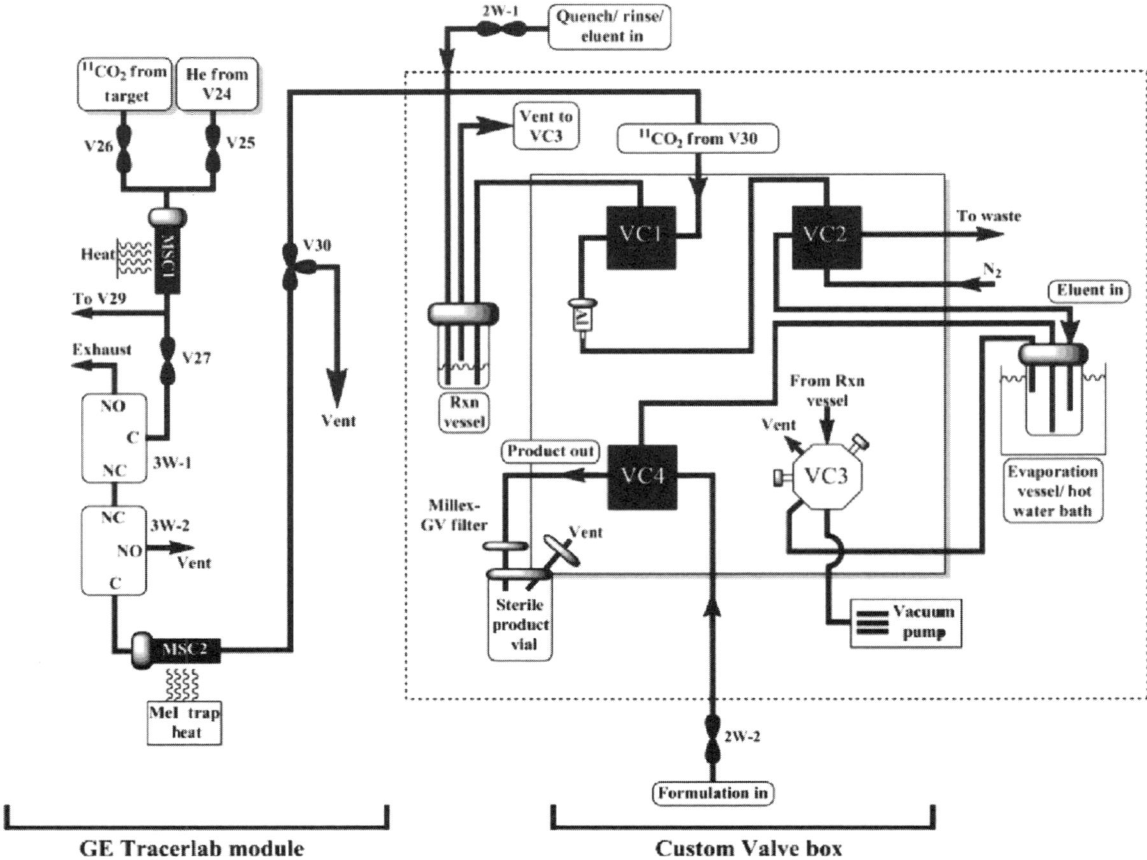

FIGURE 1 Schematic of the custom valve box used for the synthesis. The box was placed in a separate hot cell from the reaction module (dashed line).

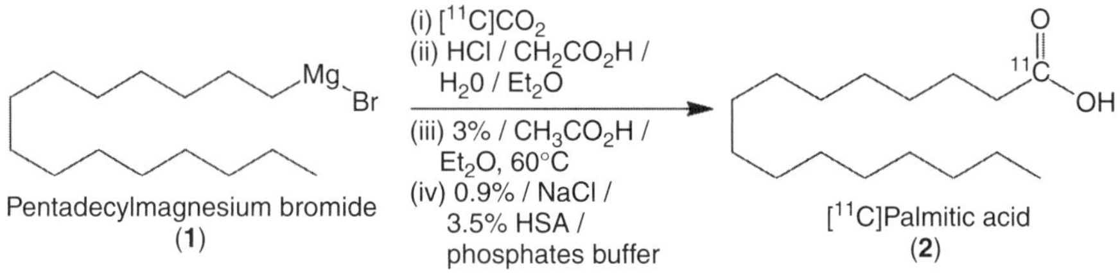

FIGURE 2 Synthesis of [^{11}C]palmitic acid.

The reaction vessel was placed in a standard clamp and precharged with the 0.5 M pentadecylmagnesium bromide (**1**, Fig. 2) and ethyl ether. Three needles connected to **2W-1, VC1**, and **VC3** were inserted into the reaction vessel septum to allow for the introduction of the quench, rinse, and eluent solutions; receipt of the [^{11}C]CO_2; and venting of the reaction vessel, respectively. Similarly, the evaporation vessel was held by a standard clamp and approximately 80% submerged in a hot water bath contained in a large sonicator [21]. Three needles were inserted into the evaporation vessel septum and attached to **VC2, VC3**, and **VC4** to function as the eluent inlet and nitrogen source, the vacuum line, and the formulation inlet and product outlet, respectively. **VC4** allowed for the transfer of the product into the sterile dose vial through a 0.22-μm sterile filter [22].

As mentioned above, each valve was controlled via an external electronic switch [23]; however, it was necessary to insert two manually controlled two-way valves (Fig. 1) by which the quench, rinse, and eluent (**2W-1**), as well as the tracer formulation (**2W-2**) were introduced into the system. The addition of each of these solutions was controlled by the production chemist manually opening and closing the two-way valves attached to separate sterile polyethylene lines that originated outside of the hot cell. A sterile syringe containing the appropriate solution was attached to the proper two-way valve, and the chemist manually injected the solution into the system.

2.2 Production of [^{11}C]CO_2

The [^{11}C]CO_2 required for the synthesis was produced by a GEMS PETTrace cyclotron [15] by bombarding zero-grade nitrogen in 1% oxygen (according to the nuclear reaction $^{14}N(p, \alpha)^{11}C$) at 40 μA for 30 min. Approximately 3 Ci (~111 GBq) of [^{11}C]CO_2 was generated and delivered directly to the $FX_{C\text{-}Pro}$ module. Since the target and all transferring lines were kept under positive pressure, there is no preconditioning procedure needed.

2.3 Reagent Preparation

As previously mentioned, a column containing molecular sieve, **MSC2**, was placed into the **MeI Trap** position to capture [^{11}C]CO_2 while allowing any possible contaminants from the cyclotron to be expelled. **MSC2** was precharged with 1 mL of [^{12}C]CO_2 and was determined to be necessary to reduce impurities and prevent precipitation formation [13,24]. Precipitates were never observed during preliminary production trials for [^{11}C]palmitic acid, and peralkylated contaminants were of no concern. However, considerable precautions were taken when designing the custom valve box for the synthesis, which allowed easy and thorough cleaning of all the lines after each run.

Once the reaction vessel was evacuated of all atmospheric gases as described above, it was precharged with 650 μL of 0.5 M $CH_3(CH_2)_{14}MgBr/Et_2O$ [19] and 3 mL of Et_2O [25]. Similarly, 10 mL of 0.15% CH_3CO_2H/Et_2O [25] was used to condition the Light Alumina-N Sep-Pak® (**Al**, Fig. 1) [26]. As mentioned above, the quench, rinse, eluent, and tracer formulations were prepared and stored in sterile syringes and consisted of the following:

- Quench—50 μL of 6 M HCl [25], 150 μL of sterile H_2O [27], 250 μL of 70% EtOH, and 8 mL of Et_2O
- Rinse—7 mL of Et_2O was used to rinse residual impurities and contaminants from the Al Sep-Pak before elution
- Eluent—8 mL of 3% CH_3CO_2H/Et_2O
- Tracer formulation—8.3 mL of NaCl, 0.9% USP; 300 μL of sodium phosphates, Injection, USP [27]; and 1.4 mL of 3.5% human serum albumin (HSA) [28] prepared the dose for injection.

2.4 Synthesis of [^{11}C]Palmitic Acid

Activity from the cyclotron was trapped by both **MSC1** and **MSC2** to ensure that any potential contaminants from the cyclotron had been vented off. The [^{11}C]CO_2 was released from **MSC2** by heating to 250°C, then received by **V30** and directed to the custom valve box. **VC1** directed [^{11}C]CO_2 to the reaction vessel (precharged with 650 μL of 0.5 M $CH_3(CH_2)_{14}MgBr$ and 3 mL of Et_2O), where it was bubbled through the solution for 4 min at 10 mL/min. The quench (described in Section 2.3) was added via **2W-1** to kill all reactions. Once the quench was added, the solution in the reaction vessel was passed over the Al Sep-Pak for 2 min to retain [^{11}C]palmitic acid. Very little activity was observed in the waste from the Al Sep-Pak when rinsed, which indicated that the activity was binding efficiently. The product was eluted with 8 mL of 3% CH_3CO_2H/Et_2O, and the eluent was collected in the evaporation vessel. The hot water bath was heated to 60°C (measured at the beginning of the synthesis before activity was delivered to the $FX_{C\text{-}Pro}$) and poured into the large sonicator. The eluent was placed under vacuum, while nitrogen was bubbled through the solution for 7 min to encourage efficient evaporation. After evaporation, the anatomical formulation was added via **2W-2**, and **VC4** directed it into the evaporation vessel. Once the formulation had been added to the evaporation vessel, the solution was sonicated for 5 min before the final dose was transferred into the sterile product vial through a 0.22-μm filter [22]. Approximately 0.5 mL of the dose was collected and released for quality control analyses. Total synthesis time was approximately 36 min.

2.5 Production Considerations

The production of [^{11}C]palmitic acid presented a number of unique obstacles, with the primary obstacle being the necessity of building a custom valve box as well as the utilization of an immediately adjacent hot cell. Consequently, each step performed on the custom valve box was controlled by an external switch, and each solution had to be prepared and stored in a syringe. Although this may have introduced some error into the synthesis, that error was negligible and did not affect the quality of the final dose. Nonetheless, a consistent difficulty observed was trying to achieve a stable pH for the final dose.

The synthesis utilized a phosphate buffer, USP, for the injection solution, which was incorporated into the anatomical formulation to prepare the dose for injection. However, after the first couple of trials, it was realized that the pH seemed to be variable and nearly failing release criteria for dose pH [12]. Therefore, it must be noted that in order to prevent dose failure due to low pH, it was absolutely imperative that the solution be evaporated for a minimum of 7 min with nitrogen bubbling through the solution while the evaporation vessel was placed under a vacuum and submerged in a hot water bath. A final dose pH ~6.0 was characteristic of the final product after thorough evaporation, and the dried product was reconstituted with the final reformulation solution.

Lastly, the reactivity of pentadecylmagnesium bromide made it susceptible to degradation over time and sensitive to temperature. In order to minimize the degradation of the Grignard reagent, it was refrigerated until it was required. Special efforts to prevent exposing the stock reagent to room temperature significantly improved the success of the synthesis. If the reagent was brought to room temperature and then returned to the refrigerator, or simply left out, the synthesis resulted in low dose concentrations as well as a variety of residual contaminants.

3 QUALITY CONTROL PROCEDURES

CAUTION: All radiopharmaceuticals produced for clinical use must have local regulatory approval before use in humans. Trained personnel must carry out quality control procedures, and each dose must meet all established quality control criteria before release to the clinic.

Although the US Pharmacopeia (USP) [29] sets forth specific quality control guidelines for radiopharmaceuticals that are to be injected, a quality control process for the production of [^{11}C]palmitic acid has not been published. However, the guidelines described below are in agreement with those for [^{11}C]acetic acid [30], as set forth by the USP [30–32]. The results for three production validation trials are presented herein (Table 1).

3.1 Visual Inspection

The dose was examined behind a PET L-block. [^{11}C]Palmitic acid was only required to be free of particulates since the color should be slightly yellow because of the addition of the HSA.

TABLE 1 Quality Control Data for Three Site Qualification Trials ([11C]Palmitic acid[a])

QC Process	Release Criteria	Trial 1	Trial 2	Trial 3
Radiochemical purity	≥ 95%	>99.9	>99.9	>99.9
pH	4.5–7.5	5.0	5.5	4.5
Visual inspection	Clear, no precipitate	Pass	Pass	Pass
Residual solvent analysis				
Acetone (μg/mL)	≤5000	3	2	<1
Ethyl ether (μg/mL)	≤5000	276	2	71
Ethanol (μg/mL)	≤5000	82	24	54
Total (μg/mL)		361	28	126
Half-life (min)	18.4–22.4	21.13	20.04	21.16
Endotoxin (EU[b]/mL)	<25.0	<2.00	<2.00	<2.00
Radiochemical identity (RRT[c])	0.9–1.10	1.00	0.99	0.99
Activity (mCi)		49	117	95
Sterile filter integrity test (psi)	>45	Pass	Pass	Pass
Sterility testing	No colony growth after 14 d	Pass	Pass	Pass

[a]USP quality control guidelines not specified.
[b]EU = endotoxin units.
[c]RRT = relative retention time.

3.2 Dose pH

The pH for [^{11}C]palmitic acid was determined by applying a small volume of the dose to a colorpHast® pH 2.0–9.0 nonbleeding pH-indicator strip [33] and determined by visual comparison to the color scale provided by the manufacturer. Acceptable doses were between pH 4.5–7.5.

3.3 Radiochemical Purity and Radiochemical Identity

Radiochemical purity and identity were determined by using a Shimadzu LC-2010A$_{HT}$ Liquid Chromatograph [34] equipped with a UV detector and a Bioscan FC3300 γ-detector [35] (column: Phenomenex Luna C8(2) [36]; mobile phase: 80% MeCN/0.1% acetic acid [37]; flow rate = 1.0 mL/min). Radiochemical purity was required to be >95%. Palmitic acid, 99% [38], was used as the nonradioactive carbon-12 reference standard, and representative chromatograms are presented in Fig. 3.

Radiochemical identity was confirmed and quantified by calculating the relative retention time (RRT = [retention time of [^{11}C]palmitic acid]/[retention time of [^{12}C]palmitic acid]), and was required to be 0.9–1.10 for dose approval. Average retention times for [^{11}C]palmitic acid and [^{12}C]palmitic acid were 5.750 and 5.783 min, respectively, and the activity peak was observed at an average time of 5.808 min.

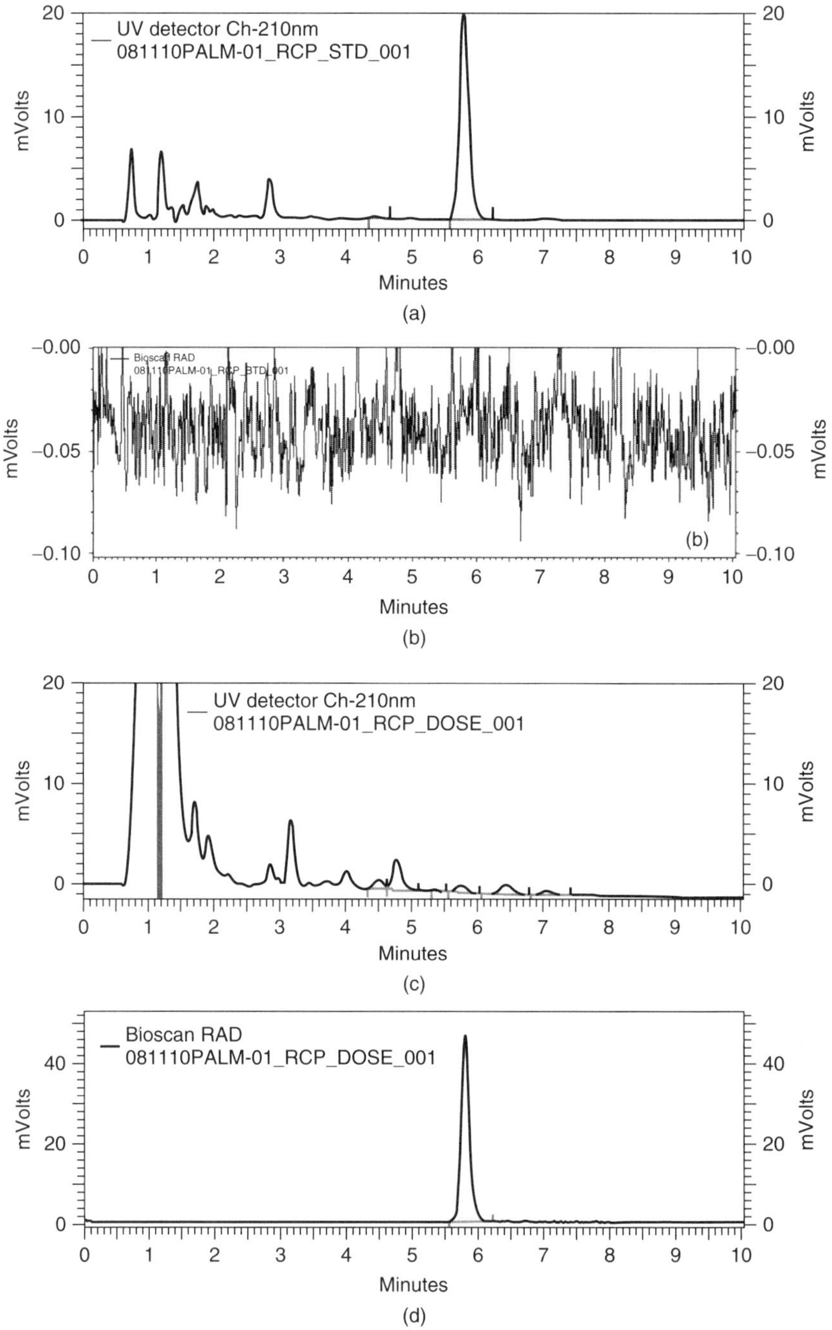

FIGURE 3 HPLC results for the UV and radioactivity signals for the [^{12}C]palmitic acid standard (a and b, respectively) and [^{11}C]palmitic acid dose (c and d, respectively).

3.4 Radionuclide Purity

Radionuclide purity was analyzed by γ-ray spectrometry, and doses were allowed to decay, and then sent out to a contract laboratory [39] to test for the presence of long-lived radioactive contaminants.

3.5 Radionuclide Identity

The radionuclide identity was determined by measuring the half-life of [^{11}C]palmitic acid and comparing it to the established and accepted half-life of carbon-11. Radionuclide activities were determined by a Capintec CRC®-15 Radioisotope Dose Calibrator [40], and half-life ($t_{1/2}$) was calculated according to Equation 1.

$$t_{1/2} = -\ln 2 \left[\frac{\Delta \text{time}}{\ln \frac{(\text{ending activity})}{(\text{initial activity})}} \right] \tag{1}$$

3.6 Bacterial Endotoxins

Endotoxin content in doses was determined using a Charles River Laboratories EndoSafe® Portable Testing System [41] and in accordance with the USP [31]. Maximum recommended total dose was required to contain <175 endotoxin units (EU).

3.7 Sterility

Culture tubes of fluid thioglycolate media (FTM) and tryptic soy broth (TSB) were inoculated with samples of the [^{11}C]palmitic acid dose and incubated for 14 days. Negative controls were included in the analysis. FTM cultures allowed for the determination of anaerobes, aerobes, and microaerophiles. Conversely, soybean casein digest media (SCDM) cultures determine the presence or absence of nonfastidious and fastidious microorganisms [42]. Cultures were visually inspected on the third, eighth, and fourteenth days of the incubation period and compared to the positive and negative controls. Positive cultures must show growth while the [^{11}C]palmitic acid samples and negative controls should be clear of any growth at the conclusion of the incubation period to verify sterility. All incubations were in accordance with current USP standards for sterility [32].

CHEMICAL ABSTRACTS NOMENCLATURE (REGISTRY NUMBER)

Acetic acid (64-19-7)
Acetonitrile (75-05-8)
Ethanol (64-17-5)
Ethyl ether (60-29-7)
Hydrochloric acid (7647-01-0)
Pentadecylmagnesium bromide (78887-70-4)

REFERENCES AND NOTES

1. Weiss ES, Ahmed SA, Welch MJ, Williamson JR, Ter-Pogossian MM, Sobel BE. Circulation 1977;55:66–73.
2. Klein MS, Goldstein RA, Welch MJ, Sobel BE. Am J Physiol Heart Circ Physiol 1979;237: H51–H58.
3. Goldstein RA, Klein MS, Welch MJ, Sobel BE. J Nucl Med 1980;21:342–348.
4. Neely JR, Morgan HE. Annu Rev Physiol 1974;36:413–459.
5. Kong Y, Friedberg SJ. Metabolism 1971;20:681–689.
6. Weiss ES, Hoffman EJ, Phelps ME, Welch MJ, Henry PD, Ter-Pogossian MM, Sobel BE. Circ Res 1976;39:24–32.
7. Machulla HJ, Stocklin G, Kupfernagel C, Freundlieb C, Hock A, Vyska K, Feinendegen LE. J Nucl Med 1978;19:298–302.
8. Lerch RA, Ambos HD, Bergmann SR, Welch MJ, Ter-Pogossian MM, Sobel BE. Circulation 1981;64:689–699.
9. Welch MJ, Dence CS, Marshall DR, Kilbourn MR. Remote system for production of carbon-11 labeled palmitic acid. J Labelled Comp Radiopharm 1983;20:1087–1095.
10. Padgett HC, Robinson GD, Barrio JR. [1-(11)C]palmitic acid: improved radiopharmaceutical preparation. Int J Appl Radiat Isot 1982;33:1471–1472.
11. Jewett DM, Ehrenkaufer RL, Ram S. A captive solvent method for rapid radiosynthesis: application to the synthesis of [1-(11)C]palmitic acid. Int J Appl Radiat Isot 1985;36:672–674.
12. Runkle AC, Shao X, Tluczek LJM, Henderson BD, Hockley BG, Scott PJH. Appl Radiat Isot. 2011;69:691–698.
13. Shao X, Kilbourn MR. Appl Radiat Isot 2009;67:602–605.
14. Iwata R, Ido T, Tada M. Appl Radiat Isot 1995;46:117–121.
15. The Tracerlab $FX_{C\text{-}Pro}$ and PETTrace cyclotron are products of GE Healthcare, USA.
16. The radiation detector for the MeI Trap was used to reference the amount of activity received by the GEMS Tracerlab $FX_{C\text{-}Pro}$.
17. Molecular sieve (part no. 5123116) purchased from Alltech, Inc., State College, PA, USA.
18. Ten-milliliter sterile vials were purchased from Hollister-Stier (part no. 7515ZA) and vented with a sterile Millex-FG vent filter (part no. SLFG025LS).
19. Pentadecylmagnesium bromide 0.5 M in ether was purchased from Sigma-Aldrich (part no. 68196DJ).
20. Standard operating procedure (SOP) for the preparation of a sterile dose vial: clean stainless steel ventilation hood with 70% EtOH, clean septum of dose vial with 70% EtOH before placing in rabbit, and perform sterile technique when opening syringes and filters and insert syringes into the vial septum.
21. Branson 1501R-MTH (part no. CPN-952-117) was purchased from Branson Ultrasonics Corp., CT, USA.
22. Millex-GS sterile 0.22-μm filters were purchased from Millipore (part no. SLGSV255F) and used as received.
23. Each valve was electronically controlled by a dedicated switch located on the exterior of the hot cell, which was activated and deactivated by the production chemist. VC1, VC2, and VC4 were four-way valves, and VC3 was an eight-way valve.
24. Precipitation was previously reported by Runkle et al. [12] in the production of [^{11}C]acetic acid, and because of the greater length of the alkylating group of pentadecylmagnesium bromide than methylmagnesium bromide and the similarity between reaction mechanisms, to preemptively counteract precipitate formation in the synthesis of [^{11}C]palmitic acid, precharging the MSC2 was determined to be necessary.
25. Ethyl ether, glacial acetic acid, and 6 N HCl were purchased from Fischer Scientific, Inc. A 3% acetic acid in ether solution was the stock solution for subsequent dilutions.
26. Light Alumina-N Al Sep-Pak cartridges were purchased from Waters Corp. (part no. WAT023561) and conditioned with 10 mL of 0.15% CH_3CO_2H/Et_2O before use.

27. Sterile water for injection (part no. 0409-4887-50), NaCl, 0.9% USP (part no. 0409-4888-50), and sodium phosphates, Injection, USP (part no. 0409-7391-72) were purchased from Hospira and used as received.
28. Human serum albumin was obtained from the University of Michigan health stand and diluted to the final concentration of 3.5%. It was purchased from Baxter Healthcare Corp., West Lake Village, CA, USA.
29. U. S. Pharmacopeia, <823>. Radiopharmaceuticals for positron emission tomography-compounding. USP 32–NF 27, 2009.
30. U. S. Pharmacopeia. Sodium acetate C 11 injection. USP 32–NF 27, 2009.
31. U. S. Pharmacopeia, <85>. Bacterial endotoxins test. USP 32–NF 27, 2009.
32. U. S. Pharmacopeia <71>. Sterility Tests. USP 32–NF 27, 2009.
33. EMD Chemicals, Inc., USA (part no. 9578-3).
34. Shimadzu Corporation, USA.
35. Bioscan, Inc., USA.
36. Analytical HPLC column: Luna C8(2) 5 μm, 100 × 4.6 mm, purchased from Phenomenex (part no. 00D-4249-E0); flow rate = 1.0mL/min; λ = 210 nm; analytical column was equilibrated with the analytical mobile phase for 30–60 min before use.
37. Analytical mobile phase: 80% MeCN in 0.1% HOAc.
38. Palmitic acid, >99% reference standard (part no. 258725) purchased from Sigma-Aldrich.
39. Dade Moeller & Associates, Gaithersburg, MD, USA.
40. Capintec, Inc., USA.
41. Charles River Laboratories, USA.
42. Becton, Dickinson and Company, USA.

CHAPTER 24

SYNTHESIS OF O-methyl-^{11}C]METOMIDATE ([^{11}C]-MTO)

István Boros and Franklin I. Aigbirhio

Wolfson Brain Imaging Centre, Department of Clinical Neurosciences, University of Cambridge, Addenbrooke's Hospital, Cambridge, UK

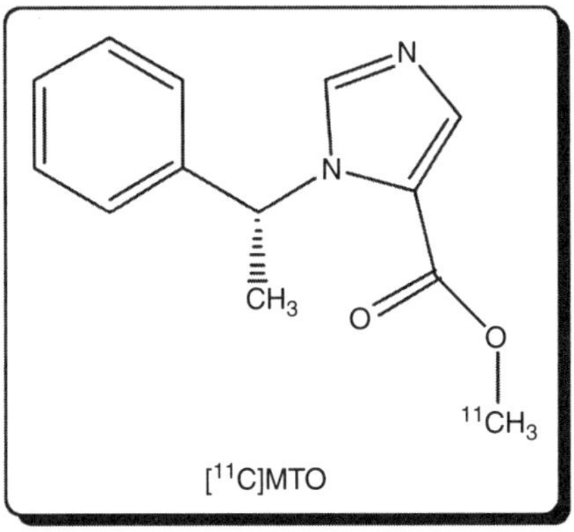

1 INTRODUCTION

Nonradiolabeled metomidate is also known as Hypnodil and was marketed by Janssen-Cilag AG as an animal anesthetic with hypnotic effect. It was widely used as an injection hypnotic in small animals. Its nearest chemical analog etomidate ((*R*)-methyl 1-(1-phenylethyl)-1*H*-imidazole-5-carboxylic acid ethyl ester) is used as an effective hypnotic in humans and marketed by Janssen-Cilag as Hypnomidate [1].

Its primary pharmacological action is inhibition of 11β-hydroxylase, a key enzyme in the biosynthesis of cortisol and aldosterone in the adrenocortex. This enzyme can be overexpressed in adrenocortical carcinomas, which secrete excessive amounts of catecholamines and cortical hormone.

[^{11}C]Metomidate was introduced as an adrenocortical imaging tracer by Uppsala University positron emission tomography (PET) center [2] and has proved to

Radiochemical Syntheses, Volume 1: Radiopharmaceuticals for Positron Emission Tomography, First Edition. Edited by Peter J. H. Scott, Brian G. Hockley.

be useful in distinguishing adrenocortical from noncortical lesions and in identifying incidentalomas of adrenocortical origin [3,4].

Since its introduction in late nineties, [^{11}C]MTO was widely used in human studies by several groups in Europe and the United States for evaluation of adrenal masses [5–7].

2 SYNTHESIS PROCEDURE

CAUTION: All radiochemical syntheses must be carried out using appropriate equipment in a facility authorized for the use of radioactive materials. Personal protective equipment must be worn and all local radiation safety laws followed.

An optimized production method of [^{11}C]MTO (Fig. 1) is based on a standard methylation of the carboxylic acid precursor with [^{11}C]iodomethane in dimethylformamide in the presence of a base catalyst (tetrabutylammonium hydroxide, TBA) [8].

A fast and reliable method for the production of carbon-11 radiopharmaceuticals on the GE TRACERlab FX$_C$ module has been developed, which allows it to perform [^{11}C]methylation reactions inside the built-in injector loop of the preparative high performance liquid chromatography (HPLC) system [9]. Purification of the crude reaction mixture is achieved in aqueous ethanol solution modified with physiologic phosphate buffer. The product fraction is collected from HPLC column directly in the final formulation solution (physiologic saline) and is subjected to the final sterilizing filtration by means of a 0.22-μm sterile filter.

2.1 Presetup for Radiochemistry

2.1.1 Cyclotron

The [^{11}C]CO_2 is produced by the GE PETtrace™ cyclotron via the $^{14}N(p,\alpha)^{11}C$ nuclear reaction; used gases are the ion source gas, hydrogen, helium, and oxygen/nitrogen mixture gas. Production of [^{11}C]CO_2 is scheduled so that the end of bombardment (EOB) occurs after setup of the [^{11}C]MeI and microreactor systems. The standard parameters for production are 40 μA for 40 min. This will produce approximately 2.3 Ci (85 GBq) of [^{11}C]carbon dioxide. The production parameters could be modified on the basis of minimum activity required for the study.

2.1.2 Preparative HPLC System

The TRACERlab FX$_C$ module has an integrated HPLC system for the purification of the radiotracer. This consists of an HPLC pump, a 1-mL injection valve,

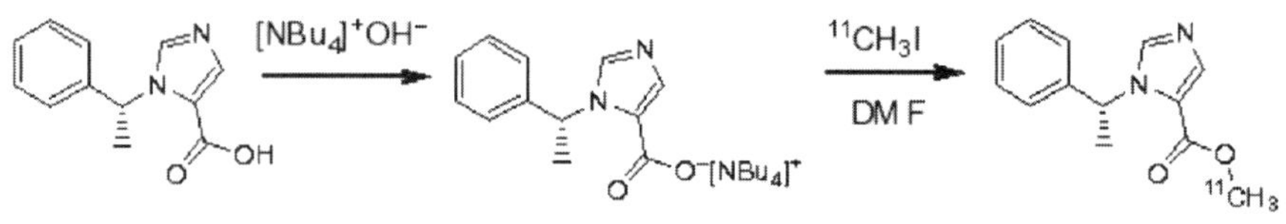

FIGURE 1 Radiosynthesis of [^{11}C]MTO.

a γ-ray radiation detector, and a UV detector to be set at 254 nm for MTO production. The flow rate of the HPLC pump using a mobile phase of 0.05 M sodium phosphate buffer, 24.5% ethanol in water solution is to be set at 1 mL/min flow rate. The preparative HPLC column used is a Phenomenex Synergi 4 μm MAX-RP 80A 50 × 3 mm column. The system is switched on to equilibrate automatically at the start of synthesis.

The product fraction after preparative HPLC is collected directly into the product collection vial. Therefore, aseptic conditions of the mobile phase preparation are of a major concern. The stock solution of preparative mobile phase (phosphate–ethanol buffer solution) is constituted for one week using only sterile ingredients. The mobile phase is prepared by dissolving 3.9 g sodium dihydrogen phosphate dihydrate and 3.6 g disodium hydrogen phosphate in 650 mL of sterile water and mixing the obtained solution with 350 mL of 70% sterile ethanol. The solution is degassed in a clean and depyrogenized 1-L glass bottle by applying ultrasound and vacuum for at least 15 min. On the day of MTO preparation, 250 mL of the preparative mobile phase is filtered through a 250-mL CORNING® sterilizing filter unit (0.22-μm, sterile, polyethersulfone (PES) membrane single-use filter).

2.1.3 GE MeI Microlab MeI System

[^{11}C]MeI is produced by the dry chemistry method in a GE MeI Microlab system. Production of [^{11}C]MeI is achieved in the gas phase by free radical iodination with recirculated iodine vapor at elevated temperatures (700°C). Before starting each synthesis, a leak test is performed by applying helium pressure into the system. The preparation for each run also includes a conditioning sequence that is run automatically from the MeI module control box during or before the cyclotron run. The run sequence in the Microlab MeI system must be set to optimal transfer values between the MeI Microlab module and the FX-C synthesizer.

For productions for human use, the iodine column is normally changed after 5 or 6 runs; for research or testing applications, up to 10 syntheses are acceptable.

2.2 Setup of the GE TRACERlab FX$_C$ System for Radiochemistry

The standard configuration of the module as supplied by the manufacturer (GE) was modified by installing additional valves, feed-back digital input, and rearranging capillary line interconnections. The flow path of the [^{11}C]iodomethane is rerouted to enable "captive solvent" [^{11}C]methylations followed by direct injection onto the HPLC preparative column [9].

The TRACERlab FX$_C$ system is first set up to the initial settings by loading the appropriate synthesis sequence, after which the software will present a list of preparation steps to be performed and checked before it will proceed to the next synthesis phase. Each step should be confirmed manually in the reference field of the production preparation window. Example of the screen is shown in Fig. 2. Before emptying the [^{11}C]CO_2 target, the precursor solution (consisting

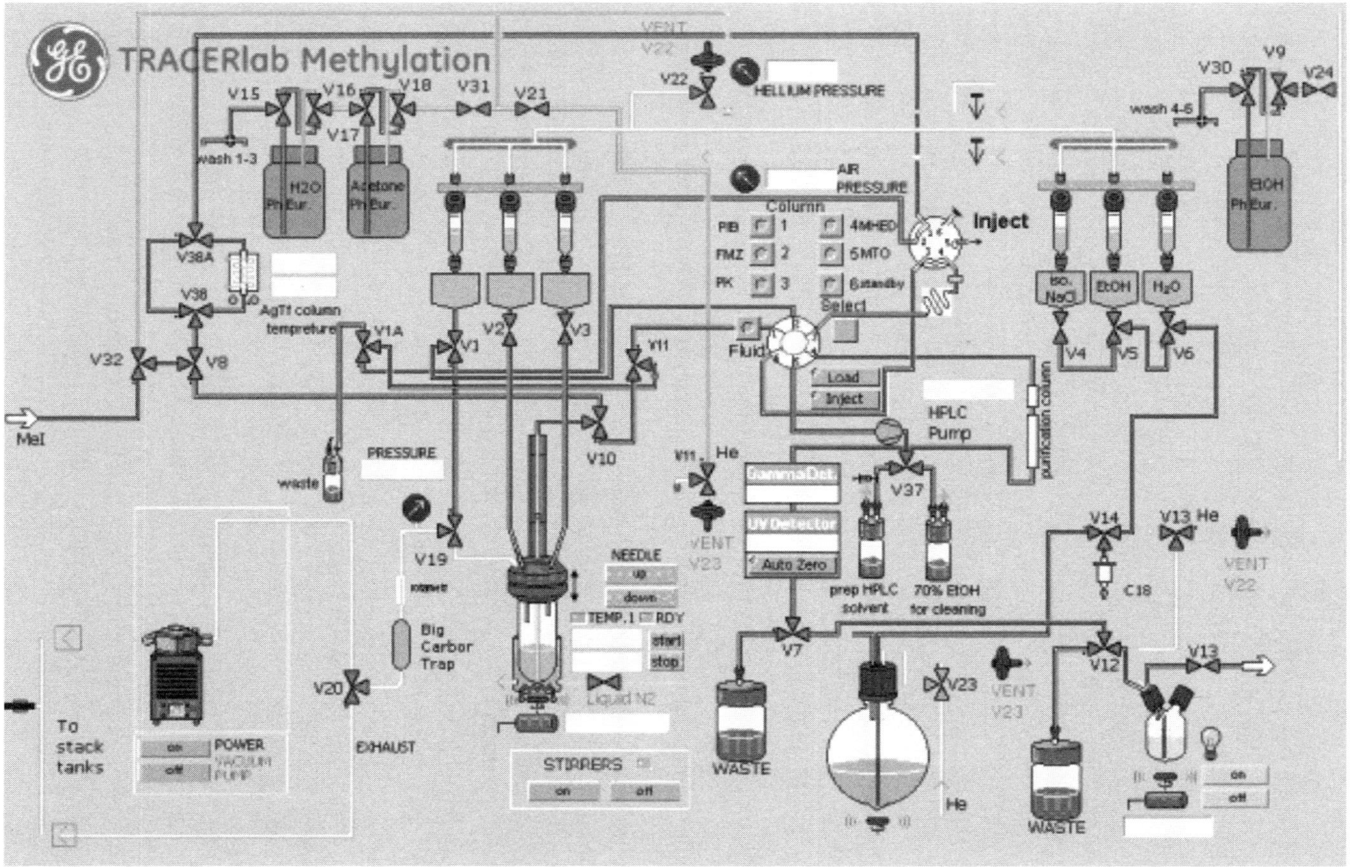

FIGURE 2 Example of a screen-shot “LOOP-methylation.bmp”.

of 6 μL of TBA 55–60% solution in water, added to 3 mg of precursor dissolved in 100 μL of anhydrous *N,N*-dimethyl formamide (DMF) mixture) is loaded in the reaction loop.

When all setup procedures have been checked, the program will place the TRACERlab FX_C system in a standby mode waiting for the synthesis.

2.3 Radiochemistry Synthesis

When the system is ready for radiosynthesis, the target is emptied and the $[^{11}C]$MeI production and consequent MTO production is initiated. The production and delivery of $[^{11}C]$MeI is completed in about 15 min. The flow of the prepared $[^{11}C]$MeI is directed through the injection loop loaded with the precursor solution. When the release of $[^{11}C]$MeI is finished, the injection loop system is closed for 1.5 min to complete the methylation reaction. Then the HPLC system is automatically switched to the inject mode, and the contents of the loop are transferred to the HPLC column for purification.

In the HPLC traces of the TracerLabFX_C software, the $[^{11}C]$MTO fraction is collected manually by pressing the "Start Peak" and the "End Peak" tabs on the top of the HPLC trace display (alternatively, the HPLC outlet can be manually deviated to the product vial). $[^{11}C]$MTO is eluted at about 11–12 min.

After the collection of the $[^{11}C]$MTO fraction in the product collect vial, the program will automatically start product transfer to the product vial through a final sterilizing filter. The transfer is usually complete after 5–6 min.

2.4 Purification and Formulation of $[^{11}C]$MTO

A typical preparative HPLC chromatogram for the purification of $[^{11}C]$MTO is shown in Fig. 3. The chromatogram on top represents the radioactive profile, whereas the bottom one represents the UV profile.

The final reaction product consists of [*O-methyl*-^{11}C]metomidate in a mixture of 10 mL of saline and ~2–4 mL of 15% ethanol (the final solution containing <10 vol% ethanol).

2.5 Aseptic Cleaning of TRACERLab $FX_{F\text{-}C}$ Unit System

After each synthesis, the production system is cleaned with acetone, water, and 70% ethanol using an automated cleaning sequence preprogrammed in the TRACERlab software [9]. After the cleaning sequence is finished, normally in <30 min, the module is clean and dry and ready for the next run. The preparative HPLC column is also rinsed with 70% ethanol solution by this sequence.

3 QUALITY CONTROL PROCEDURES

CAUTION: All radiopharmaceuticals produced for clinical use must have local regulatory approval before use in humans. Trained personnel must carry out quality control (QC) procedures, and each dose must meet all established QC criteria before release to the clinic.

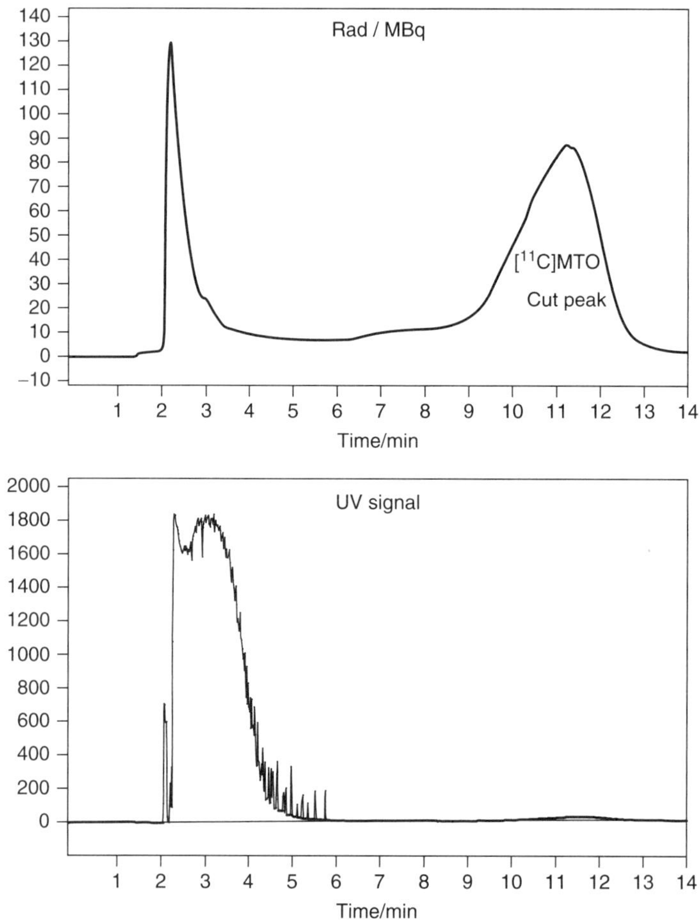

FIGURE 3 Purification of [^{11}C]MTO using a semipreparative HPLC.

Owing to the short half-life of carbon-11 (20.4 min) and high levels of radioactivity, the QC method is divided into prerelease tests and postrelease tests (Table 1). Prerelease tests have to be performed before release, and all tests have to be passed for release of the production batch to be allowed. These postrelease tests are performed after the release of the product; however, if any of these tests fails, the production of [^{11}C]MTO has to stop immediately and the reason for the failure has to be investigated. The production procedure has to be revalidated before production of [^{11}C]MTO for human use is allowed again.

3.1 Prerelease Tests

3.1.1 Visual Inspection

A visual check of the product is performed behind a lead window; the product must be clear, colorless, and free of particulate matter.

TABLE 1 QC Data for Three Repeat Runs of [^{11}C]MTO

QC Test	Release Criteria	Run 1	Run 2	Run 3
Yield mCi (GBq)	≥13.5 mCi (0.5 GBq) in 10 mL	44 (1.64)	31 (1.16)	22 (0.83)
Visual inspection	Clear colorless	Pass	Pass	Pass
Radiochemical identity	$RT_{[^{11}C]MTO} = (RT_{STD} + 15'') \pm 10''$	5	5	6
Radiochemical purity (%)	≥95	99.93	100	99.8
Chemical amount (μg/dose)	nor-MTO ≤10	ND	4.0	9.8
	MTO ≤10	3.5	2.1	0.8
Specific activity	NA	55.2	109	181.9
Residual solvent analysis	DMF ≤0.88 mg/mL	ND	ND	ND
	Ethanol ≤10%	7.1	4.3	
Residual TBA	TBA ≤0.27 mg/mL	nd	nd	nd
Dose pH	5–8	6.5	7	7
Sterile filter integrity test (bar)	≥3.4	4.3	4.2	4.4
Radionuclidic identity ($t_{1/2}$) (min)	20.4 ± 0.5	nd	nd	nd
Endotoxin analysis (EU/mL)	LAL <17.5	2.5	2.5	2.5
Sterility testing	Sterile	Sterile	Sterile	Sterile

NA, not applicable; ND, not detected; nd, not determined; RT, retention time.

3.1.2 pH Measurement

The pH of the product is determined using pH indicator sticks and should be within the range 4.5–8.5 (ex. Ph. Eur. monograph 1325).

Note: The sensitivity of pH paper does not allow measuring the pH precisely; therefore, the reduced range of pH (5–8) is used in the product specifications.

3.1.3 Radiochemical Purity

The radiochemical purity should be ≥95% and is determined using the following HPLC method:

Column: Phenomenex Gemini 5 μm C18 110A 250 × 4.6 mm
Injection loop: 20 μL
Injection volume: 25 μL
Eluent: ethanol : water : triflouracetic acid (30:69.95:0.05 v/v)
Pump settings: Flow rate: 1 mL/min
Radioactivity detector: EG&G Ortec Amplifier: 0.70 kV
UV detector: Gilson 118 UV/vis: λ = 254 nm; absorbance units full scale (AUFS) = 0.001

This system should be set up at least 30 min before analysis to allow the baselines to equilibrate.

3.1.4 *Radiochemical Identity*

The radiochemical identity test is performed by coeluting [^{11}C]MTO and the reference compound on HPLC system used for analysis (see above). The chromatograms are to be filed with the production documentation. [^{11}C]MTO must have the same elution properties (retention time) as the reference compound.

Reference standards of *R*-nor-Metomidate (precursor) and *R*-Metomidate (product) are purchased from ABX. These compounds have good solubility in water and other solvents.

3.1.5 *Chemical Purity (Chemical Amount of MTO and nor-MTO)*

The chemical amount of MTO (product) and nor-MTO (precursor) is determined by the corresponding HPLC method (see above). The acceptance limit of 10 μg of *R*-MTO and *R*-nor-MTO in the injection dose has been established on the basis of toxicological data available.

3.1.6 *Maximum Allowable Injection Volume*

The maximum allowable injection volume, expressed in milliliters, is estimated based on the previously calculated chemical amounts of MTO and nor-MTO relative to their specifications.

Formula to calculate the maximum allowed injection volume:

Maximum allowable injection volume = Specified limit of critical component (μg)/concentration of the critical component found in a QC sample (μg/mL).

By default, the maximum allowable injection volume of Wolfson Brain Imaging Centre (WBIC) products is limited to 10 mL; therefore, the specified limit of a critical component in product specifications is normally expressed in micrograms per 10 mL, where 10 mL represents the maximum injected dose, *V*.

3.1.7 *Specific Activity*

The specific activity is calculated using a validated QC Excel sheet.

3.2 Postrelease Tests

3.2.1 *Filter Integrity Test*

The integrity of the filter is tested using the bubble point test; to pass the test, the bubble point has to be $\geq$3.4 bar (50 psi).

3.2.2 *Sterility Tests*

Sterility test is performed by an external contractor upon decay of the radioactivity in the product. The sterility is assessed by membrane filtration method (Ph. Eur. method 2.6.1, in European Pharmacopoeia, 7th Edition, 2010, published by the European Pharmacopoeia Commission).

3.2.3 *Bacterial Endotoxins (LAL Tests)*

Bacterial endotoxins are tested using the Endosafe limulus amebocyte lysate (LAL) test system by the portable test system (PTS) kinetic reader. The LAL test is performed immediately after release of the product, and for MTO, a limit of <17.5 EU/mL has been established.

3.2.4 Analysis of Formulation and Residual Solvents

The presence of residual solvents is analyzed using the following gas chromatographic (GC) method:

PerkinElmer AutoSystem XL Gas Chromatograph
Column: Alltech Gas Chrom 254 80/100, 6′ × 1/8″ × 0.85″
Detector: flame ionization detector (FID)
Gas flows:
Helium (carrier gas): ~25 mL/min
Hydrogen (FID): ~45 mL/min
Air (FID): ~350 mL/min

This is a spot-check test conducted on every first production batch of a given month. The specified limits are

Residual solvent DMF $\leq$0.88 mg/mL
Formulation solvent ethanol $\leq$10%

3.2.5 Analysis of Tetrabutylammonium Hydroxide

For TBA analysis present in the injection dose, a thin layer chromatography (TLC) method is used: strips of Alugram SIL G chromatography plate (Machery-Nagel, cat. 818161) are placed for 20 min in a 15-mL polypropylene (Corning, cat. 430791) falcon tube containing 10 mg of iodine leaflets (Sigma, cat. 37,655-8), with the cap securely fitted. A 2.5-μL droplet of the QC sample and 2.5-μL droplet of standard solution at the limit concentration (2.5 mg/dose) are added to the strip, which is then placed back in the falcon tube containing the iodine. After 3 min, the developed spots are compared for intensity.

The specification limit of TBA in samples is 0.275 mg/mL.

CHEMICAL ABSTRACTS NOMENCLATURE (REGISTRY NUMBER)

Acetonitrile [75-05-8]
Disodium hydrogen phosphate [7558-79-4]
DMF anhydrous [68-12-2]
Ethanol [64-17-5]
1*H*-Imidazole-5-carboxylic acid, 1-[(1*R*)-1-phenylethyl]- [56649-48-0]
1*H*-Imidazole-5-carboxylic acid, 1-[(1*R*)-1-phenylethyl]-, methyl ester[5377-20-8]
Sodium dihydrogen phosphate dihydrate [13472-35-0]
Tetrabutylammonium hydroxide [2052-49-5]

For detailed supplier information, see appendix (below).

APPENDIX Reagents and Consumables for Production of [^{11}C]MTO

Item	Specification	Supplier/Catalogue No.	Storage
(*R*)-methyl 1-(1-phenylethyl)-1*H*-imidazole-5-carboxylic acid (precursor)	>95%	3 mg per vial ABX/1780.0003	Desiccated at 2–8°C, protected from light
DMF anhydrous	>99.5%	Sigma-Aldrich 40228-250 ML	Store at RT over dried molecular sieves.
Tetrabutylammonium hydroxide solution	55–60% aqueous solution	Fluka 86863-100 ML	Store at RT
Water for injections	For i.v. injection	Severn Biotech 20-110-10	Store at RT
Disodium hydrogen phosphate	Ph. Eur.	Sigma-Aldrich56814-1 KG-F	Keep tightly closed at room temperature
Sodium dihydrogen phosphate dihydrate	Meets analytical specification of Ph. Eur., BP, USP, FCC, E 339	Sigma-Aldrich 04269-1 KG	Keep tightly closed at room temperature
Ethanol 70%, sterile solution, Ph. Helv.	Ph. Helv.	FLUKA/1L 02877	Store at RT
Prep HPLC column: Synergi 4 μm MAX-RP 80A 50 × 3 mm		Phenomenex 00B-4337-Y0	Store at RT with 70% EtOH
NaCl solution for injection, BP, 0.9% w/w	Sterile	Fannin, Noridem Enterprises PL 24598/0002	Store at RT
Acetone Ph. Eur.	Puriss.P.A., Ph. Eur.	Fluka, Riedel de Haën 00570–32201	Store at RT
11 mL TechneVial or 10 mL nitrogen filled	Sterile Sterile	Mallinckrodt Amersham	Store at RT Store at RT
Millex GV	Sterile, 0.22-μm MF-Durapore membrane	Millipore SLGV013SL	Store at RT
Mobile phase filter	Sterile, 0.22 μm, PES membrane	Corning 431097	Store at RT

REFERENCES AND NOTES

1. Janssen-Cilag Hypnomidate™ 2 mg/ml Injection. Etomidate. Etomidate Patient Leaflet.
2. Bergstrom M, Bonasera TA, et al. J Nucl Med 1998;39(6):982–989.
3. Khan TS, Sundin A, et al. Eur J Nucl Med Mol Imaging 2003;30(3):403–410.
4. Zettinig G, Mitterhauser M, et al. Eur J Nucl Med Mol Imaging 2004;31(9):1224–1230.

5. Hennings J, Lindhe O, et al. J Clin Endocrinol Metab 2006;91(4):1410–1414.
6. Minn H, Salonen A, et al. J Nucl Med 2004;45(6):972–979.
7. Mitterhauser M, Dobrozemsky G, et al. Eur J Nucl Med Mol Imaging 2006;33(8):974.
8. Mitterhauser M, Wadsak W, et al. Appl Radiat Isot 2003;59(2–3):125–128.
9. Kuznetsov M, Soloviev D, et al. J Labelled Comp Radiopharm 2009;52(S1):S313.

CHAPTER 25

SYNTHESIS OF [^{11}C]CARFENTANIL ([^{11}C]CFN)

LOUIS TLUCZEK AND XIA SHAO

Division of Nuclear Medicine, Department of Radiology, University of Michigan, Ann Arbor, Michigan

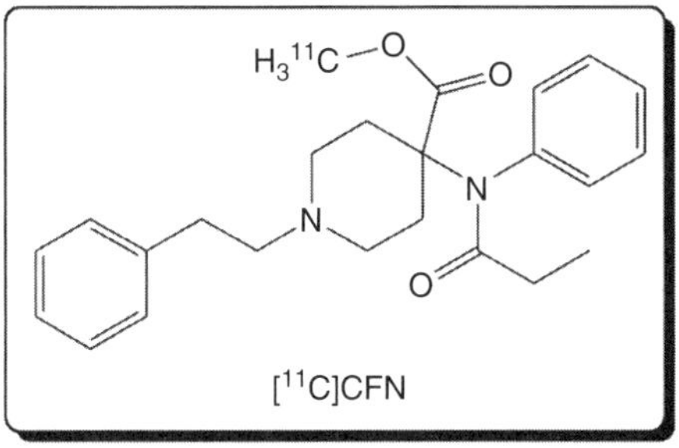

1 INTRODUCTION

The μ-opioid receptors are a class of receptors that are important in mediating pain and feelings of well-being. They lie at the core of the unique neurological mechanisms that make us human. During the mid-1960s, binding studies using brain homogenates made it apparent that opioid agents likely act through interaction with specific receptors, the first to be identified were the μ-opioid receptors [1,2]. These belong to a group of G-protein-coupled pre- and postsynaptic receptors with opioids as ligands.

Carfentanil, one of the most potent opioids known, has been shown to bind very specifically with μ-opioid receptors as a full agonist.

Carbon-11-labeled carfentanil has found widespread use as a positron emission tomographic (PET) tracer to measure and map μ-opioid receptors in the in vivo brain. Many investigations have been performed using [^{11}C]carfentanil, which relates in vivo μ-opioid distribution to many normal and altered neurophysiological and psychic phenomena [3,4].

Radiochemical Syntheses, Volume 1: Radiopharmaceuticals for Positron Emission Tomography, First Edition. Edited by Peter J. H. Scott, Brian G. Hockley.

2 SYNTHESIS PROCEDURES

CAUTION: All radiochemical synthesis must be carried out using appropriate equipment in a facility authorized for the use of radioactive materials. Personal protective equipment must be worn and all local radiation safety laws followed.

2.1 Production of [^{11}C]CO_2

[^{11}C]CO_2 was produced by General Electric Medical Systems (GEMS) PETTrace cyclotron. The target [5] was loaded with [^{14}N]N_2 gas [6] and bombarded with a proton beam (40 μA beam for 30 min) to generate ~3 Ci (111 GBq) of [^{11}C]CO_2 by the $^{14}N(p,\alpha)^{11}C$ nuclear reaction. Ultra high purity nitrogen gas (99.9999%) plus 1% of high purity oxygen gas (99.9999%) was used to achieve high specific activity.

2.2 Production of [^{11}C]Carfentanil

[^{11}C]CO_2 was delivered from the target by nitrogen gas pressure to a column filled with a mixture of molecular sieve (0.3 g) [7] and Shimalite-Ni catalyst (0.2 g) [8] at room temperature in a GE TRACERlab $FX_{C\text{-}Pro}$ synthesis module. The column was sealed under H_2 and then heated to 350°C for 20 s to reduce the [^{11}C]CO_2 to [^{11}C]CH_4. [^{11}C]CH_4 was passed through a phosphorus pentoxide desiccant [9] column and trapped in a carbosphere [10] column at −78°C. Gaseous [^{11}C]CH_4 was passed into a circulation loop system including a gas pump, an iodine [11] column at 100°C, a reactor tube at 720°C, two adjacent Ascarite II [12] columns, and a Porapak [13] column at room temperature. The gaseous mixture was then circulated for 5 min, while [^{11}C]CH_3I accumulated on the Porapak column.

[^{11}C]CH_3I was converted to [^{11}C]CH_3OTf by passing it through a silver triflate-Graphpac [14] column heated to 190°C. This was conveyed to and bubbled, at 15 mL/min flow rate, into a reaction vessel containing 0.4 mg desmethyl carfentanil tetrabutyl ammonium (TBA) salt [15] dissolved in 100 μL dimethyl sulfoxide, whereby [^{11}C]carfentanil was produced at room temperature (Fig. 1).

The modified configurations of the GE TRACERlab $FX_{C\text{-}Pro}$ system are shown in Fig. 2.

2.3 Purification and Formulation of [^{11}C]Carfentanil

After production, 1 mL of 1% ammonium hydroxide was added to the reaction vessel. This was conveyed via helium gas to a dilution vessel containing 5 mL

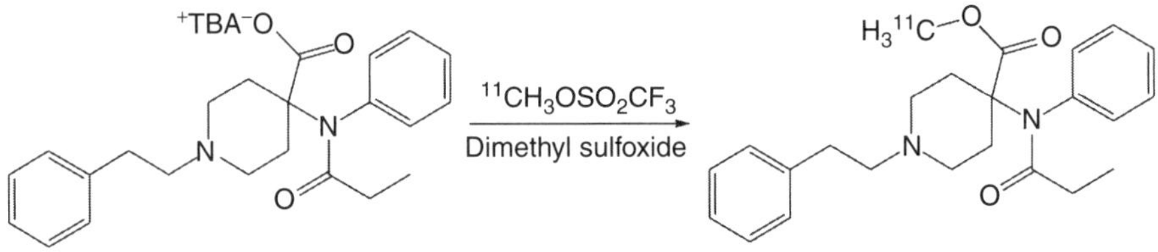

FIGURE 1 Synthesis of [^{11}C]carfentanil.

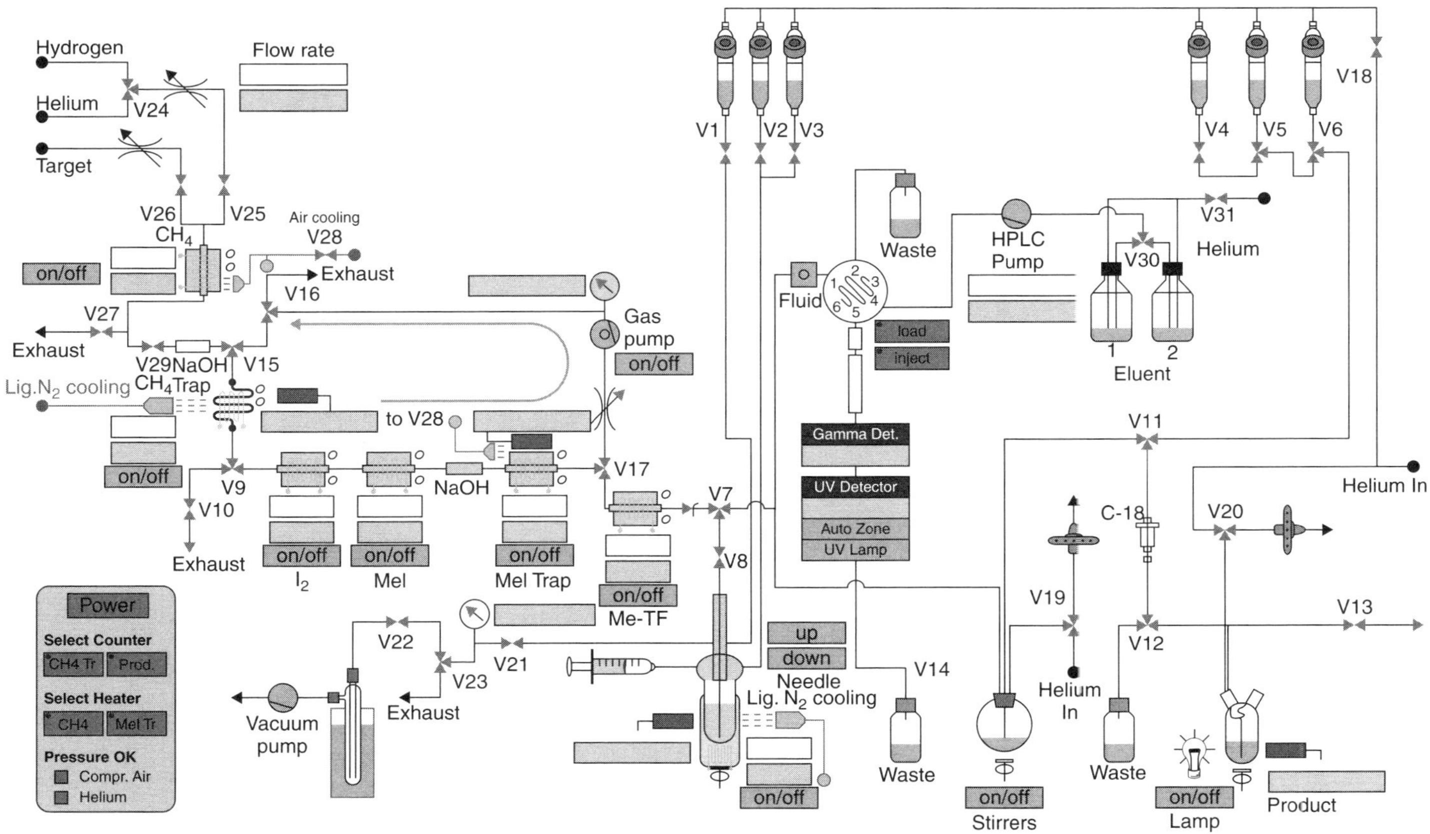

FIGURE 2 Modified TRACERlab $FX_{C\text{-}Pro}$ for preparation of [^{11}C]carfentanil.

of 1% ammonium hydroxide. This mixture was passed through a 3M Empore C2 extraction disk [16] where the [^{11}C]carfentanil was trapped. Three milliliters of 10% *n*-propanol followed by 10 mL Millique water was then passed through to remove impurities from the disk.

The disk was then dried by passing helium gas through for 1.0 min, and the [^{11}C]carfentanil was eluted off using 0.5 mL ethanol followed by 9.5 mL water. This was passed through a Millipore Millex-GV filter [17] into a sterile dose vial to produce the final formulated [^{11}C]carfentanil product.

3 QUALITY CONTROL PROCEDURES

CAUTION: All radiopharmaceuticals produced for clinical use must have local regulatory approval before use in humans. Trained personnel must carry out quality control (QC) procedures, and each dose must meet all established QC criteria before release to the clinic.

QC procedures for [^{11}C]carfentanil, based on the current requirements for radiopharmaceuticals defined in the United States Pharmacopeia (USP) [18], are summarized below. Complete QC data for three repeat batches of [^{11}C]carfentanil produced using the methods disclosed herein are summarized in Table 1. Each of the three doses met all of the established QC criteria.

3.1 Visual Inspection

The [^{11}C]carfentanil QC samples were examined behind a PET L-block and had to be clear, colorless, and free of any particulate matter.

3.2 Dose pH

The pH of the [^{11}C]carfentanil doses were determined by applying small amounts of the QC sample to colorpHast® pH 2.0–9.0 nonbleeding pH-indicator strips [19] and visual comparison to the scale provided. Release stipulate of dose pH had to be 4.5–7.5.

3.3 Radiochemical Identity

High performance liquid chromatographic (HPLC) analysis of radiochemical identity was conducted using a Shimadzu LC 2010 AHT Liquid Chromatograph fitted with UV detectors and Bioscan γ-detectors. Conditions: column was Luna C8 (2) 5 µm 100 × 2 mm; solvent was 35% MeOH/20 mM NH_4OAc, pH = 4.5; flow rate was 0.8 mL/min; oven set to 40°C; injection volume was 20 µL; 218 nM. The retention time of [^{11}C]carfentanil (~4.5 min) was compared to that of the carfentanil reference standard and had to be within 10% (relative retention time (RRT) had to be 0.9–1.1).

$$\mathrm{RRT} = t_{\mathrm{r}(^{11}\mathrm{C\text{-}raclopride})}/t_{\mathrm{r}(^{12}\mathrm{C\text{-}raclopride})}$$

TABLE 1 QC Data for Three Repeat Runs of [^{11}C]Carfentanil

QC Test	Release Criteria	Run 1	Run 2	Run 3
Yield mCi (GBq)	N/A	207 (7.7)	250 (9.3)	217 (8.0)
Visual inspection	Clear, colorless	Pass	Pass	Pass
Radiochemical identity	RRT = 0.9–1.1	1.0	1.0	1.0
Radiochemical purity	≥95%	97.1%	97.2%	97.2%
Specific activity (mCi/μmol)	N/A	57,202	79,569	72,504
Dose pH	4.5–7.5	5.0	5.0	5.0
Radionuclidic identity (min)	18.5–22.5	20	20	20
Sterile filter integrity test (psi)	>45	Pass	Pass	Pass
Residual solvent analysis (ppm)	Acetone (<5000) *n*-Propanol (<5000) DMSO (<5000)	142 4 3	Not determined	Not determined
Endotoxin analysis (EU/mL)	≤17.5	≤2	≤2	≤2
Sterility testing	No colony growth after 14 d	Pass	Pass	Pass

3.4 Radiochemical Purity

HPLC analysis of radiochemical purity was conducted simultaneously with the radiochemical identity determination. Radiochemical purity of [^{11}C]carfentanil had to be >93.0% (Fig. 3).

3.5 Specific Activity

The concentration of [^{11}C]carfentanil can be determined using the HPLC system described in Section 3.3, by integrating the area under the UV peak of [^{11}C] carfentanil and comparing with the area of known reference standard peak. The specific activity (As) was calculated as the division of radioactivity (millicurie) by the amount of raclopride (micromoles).

$$\mathrm{As(mCi/\mu mol)} = A\mathrm{(mCi)}/[C_{\mathrm{CFN}}(\mu\mathrm{g/mL}) \times V\mathrm{(mL)}/394.5(\mu\mathrm{g}/\mu\mathrm{mol})]$$

A is the measured radioactivity; V is the volume of sample measured; C is the calculated concentration of carfentanil.

There is no specific activity release limit for [^{11}C]carfentanil in place at our institution. The injection volume was based on the injectable mass limit for carfentanil and was calculated on a dose-by-dose basis by the administering study team.

3.6 Radionuclidic Identity

Activities were measured using a Capintec CRC® -15R Radioisotope Dose Calibrator, and half-life was calculated using the equation below. Calculated half-life must be 18.5–22.4 min.

$$t_{1/2} = -\ln 2\{\text{time difference}/[\ln(\text{ending activity/starting activity})]\}$$

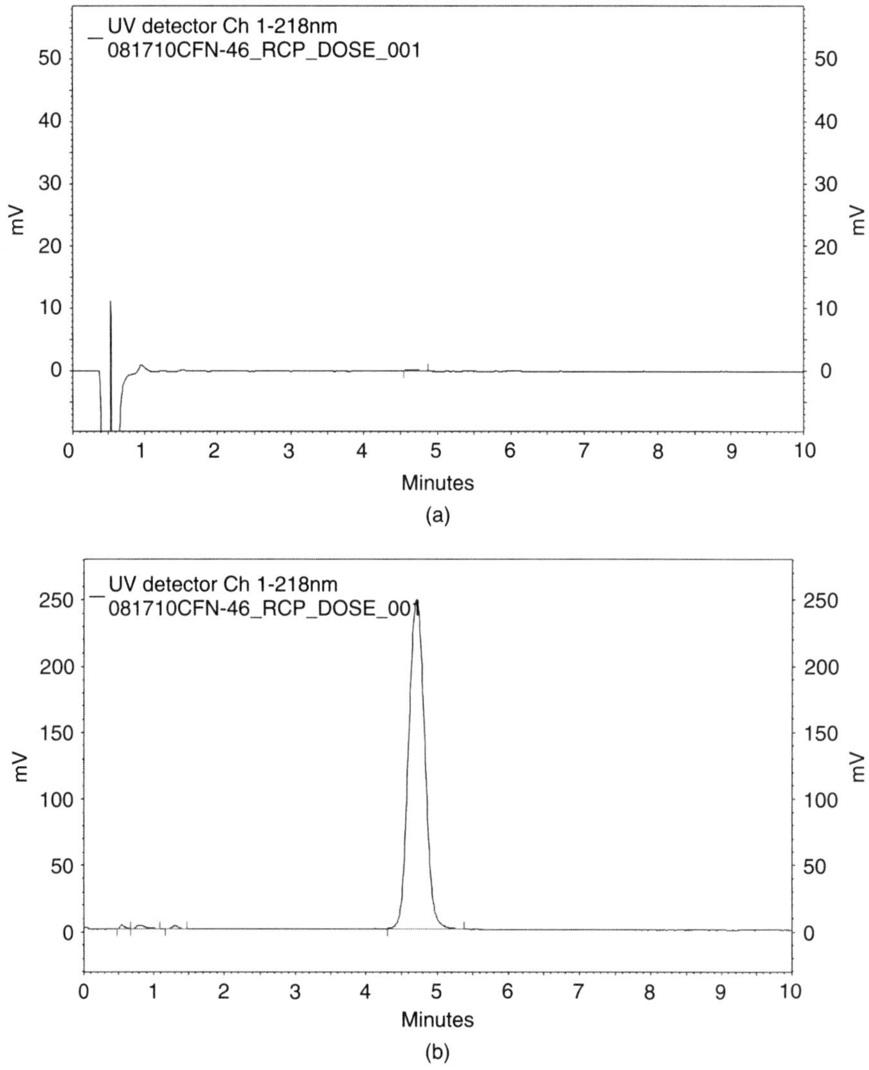

FIGURE 3 Analytical UV (a) and radioactive (b) HPLC trace of [^{11}C]carfentanil.

3.7 Sterile Filter Integrity Test

The sterile filter with the needle that was used to convey the dose into the dose vial was connected to a nitrogen supply via a regulator. The filter was immersed in water, and the pressure was gradually increased while observing for a steady stream of bubbles emanating from the needle. If this happened at a pressure greater than the filter acceptance level then the filter was considered to have been intact and thus passed.

3.8 Residual Solvent Analysis

Gas chromatography (GC) is used to determine the residual solvents. The GC system used was a Shimadzu GC-2010 with an AOC-20i autoinjector, split/splitless inlet, a flame ionization detector (FID) and a Restek column (Stabilwax 30 m × 0.25 mm, 0.25 μm G16 stationary phase). Chromatographic conditions are outlined below.

Injection volume: 0.25 μL
Split ratio: 1:15
Injector temperature: 180°C
Linear velocity: 11.7 cm/s
Detector temperature: 250°C

Oven temperature:
Initial temperature: 55°C
Hold at 55°C for 6.5 min
Heat to 180°C at 80°C/min
Hold at 180°C for 8.5 min

3.9 Endotoxin Analysis

Endotoxin content in doses of [^{11}C]carfentanil was analyzed by a Charles River Laboratories EndoSafe™ Portable Testing System as per the USP [20]. Doses had to contain ≤175 endotoxin units (EU), which was adjusted to 17.5 EU/mL, as our formulation volume was 10 mL.

3.10 Sterility Testing

Samples of the [^{11}C]carfentanil dose was added to tubed fluid thioglycolate media (FTM) and to tubed tryptic soy broth (TSB) media. FTM was used to test for anaerobes, aerobes, and microaerophiles, TSB was used to test for fastidious and nonfastidious organisms [21]. These media tubes as well as positive and negative controls were incubated at 30–35°C (FTM) and 22°C (TSB) for 14 days according to current USP guidelines [22]. Tubes were visually inspected on the third, eighth, and fourteenth days after inoculation and were compared to positive and negative controls. Positive controls had to show growth (turbidity), and the test article samples ([^{11}C]carfentanil) and negative controls had to have no growth after 14 days in order to pass the test for sterility. Positive controls could be established separately as growth promotion tests according to USP guidelines [22].

CHEMICAL ABSTRACTS NOMENCLATURE (REGISTRY NUMBER)

Acetonitrile (75-05-8)

1-Butanaminium, *N,N,N*-tributyl-, 4-[(1-oxopropyl)phenylamino]-1-(2-phenylethyl)-4-piperidinecarboxylate (1:1) (872043-55-5)

Ethanol (64-17-5)

Methane, 1,1′-sulfinylbis- (67-68-5)

4-Piperidinecarboxylic acid, 4-[(1-oxopropyl)phenylamino]-1-(2-phenylethyl)-, methyl-^{11}C ester (98598-83-5)

REFERENCES AND NOTES

1. Martin W. Opioid antagonists. Pharmacol Rev 1967;19(4):463–521.
2. Pert CB, Snyder SH. Opioid receptor: demonstration in nervous tissue. Science 1973;179(77):1011–1014.
3. Zubieta J-K, Stohler CS. Neurobiological mechanisms of placebo responses. Ann N Y Acad Sci Rev 2009;1156:198–210.
4. Sprenger T, Berthele A, Platzer S, Henning T, Thomas R. What to learn from in vivo opioidergic brain imaging? Eur J Pain Rev 2005;9(2):117–121.
5. GEMS standard carbon-11 target with a 75 μm Havar target foil and GEHC standard vacuum foil.
6. Ultra high purity nitrogen gas with 1% of oxygen was purchased from Purity Plus.
7. Molecular sieve (4A, 80/100 mesh) was purchased from Alltech, pn: 5624.
8. Shimalite-Ni reduced (80/100 mesh) was purchased from Shimadzu (Kyoto, Japan), pn: 22127719.
9. Phosphorus pentoxide (with moisture indicator) was purchased from Fluka, pn: 79610.
10. Carbosphere (60/80 mesh) was purchased from Alltech (Grace), pn: 5680.
11. Iodine (crystals) was purchased from EMD (Germany), pn: IX01203.
12. Ascarite II (8–20 mesh, 3 in in TRACERlab standard column) from Thomas Scientific, pn: C049H40.
13. Porapak type Q (0.3 g, 50-80 mesh) was purchased from Alltech, pn: 2700.
14. Silver triflate-Graphpac was prepared by dissolving 5 g of silver trifluoromethanesulfonate (Sigma Aldrich) in 100 mL of acetonitrile (HPLC grade, Fisher) and then mixed with 10 g of Graphpac (80/100 mesh, GC grade, Altech). The solvent was evaporated by rotary vapor under vacuum, and 0.35 g of dried powder was used to pack a TRACERlab standard column.
15. Desmethyl carfentanil tetrabutyl ammonium salt was prepared by dissolving 1 mg of desmethyl carfentanil (Sigma-Aldrich) in 100 μL of water and then adding 10 μL of 1 M tetrabutylammonium hydroxide in MeOH (Aldrich). The precursor solution was diluted with 6 mL of water and loaded onto a preconditioned C-18 extraction disk (3M Empore). The C-18 extraction disk was washed with 2 × 5 mL of water and dried. The product was eluted with 0.2 mL of ethanol into a clean vial and dried under vacuum.
16. C2 extraction disk was purchased from 3M Empore, pn: 4311 (HD).
17. Millex-GV 0.22 μm filter was purchased from Millipore (USA), pn:SLGV 013SL.
18. U. S. Pharmacopeia <823>. Radiopharmaceuticals for Positron Emission Tomography Compounding. 2009, USP 32-NF 27.
19. USA, E.C.I., (part no. 9578-3).
20. U. S. Pharmacopeia <85>. Bacterial Endotoxins Test. 2009, USP 32-NF 27.
21. Becton, D.a.C.U.
22. U. S. Pharmacopeia <71>. Sterility Test; 2009, USP 32-NF 27.

CHAPTER 26

SYNTHESIS OF [carbonyl-^{11}C]WAY-100635

JAN D. ANDERSSON AND SANGRAM NAG

Department of Clinical Neuroscience, Psychiatry Section, Karolinska Institutet, Stockholm, Sweden

RAISA N. KRASIKOVA

Department of Clinical Neuroscience, Psychiatry Section, Karolinska Institutet, Stockholm, Sweden; Institute of Human Brain, Russian Academy of Sciences, St. Petersburg, Russia

VICTOR W. PIKE

Molecular Imaging Branch, National Institute of Mental Health, National Institutes of Health, Bethesda, Maryland

CHRISTER HALLDIN

Department of Clinical Neuroscience, Psychiatry Section, Karolinska Institutet, Stockholm, Sweden

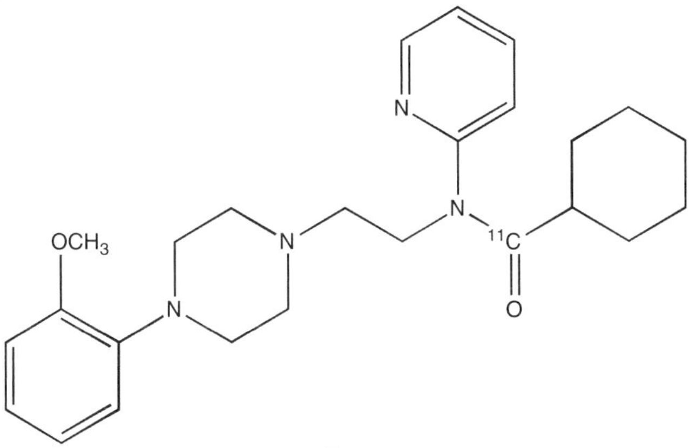

[*Carbonyl*-^{11}C]WAY-100635

Radiochemical Syntheses, Volume 1: Radiopharmaceuticals for Positron Emission Tomography, First Edition. Edited by Peter J. H. Scott, Brian G. Hockley.

1 INTRODUCTION

The central serotonin subtype 1A (5-HT$_{1A}$) receptor is strongly implicated in a variety of neuropsychiatric disorders such as depression, schizophrenia, and anxiety. WAY-100635 is a potent and selective 5-HT$_{1A}$ antagonist, and [*O-methyl*-^{11}C]WAY-100635 was the first radioligand developed for the study of brain 5-HT$_{1A}$ receptors in humans with positron emission tomography (PET) [1,2]. WAY-100635 can easily be labeled with carbon-11 in the 2-methoxy position; however, human liver metabolism cleaves the amide bond and generates a radiometabolite, [^{11}C]WAY-100634, which crosses the blood–brain barrier and has affinity for 5-HT$_{1A}$ receptors and α_1-adrenoceptors [3]. [^{11}C]-Acylation of WAY-100634 with [^{11}C]cyclohexanecarbonyl chloride yields [*carbonyl*-^{11}C]WAY-100635, which in vivo is hydrolyzed to generate [^{11}C]cyclohexanecarboxylic acid as the major radiometabolite. This acid accumulates only transiently in brain, and hence nonspecific binding is greatly reduced and image contrast greatly enhanced. For this reason, [*carbonyl*-^{11}C]WAY-100635 has become the most widely used antagonist radioligand for studying human brain 5-HT$_{1A}$ receptors [4]. The radiosynthesis of [*carbonyl*-^{11}C]WAY-100635 and especially its automation are demanding but practicable [5,6].

2 SYNTHESIS PROCEDURES

CAUTION: All radiosyntheses must be carried out using appropriate equipment in a facility authorized for the use of radioactive materials. Personal protective equipment must be worn and all local radiation safety laws followed.

The following description is based on the production of [*carbonyl*-^{11}C]WAY-100635 (Fig. 1) at the Karolinska Institutet.

2.1 Production of [^{11}C]Carbon Dioxide ([^{11}C]CO_2)

[^{11}C]CO_2 was produced in a General Electric Medical Systems PETtrace cyclotron [7] (GEMS, Uppsala, Sweden) via bombardment of a nitrogen/oxygen gas target [8] using 16.4 MeV protons. Irradiation was performed for 40 min with a beam intensity of 50 μA. At the end of irradiation, radioactivity was vented from the target in a nitrogen stream and trapped on molecular sieves [9].

2.2 Synthesis of [carbonyl-^{11}C]Cyclohexanecarboxylic Acid Chloride

The molecular sieve trap was heated to 360°C and the [^{11}C]CO_2 delivered in a stream of nitrogen to a reaction vial [10] containing cyclohexylmagnesium chloride [11] (0.05 mL) in tetrahydrofuran [12] (THF, 0.35 mL). A solution of thionyl chloride [13] ($SOCl_2$, 0.02 mL) in THF [14] (0.28 mL) was added to the reaction vial, which was then heated at 70°C for 2 min before evaporation of solvent under nitrogen gas flow.

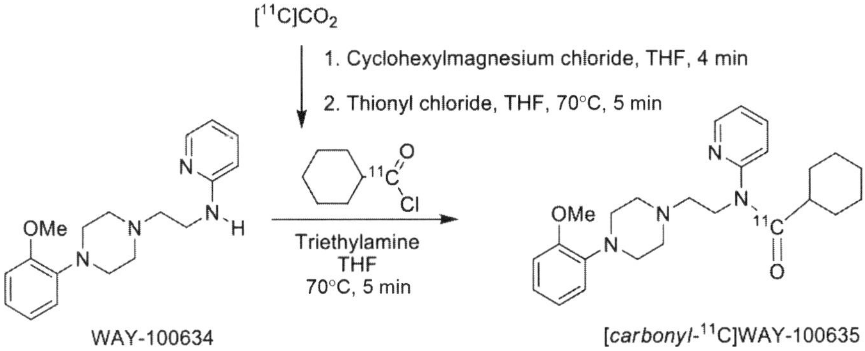

FIGURE 1 Synthesis of [*carbonyl*-^{11}C]WAY-100635.

2.3 Synthesis of [carbonyl-^{11}C]WAY-100635

A solution of amino precursor (WAY-100634 [15], 3.5 mg) and triethylamine (TEA) [16] (0.1 mL in THF (0.3 mL)) was then transferred to the reaction vial containing the [*carbonyl*-^{11}C]cyclohexanecarboxylic acid chloride. The acylation reaction was allowed to take place as THF was evaporated off at 70°C under nitrogen gas flow for 5 min.

2.4 Purification and Formulation of [carbonyl-^{11}C]WAY-100635

The residue of radioactive product was diluted with acetonitrile (MeCN) [17] : water : trifluoroacetic acid (TFA) [18] (40:60:1 by volume) (4 mL), and the solution was introduced into a 5-mL high performance liquid chromatographic (HPLC) injection loop. The purification was performed using semipreparative HPLC (ACE column [19]) starting with TFA (0.1%) in water as mobile phase for the first 5 min and then changing to the MeCN/water/TFA (40:60:1 by volume) as mobile phase for the remainder of the chromatography (Fig. 2).

The fraction corresponding to [*carbonyl*-^{11}C]WAY-100635 (typically eluting between 16 and 20 min) was collected for 1 min into a vial containing sterile water (50 mL). The mobile phase was removed by pushing the diluted product onto an Oasis® HLB cartridge (Waters, Milford, MA, USA) and rinsing with water (2 × 5 mL). The product was eluted with ethanol (1 mL) into a sterile vial containing phosphate-buffered saline [20] (9 mL). The final formulation was then passed through a sterile filter [21] into a sterile vial [22] to provide [*carbonyl*-^{11}C]WAY-100635 (typically 500–1000 MBq) in an isotonic solution and then released for quality control (QC).

After synthesis was complete, the chemistry system was cleaned with acetone (Puriss. p.a., ACS reagent, reag. ISO, reag. Ph. Eur., ≥99.5% (GC), Sigma-Aldrich), and flushed extensively with nitrogen gas. The SepPak module and semipreparative HPLC column was washed with ethanol (99.5%, Kemetyl A®, Kemetyl, Sweden).

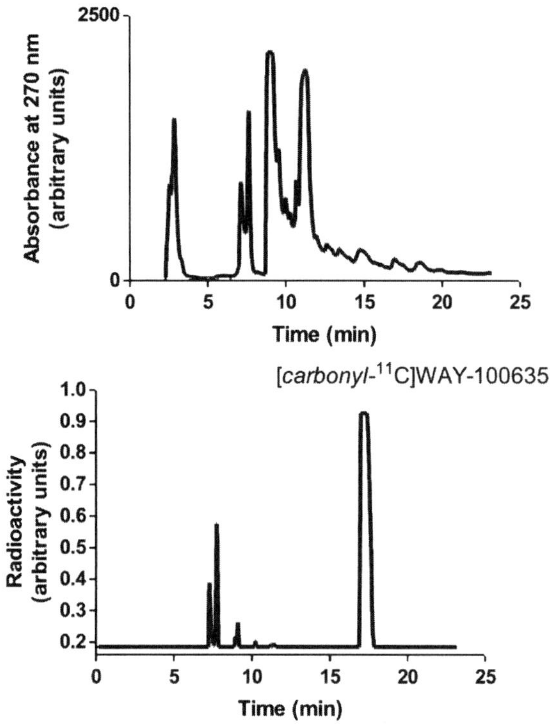

FIGURE 2 Absorbance and radioactivity HPLC traces for the purification of [*carbonyl*-^{11}C]WAY-100635.

3 QUALITY CONTROL PROCEDURES

CAUTION: All radiopharmaceuticals produced for clinical use must have local regulatory approval before use in humans. Trained personnel must carry out QC procedures, and each dose must meet all established QC criteria before release to the clinic.

All quality control procedures and the results from three batches of [carbonyl-11C]WAY-100635 is presented in Table 1.

3.1 Visual Inspection

The [*carbonyl*-^{11}C]WAY-100635 dose must be clear and free of particulate matter.

3.2 Radiochemical Identity and Purity

The analytical HPLC system includes a Waters μ-Bondapak C18 column (300 mm × 3.9 mm, 10 μm particle size), Merck-Hitachi L-7100 Pump, L-7400 UV detector (VWR international, Stockholm, Sweden), and a β-flow detector (Beckman, model 170) for radioactivity detection. A wavelength of 270 nm and a mobile phase of 35% MeCN in H_3PO_4 (0.01 M) at 3 mL/min was used for the analysis (Fig. 3a). The retention time of [*carbonyl*-^{11}C]WAY-100635

TABLE 1 QC Data for Three Repeat Runs of [*carbonyl*-^{11}C]WAY-100635

QC Test	Release Criteria	Run 1	Run 2	Run 3
Yield mCi (MBq)	N/A	21 (783)	12 (458)	18 (656)
Visual inspection	Clear	Clear	Clear	Clear
Radiochemical identity	RRT = 0.09–0.11	1.0	1.0	1.0
Radiochemical purity (%)	≥95	>99	>99	>99
Specific activity (GBq/μmol)	N/A	30	10	15
Chemical mass of WAY-100635 (μg)	Total mass ≤14 [23]	4.62	12.12	8.46
Chemical mass of WAY-100634 (μg)		2.67	0.75	1.14
Chemical identity	*m/z* conforms to reference	Conforms	Conforms	Conforms
Residual solvent analysis (ppm)	TEA <320	Not detected	Not detected	Not detected
	THF <410	0.2	0.1	0.2
	MeCN <410	72	1.5	27
	Acetone <5000	Not detected	Not detected	Not detected
Dose pH	4.5–8	7.5	7.5	7.5
Sterile filter integrity test (bar)	>3.5	>3.5	>3.5	>3.5
Radionuclidic identity ($t_{1/2}$) (min)	19–22	20.25	20.03	20.52
Endotoxin analysis (EU/mL)	≤17.5	<1.25	<1.25	<1.25
Sterility testing	No colony growth (14 d)	Pass	Pass	Pass

is compared to that of the WAY-100635 reference standard [15] and must be ±10% (relative retention time (RRT) must be 0.9–1.1). Radiochemical purity must be >95%.

3.3 Injectable Mass (Specific Radioactivity)

For measurements of chemical amounts of WAY-100635 and WAY-100634 in the final formulation, an analytical HPLC system from Merck-Hitachi (VWR International) with D-6200A pump, L-4000 UV detector, and D-6000 interface was used. A Waters μ-Bondapak C18 column (300 mm × 3.9 mm, 10 μm particle size) was eluted with 50% MeCN in HCO_2NH_4 (0.1 M) at 3 mL/min, with eluate monitored for absorbance at 270 nm (Fig. 3b). The specific radioactivity and the mass were calculated by comparing the UV absorbance area counts for the formulation to a reference standard with known concentrations. In our institute, the release criteria for the maximum amount of WAY-100635 plus WAY-100634 in an injected dose is 14 μg [23–25], which is calculated on a dose-by-dose basis.

3.4 Chemical Identity

The chemical identity of [*carbonyl*-^{11}C]WAY-100635 was verified by liquid chromatography–mass spectrometry (LC-MS). A Waters Acquity™ ultra performance LC system (Waters, Milford, MA, USA) was used for pumping the

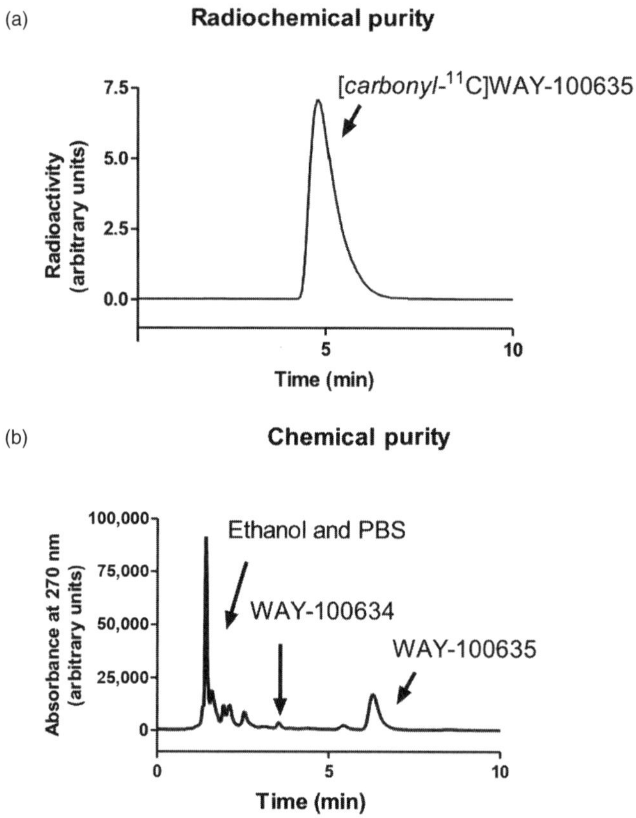

FIGURE 3 Analytical radioactivity (a) and UV absorbance (b) traces for [*carbonyl*-^{11}C]WAY-100635 using HPLC methods described in the text.

mobile phase, injection of the samples, and heating of the column. The mobile phase consisted of 0.1% formic acid plus 5% MeCN in water (A) and 0.1% formic acid in MeCN (B). The pumps were programmed to deliver a linear gradient running from 0% to 50% B from 0 to 4.00 min, 50% to 100% B from 4.00 to 4.50 min, at a flow rate of 0.5 mL/min. The composition was returned to 0% B in A from 5.00 to 5.50 min. The end time of the program was set at 6.5 min [26]. The m/z and the fractionation must conform to the reference standard of WAY-100635.

3.5 Residual Solvent Analysis

Analysis of residual solvent levels (triethylamine, acetonitrile, THF, acetone, ethanol) in [*carbonyl*-^{11}C]WAY-100635 doses was conducted with gas chromatography [27]. Limits according to the ICH Harmonised Tripartite Guidelines [28] are <410 ppm for acetonitrile, <720 ppm for THF, and <5000 ppm for class 3 solvents (acetone); the limit for triethylamine is <320 ppm [29].

3.6 Dose pH

The pH of the [*carbonyl*-^{11}C]WAY-100635 dose was measured by applying a small amount of the dose to indicator paper [30]. Dose pH must be in the range 4.5–8.

3.7 Sterile Filter Integrity Test

The filter used for the sterile filtration of [*carbonyl*-^{11}C]WAY-100635 was connected to a bubble point machine [31]. Only if the filter could hold a pressure over the filter acceptance pressure (3.5 bar) without seeing a stream of bubbles, the filter was considered intact.

3.8 Radionuclidic Identity

Radioactivities were measured using a dose calibrator (Capintec, Ramsey, NJ, USA) at two time points within a time interval of 20 min, and the half-life was calculated. Calculated half-life of [*carbonyl*-^{11}C]WAY-100635 must be 19–22 min.

3.9 Endotoxin Analysis

Endotoxin content was analyzed by an EndoSafe® Portable Testing System (Charles River Laboratories, Wilmington, MA, USA). Doses must contain $\leq$17.5 endotoxin units (EU)/mL [32].

3.10 Sterility Testing

Samples were sent to Apotek Produktion & Laboratorier AB for sterility testing. [*carbonyl*-^{11}C]WAY-100635 must show no culture growth on the plates after 14 days to indicate sterility [33].

ACKNOWLEDGMENTS

We thank Drs. Ryuji Nakao, Anna Sumic, Arsalan Amir, and Guennadi Jogolev for their valuable support. V.W.P. is supported by the Intramural Research Program of the National Institutes of Health (NIMH).

CHEMICAL ABSTRACTS NOMENCLATURE (REGISTRY NUMBER)

Acetone (67-64-1)
Acetonitrile (75-05-8)
Cyclohexylmagnesium chloride (931-51-1)
Ethanol (64-17-5)
Sterile water (7732-18-5)
Tetrahydrofuran (109-99-9)
Thionyl chloride (7719-09-7)
Triethylamine (121-44-8)
Trifluoroacetic acid (76-05-1))

REFERENCES AND NOTES

1. Mathis CA, Simpson NR, Mahmood K, Kinahan PE, Mintun MA. Life Sci 1994;55: PL403–PL407.
2. Pike VW, McCarron JA, Lammerstma AA, Hume SP, Poole K, Grasby PM, Malizia A, Cliffe IA, Fletcher A, Bench CJ. Eur J Pharmacol 1995;283:R1–R3.
3. Osman S, Lundkvist C, Pike VW, Halldin C, McCarron JA, Swahn CG, Ginovart N, Luthra SK, Bench CJ, Grasby PM, Wikstrom H, Barf T, Cliffe IA, Fletcher A, Farde L. Nucl Med Biol 1996;23:627–634.
4. Pike VW, McCarron JA, Lammertsma AA, Osman S, Hume SP, Sargent PA, Bench CJ, Cliffe IA, Fletcher A, Grasby PM. Eur J Pharmacol 1996;301:R5–R7.
5. Krasikova RN, Andersson J, Truong P, Nag S, Shchukin EV, Halldin C. Appl Radiat Isot 2009;67:73–78.
6. McCarron JA, Turton DR, Pike VW, Poole KG. J Labelled Comp Radiopharm 1996;38: 941–953.
7. GEMS, Uppsala, Sweden.
8. Aluminum target, 78 mL. Target gas was nitrogen with 0.5% oxygen. Before entering the target, the target gas was passed through an ALL-Pure™ Gas Purifier Moisture trap (Grace, Deerfield, IL, USA, part no. 7034).
9. ProCab chemistry system setup, GE medical systems, Uppsala, Sweden.
10. Micro-Reaction Vial, 5 mL, from Supelco (part no. 33299).
11. Cyclohexylmagnesium chloride (2.0 M solution in diethyl ether) from Aldrich (part no. 224413).
12. Tetrahydrofurane (anhydrous, $\geq$99.9%, inhibitor-free) from Aldrich (part no. 40,175-7).
13. Thionylchloride (ReagentPlus®, $\geq$99%) from Sigma-Aldrich (part no. 23,046-4).
14. Stock solution prepared on the day of production (150 µL $SOCl_2$ in 2 mL THF).
15. Synthesis described in Pike VW, McCarron JA, Hume SP, Ashworth S, Opacka-Juffry J, Osman S, Lammertsmaa AA, Poole KG, Fletcher A, White AC, Cliffe IA. Med Chem Res 1995;5:208–227. WAY-100634 and WAY-100635 are available from PharmaSynth, Estonia, 2011. Available at http://www.pharmasynth.ee/.
16. Triethylamine ($\geq$99%) from Sigma-Aldrich (part no. T0886).
17. Acetonitrile (HPLC grade) from Fischer Scientific (code A/0626/17X).
18. Trifluoroacetic acid ($\geq$98.0%) from Fluka (part no. 91700).
19. ACE C18~250 mm × 10 mm, 5 µm, Advanced Chromatography Technologies (part no. ACE-321-2510).
20. pH = 7.4, APL (Apoteket Production & Laboratories AB).
21. Medical Millex-GV Filter (0.22 µm, PVDF, 33 mm, sterilized by γ-irradiation) from Millipore AB, Solna, Sweden (part no. SLGV033RS).
22. Sterile vial 10 mL (Apoteket Production & Laboratories AB).
23. The maximum injected mass was calculated from an estimation of 5% occupancy from in-house data. Specifically bound radioligand concentration in the prefrontal cortex, C_b, was calculated from 39–63 min, as described in Ref. 18. The molar concentration of specifically bound WAY100635, B, was obtained by dividing C_b with the specific radioactivity at the time of injection. Occupancy was estimated as B divided by B_{max} with a 5-HT$_{1A}$ receptor B_{max} value of 30 pmol/g tissue as obtained in autoradiographic studies [19]. WAY-100634 has lower affinity than WAY100635 [3].
24. Farde L, Ginovart N, Ito H, Lundkvist C, Pike VW, McCarron JA, Halldin C. Psychopharmacology (Berl) 1997;133:196–202.
25. Varnas K, Halldin C, Hall H. Hum Brain Mapp 2004;22:246–260.
26. The column, a Waters Acquity UPLC™ BEH C18 (50 mm × 2.1 mm, 1.7 µm) from Waters, Milford, MA, USA (part no. 186002350) was kept in a column heater at 50°C. A Micromass Premier™ Quadrupole time-of-flight mass spectrometer (Waters, Milford, MA, USA) was used for detection. It was operated in positive ESI mode, with the following settings: capillary voltage, 3.0 kV; cone voltage, 45 V; source temperature, 120°C; desolvation temperature, 350°C; and collision energy, 10 V.

27. HP 6890 GC system (Agilent Technologies Sweden AB, Kista, Sweden). Perkin Elmer HS 40 XL Headspace sampler. Column: HP-DB-624 0.32 mm × 30 m, 1.8 μm (part no. 123–1334); temperature: oven 40°C, front inlet 140°C, front detector 250°C; inlet pressure: linear velocity 36 cm/s, column mode constant flow, split 5:1; FID: H_2 flow 30 mL/min, air flow 400 mL/min, He flow 10 mL/min (column + makeup); analysis time: 10 min.
28. ICH Topic Q3C (R4) Impurities: Guideline for Residual Solvents Step 5. February 2009, CPMP/ICH/283/95.
29. EDQM Europe. Certification of suitability of Monographs of the European Pharmacopoeia, Content of the dossier for chemical purity and microbial quality, PA/PH/CEP (04) 1 4R. 2007.
30. Spezialindikator (part no. 109584), Merck, Darmstadt, Germany.
31. The bubble point machine was custom made. Before testing, the filter was washed with 20 mL of water.
32. European Pharmacopoeia chapter 2.6.14 states that for radiopharmaceuticals the endotoxin tolerance limit is 175 EU. For 10 mL of the product, the limit is 17.5 EU/mL.
33. European Pharmacopoeia chapter 2.6.1.

CHAPTER 27

SYNTHESIS OF [^{11}C]RACLOPRIDE

XIA SHAO

Department of Radiology, University of Michigan Medical School, Ann Arbor, Michigan

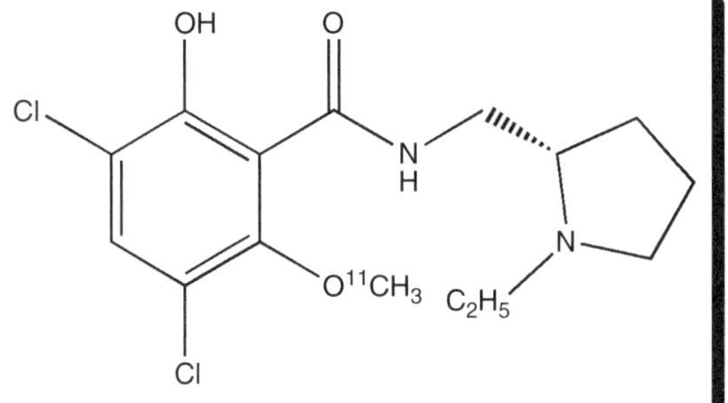

[^{11}C]Raclopride

1 INTRODUCTION

[^{11}C]Raclopride, a substituted benzamide, is a popular radiotracer for visualization and quantification of D_2-like dopamine receptors by positron emission tomography (PET) [1,2]. [^{11}C]Raclopride has been used to study neuropsychiatric disorders related to dopamine D_2 receptors for years [3–9].

Several different synthetic approaches have been reported, as well as different automated modules for production of [^{11}C]raclopride [10–14]. Both [^{11}C]methyl iodide and [^{11}C]methyl triflate were used for [^{11}C]methylation of *O*-desmethyl raclopride in the presence of base. Recently, the solvent capture (loop method) [^{11}C]methylation has become more popular because of its high efficiency [13,15–17]. With the increasing demand for this tracer, we developed a fully automated and reliable loop method to produce [^{11}C]raclopride for routine clinical application [18]. By simple modification of a GE TRACERlab $FX_{C\text{-}Pro}$ system, [^{11}C]raclopride has been prepared in high radiochemical yield and purity.

Radiochemical Syntheses, Volume 1: Radiopharmaceuticals for Positron Emission Tomography, First Edition. Edited by Peter J. H. Scott, Brian G. Hockley.

2 SYNTHESIS PROCEDURES

CAUTION: All radiochemical syntheses must be carried out using appropriate equipment in a facility authorized for the use of radioactive materials. Personal protective equipment must be worn and all local radiation safety laws followed.

2.1 Production of [^{11}C]CO_2

[^{11}C]CO_2 was produced by a General Electric Medical Systems (GEMS) PET-Trace cyclotron. The target [19] was loaded with [^{14}N]N_2 gas [20] and bombarded with a proton beam (40 μA beam for 30 min) to generate ~3 Ci (111 GBq) of [^{11}C]CO_2 by the $^{14}N(p,\alpha)^{11}C$ nuclear reaction.

2.2 Preparation of [^{11}C]Methyl Triflate

[^{11}C]CO_2 was delivered from the target by nitrogen gas pressure to a column filled with a mixture of molecular sieve (0.3 g) [21] and Shimalite-Ni catalyst (0.2 g) [22] at room temperature. The column was sealed under H_2 and then heated to 350°C for 20 s to reduce the [^{11}C]CO_2 to [^{11}C]CH_4. [^{11}C]CH_4 was passed through a phosphorus pentoxide desiccant [23] column and trapped in a carbosphere [24] column at −78°C. Gaseous [^{11}C]CH_4 was entered into a circulation loop system including a gas pump, an iodine [25] column at 100°C, a reactor tube at 720°C, two adjacent Ascarite II [26] columns, and a Porapak [27] column at room temperature. The gaseous mixture was then circulated for 5 min, while [^{11}C]CH_3I accumulated on the Porapak column.

[^{11}C]CH_3I was converted to [^{11}C]CH_3OTf by passing it through a silver triflate-Graphpac [28] column heated to 190°C.

2.3 Synthesis of [^{11}C]Raclopride

[^{11}C]Raclopride was produced by [^{11}C]methylation of *O*-desmethyl raclopride tetrabutyl ammonium salt [29] with [^{11}C]methyl triflate in high performance liquid chromatographic (HPLC) loop (Fig. 1). The modified configurations of GE TRACERlab $FX_{C\text{-}Pro}$ system are shown in Fig. 2. Modification 1 (**M1**) was to connect the outline of triflate column to the HPLC loop. Modification 2 (**M2**) was to connect the HPLC collecting line directly to product vial. *O*-desmethyl raclopride tetrabutyl ammonium salt [29] was dissolved in 100 μL of 2-butanone and slowly injected into the HPLC loop. [^{11}C]CH_3OTf described above was

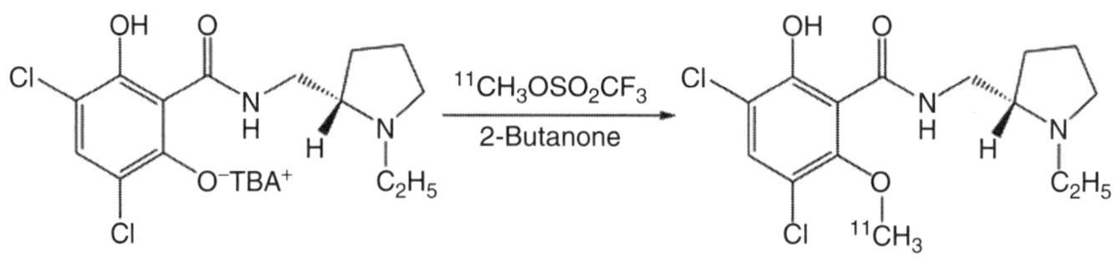

FIGURE 1 Radiosynthesis of [^{11}C]raclopride.

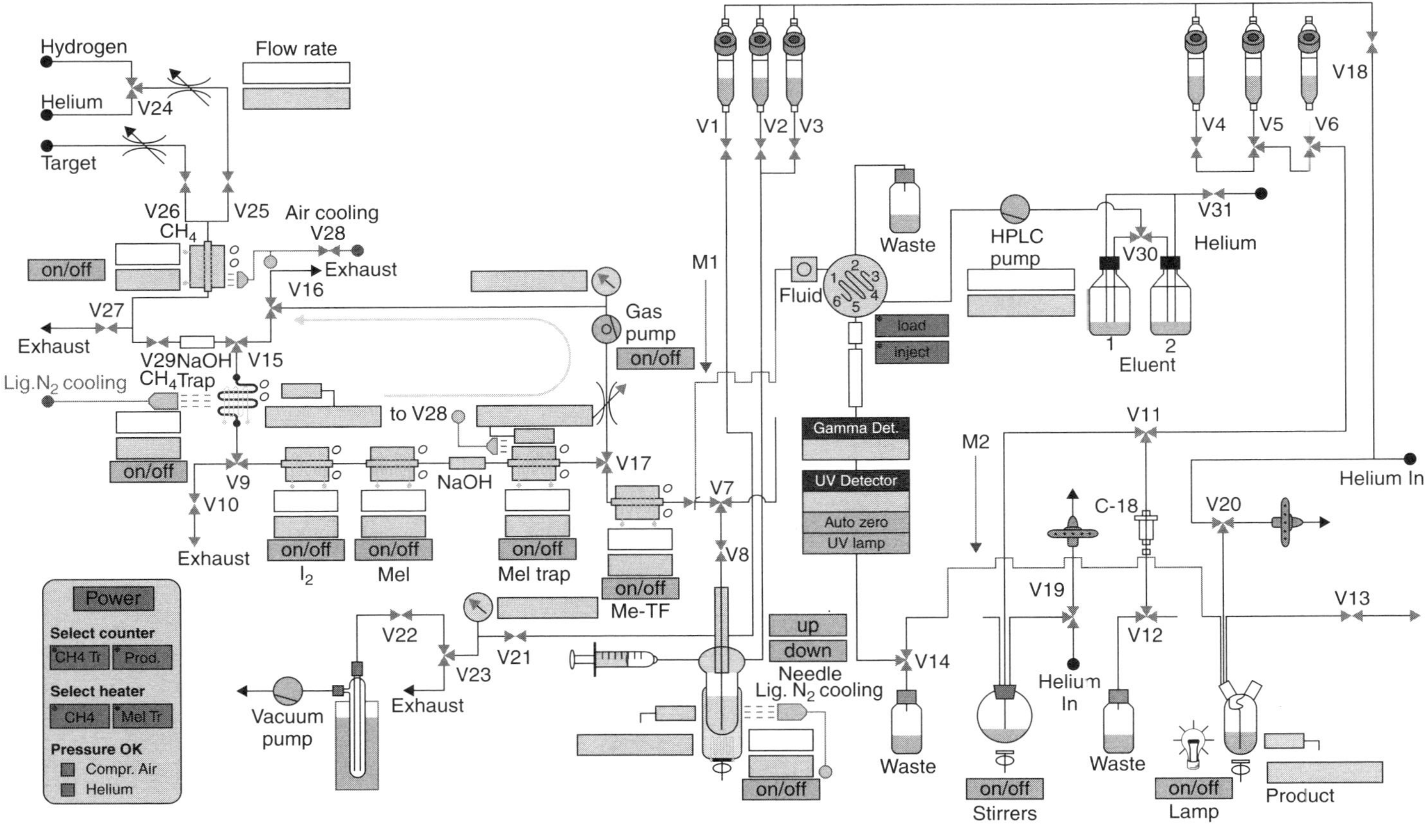

FIGURE 2 Modified TRACERlab $FX_{C\text{-}Pro}$ for preparation of [^{11}C]raclopride.

passed through the loop at a flow rate 40 mL/min. After 3 min, the reaction mixture was injected onto a HPLC column (Luna NH_2, 250 × 10 mm) and eluted using 20 mM NH_4OAc in 10% ethanol at 3 mL/min. The product peak (retention time ~8 min) was collected for 90 s (4.5 mL) into the product vial containing 5.5 mL of United States Pharmacopeia (USP) saline, and the final dose was filtered through a 0.22-μm sterile filter [30] into a dose vial.

Modification 1: connect the outline from triflate column to HPLC loop.

Modification 2: connect the HPLC collecting line to product vial.

3 QUALITY CONTROL PROCEDURES

CAUTION: All radiopharmaceuticals produced for clinical use must have local regulatory approval before use in humans. Trained personnel must carry out quality control (QC) procedures, and each dose must meet all established QC criteria before release to the clinic.

QC procedures for [^{11}C]raclopride, based on the current requirements for radiopharmaceuticals laid out in the USP [31], are summarized below. Complete QC data for three repeat batches of [^{11}C]raclopride produced using the methods described herein are summarized in Table 1. Each of the three doses met all of the established QC criteria.

3.1 Visual Inspection

The [^{11}C]raclopride QC sample was examined behind shielding lead glass in a clear vial and had to be clear, colorless, and free of visible particulate matter.

3.2 Dose pH

The pH of the [^{11}C]raclopride dose was determined by applying a small amount of the QC sample to a colorpHast® pH 2.0–9.0 nonbleeding pH-indicator strip [32] and visual comparison to the scale provided. Dose pH had to be 4.5–7.5.

TABLE 1 QC Data for Three Repeat Runs of [^{11}C]Raclopride

QC Test	Release Criteria	Run 1	Run 2	Run 3
Yield mCi (GBq)	N/A	52 (1.9)	93 (3.4)	120 (4.4)
Visual inspection	Clear, colorless	Pass	Pass	Pass
Radiochemical identity	RRT = 0.9–1.1	1.0	1.0	1.0
Radiochemical purity (%)	≥ 95	96.6	95.9	96.2
Specific activity (mCi/μmol)	N/A	8771	25465	5575
Dose pH	4.5–7.5	5.5	5.5	5.5
Radionuclidic identity (min)	18.5–22.4	20	20	20
Sterile filter integrity test (psi)	>45	Pass	Pass	Pass
Residual solvent analysis (ppm)	Acetone (<5000)	10	12	6
	2-Butanone (<5000)	Not detected	Not detected	Not detected
Endotoxin analysis (EU/mL)	≤ 17.5	≤ 2	≤ 2	≤ 2
Sterility testing	No colony growth out of 14 d	Pass	Pass	Pass

3.3 Radiochemical Identity

Radiochemical identity was determined using a Shimadzu LC 2010 AHT Liquid Chromatograph fitted with UV detectors and Bioscan γ-detectors. Conditions: column was Luna C8 (2) 5 μm100 × 2 mm; solvent was 35% MeOH/20 mM NH_4OAc (pH = 4.5); flow rate was 0.8 mL/min; oven set to 40°C; injection volume was 10 μL; UV detector operated at 218 nM. The retention time of [^{11}C]raclopride (~3.5 min) was compared to that of a known reference standard and had to be within 10% (relative retention time (RRT) had to be 0.9–1.1).

$$\mathrm{RRT} = t_{\mathrm{r}(^{11}\mathrm{C\text{-}raclopride})}/t_{\mathrm{r}(^{12}\mathrm{C\text{-}raclopride})}$$

3.4 Radiochemical Purity

Radiochemical purity of each batch was analyzed using the HPLC system described in Section 3.3. Radiochemical purity had to be greater than 95%. Representative analytical HPLC tracers are displayed in Fig. 3.

3.5 Specific Activity

The concentration of [^{11}C]raclopride can be determined using the HPLC system described in Section 3.3, by integrating the area under the UV peak of [^{11}C]raclopride and comparing with the area of known reference standard peak. The specific activity (As) was calculated by dividing radioactivity (in millicurie) by the amount of raclopride (in micromoles).

$$\mathrm{As(mCi/\mu mol)} = A\mathrm{(mCi)}/[C_{\mathbf{RAC}}(\mu\mathrm{g/mL}) \times V\mathrm{(mL)}/346(\mu\mathrm{g}/\mu\mathrm{mol})]$$

A is the measured radioactivity; V is the volume of sample measured; C is the calculated concentration of raclopride.

There is no specific activity release limit for [^{11}C]raclopride in place at our institution. The injection volume was based on the injectable mass limit for raclopride and was calculated on a dose-by-dose basis by the administering study team.

3.6 Radionuclidic Identity

Activities were measured using a Capintec CRC®-15R radioisotope dose calibrator, and half-life was calculated using the equation below. Calculated half-life had to be 18.5–22.4 min.

$$t_{1/2} = -\ln 2\{\text{time difference}/[\ln(\text{ending activity}/\text{starting activity})]\}$$

3.7 Sterile Filter Integrity Test

The sterile filter with the needle that was used to convey the dose into the dose vial was connected to a nitrogen supply via a regulator. The filter was immersed in water, and the pressure was gradually increased while observing for a steady stream of bubbles emanating from the needle. If this happened at a pressure

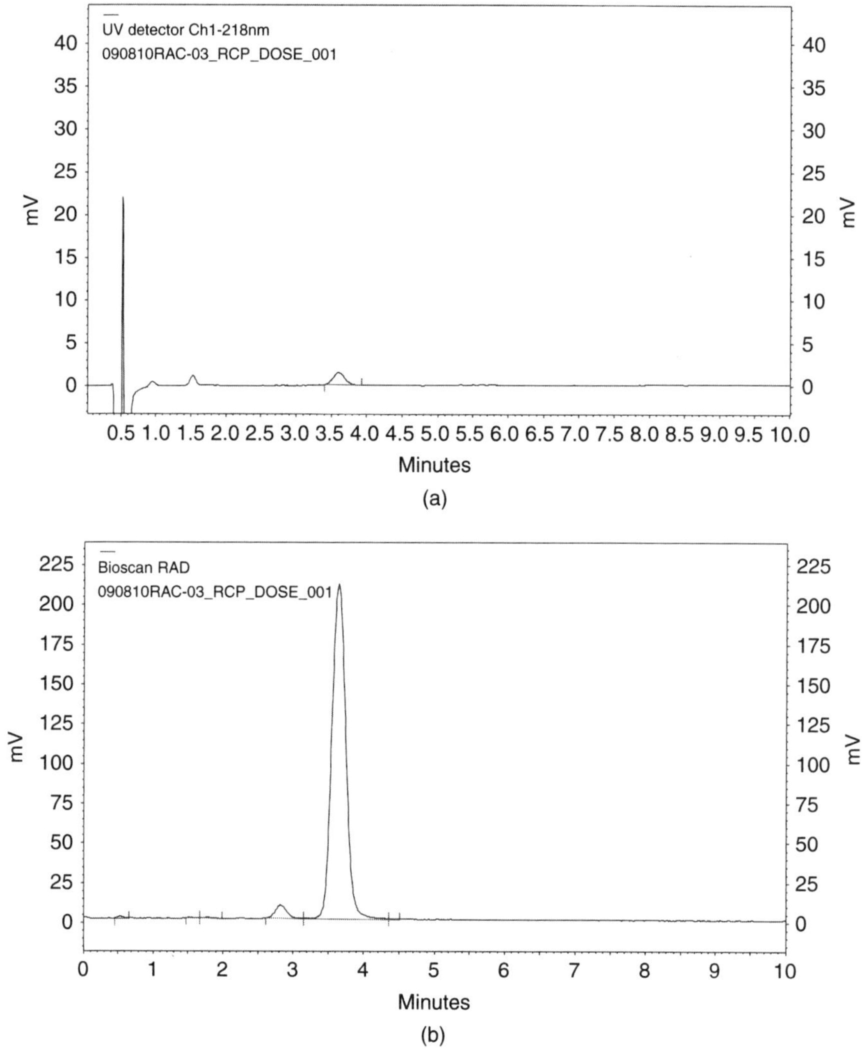

FIGURE 3 Analytical UV (a) and radioactive (b) HPLC trace of [^{11}C]raclopride.

greater than the filter acceptance level then the filter was considered to have been intact and thus passed.

3.8 Residual Solvent Analysis

Gas chromatography (GC) is used to determine the residual solvents. The GC system used was a Shimadzu GC-2010 with an AOC-20i autoinjector, split/splitless inlet, a flame ionization detector (FID), and a Restek column (Stabilwax 30 m × 0.25 mm, 0.25 μm G16 stationary phase). Chromatographic conditions are outlined below.

Injection volume: 0.25 μL
Split ratio: 1:15
Injector temperature: 180°C
Linear velocity: 11.7 cm/s
Detector temperature: 250°C

Oven temperature:
Initial temperature: 55°C
Hold at 55°C for 6.5 min
Heat to 180°C at 80°C/min
Hold at 180°C for 8.5 min

3.9 Endotoxin Analysis

Endotoxin content in doses of [^{11}C]raclopride was analyzed by a Charles River Laboratories EndoSafe™ Portable Testing System as per the USP [33]. Doses had to contain $\leq$ 175 endotoxin units (EU), which was adjusted to 17.5 EU/mL, as our formulation volume was 10 mL.

3.10 Sterility Testing

Samples of the [^{11}C]raclopride dose were added to tubed fluid thioglycolate media (FTM) and to tubed tryptic soy broth (TSB) media. FTM was used to test for anaerobes, aerobes, and microaerophiles; TSB was used to test for fastidious and nonfastidious organisms. These media tubes as well as positive and negative controls were incubated at 30–35°C (FTM) and 22°C (TSB) for 14 days according to current USP guidelines [34]. Tubes were visually inspected on the third, eighth, and fourteenth days after inoculation and compared to positive and negative controls. Positive controls had to show growth (turbidity), and the test article samples ([^{11}C]raclopride) and negative controls had to have no growth after 14 days in order to pass the test for sterility. Positive controls could be established separately as growth promotion tests [34] and microbial enumeration study [35] according to USP guidelines.

CHEMICAL ABSTRACTS NOMENCLATURE (REGISTRY NUMBER)

Ammonium acetate (631-61-8)
O-Desmethyl raclopride tetrabutyl ammonium salt (119670-11-0)
Ethanol (64-17-5)
Iodomethane (74-88-4)
Methyl triflate (333-27-7)

REFERENCES AND NOTES

1. Farde L, Ehrin E, Eriksson L, Greitz T, Hall H, Hedstroem CG, Litton JE, Sedvall G. Substituted benzamides as ligands for visualization of dopamine receptor binding in the human brain by positron emission tomography. Proc Natl Acad Sci U S A 1985;82:3863–3867.

2. Fujimura Y, Ito H, Takahashi H, Yasuno F, Ikoma Y, Zhang M-R, Nanko S, Suzuki K, Suhara T. Measurement of dopamine D2 receptors in living human brain using [11C]raclopride with ultra-high specific radioactivity. Nucl Med Biol 2010;37:831–835.
3. Nordstrom AL, Farde L, Wiesel FA, Forslund K, Pauli S, Halldin C. Central D2-dopamine receptor occupancy in relation to antipsychotic drug effects: a double-blind PET study of schizophrenic patients. Biol Psychiatry 1993;33:227–235.
4. Farde L, Hall H, Ehrin E, Sedvall G. Quantitative analysis of D2 dopamine receptor binding in the living human brain by PET. Science 1986;231:258–261.
5. Rinne JO, Laihinen A, Rinne UK, Nagren K, Bergman J, Ruotsalainen U. PET study on striatal dopamine D2 receptor changes during the progression of early Parkinson's disease. Mov Disord 1993;8:134–138.
6. Yasuno F, Suhara T, Okubo Y, Sudo Y, Inoue M, Ichimiya T. Low dopamine d(2) receptor binding in subregions of the thalamus in schizophrenia. Am J Psychiatry 2004;161:1016–1022.
7. Montgomery AJ, Mehta MA, Grasby PM. Is psychological stress in man associated with increased striatal dopamine levels? A [11C] raclopride PET study. Synapse 2006;60:124–131.
8. Montgomery AJ, Stokes P, Kitamura Y, Grasby PM. Extrastriatal D(2) and striatal D(2) receptors in depressive illness: pilot PET studies using [11C]FLB 457 and [11C]raclopride. J Affect Disord 2006;101:113–122.
9. Kuroda Y, Motohashi N, Ito H, Ito S, Takano A, Nishikawa T, Suhara T. Effects of repetitive transcranial magnetic stimulation on [11C]raclopride binding and cognitive function in patients with depression. J Affect Disord 2006;95:35–42.
10. Langer O, Nagren K, Dolle F, Lundkvist C, Sandell J, Swahn C-G, Vaufrey F, Crouzel C, Maziere B, Halldin C. Precursor synthesis and radiolabelling of the dopamine D2 receptor ligand [11C]raclopride from [11C]methyl triflate. J Labelled Comp Radiopharm 1999;42(12):1183–1193.
11. Ishiwata K, Ishii S-I, Senda M. An alternative synthesis of [11C]raclopride for routine use. Ann Nucl Med 1999;13(3):195–197.
12. Ehrin E, Farde L, De Paulis T, Eriksson L, Greitz T, Johnstroem P, Litton JE, Nilsson JLG, Sedvall G. Preparation of carbon-11-labeled raclopride, a new potent dopamine receptor antagonist: preliminary PET studies of cerebral dopamine receptors in the monkey. Int J Appl Radiat Isot 1985;36(4):269–273.
13. Iwata R, Pascali C, Bogni A, Miyake Y, Yanai K, Ido T. A simple loop method for the automated preparation of [11C]raclopride from [11C]methyl triflate. Appl Radiat Isot 2001;55(1):17–22.
14. Fei X, Mock BH, DeGrado TR, Wang J-Q, Glick-Wilson BE, Sullivan ML, Hutchins GD, Zheng Q-H. An improved synthesis of PET dopamine D2 receptors radioligand [11C]raclopride. Synth Commun 2004;34(10):1897–1907.
15. Wilson AA, Garcia A, Jin L, Houle S. Radiotracer synthesis from [11C]-iodomethane: A remarkably simple captive solvent method. Nucl Med Biol 2000;27(6):529–532.
16. Wilson AA, Garcia A, Houle S, Vasdev N. Utility of commercial radiosynthetic modules in captive solvent [11C]-methylation reactions for HPLC system. J Labelled Comp Radiopharm 2009;52(11):490–492.
17. Arai T, Zhang M-R, Ogawa M, Fukumura T, Kato K, Suzuki K. Efficient and reproducible synthesis of [1-11C]acetyl chloride using the loop method. Appl Radiat Isot 2009;67(2):296–300.
18. Shao X, Kilbourn MR. A simple modification of GE TRACERlab $FX_{C\text{-}Pro}$ for rapid sequential preparation of [11C]carfentanil and [11C]raclopride. Appl Radiat Isot 2009;67(4):602–605.
19. GEMS standard carbon-11 target with a 75-µm Havar target foil and GEHC standard vacuum foil.
20. Ultra high purity nitrogen gas with 1% of oxygen was purchased from Purity Plus.
21. Molecular sieve (4A, 80/100 mesh) was purchased from Alltech, pn: 5624.
22. Shimalite-Ni reduced (80/100 mesh) was purchased from Shimadzu (Kyoto, Japan), pn: 22127719.
23. Phosphorus pentoxide (with moisture indicator) was purchased from Fluka, pn: 79610.
24. Carbosphere (60/80 mesh) was purchased from Alltech (Grace), pn: 5680.
25. Iodine (crystals) was purchased from EMD (Germany), pn: IX01203.

26. Ascarite II (8-20 mesh, 3 inches in TRACERlab standard column) was purchased from Thomas Scientific, pn: C049H40.
27. Porapak type Q (0.3 g, 50-80 mesh) was purchased from Alltech, pn: 2700.
28. Silver triflate-Graphpac was prepared by dissolving 5 g of silver trifluoromethanesulfonate (Sigma Aldrich) in 100 mL of acetonitrile (HPLC grade, Fisher) and then mixed with 10 g of Graphpac (80/100 mesh, GC grade, Altech). The solvent was evaporated by rotary vapor under vacuum, and 0.35 g of dried powder was used to pack a TRACERlab standard column.
29. *O*-desmethyl raclopride tetrabutyl ammonium salt was prepared by dissolving 1 mg of *O*-desmethyl raclopride hydrobromide in 100 μL of ethanol and 50 μL of water and then adding 10 μL of 1 M tetrabutylammonium hydroxide in MeOH (Aldrich). The precursor solution was diluted with 6 mL of water and loaded onto a preconditioned C-18 extraction disk (3M Empore). The C-18 extraction disk was washed with 2 $\times$ 5 mL of water and dried. The product was eluted with 0.2 mL of ethanol into a clean vial and dried under vacuum.
30. Millex-GV 0.22-μm filter was purchased from Millipore (USA), pn:SLGV 033RS.
31. U. S. Pharmacopeia <823>, Radiopharmaceuticals for Positron Emission Tomography Compounding; 2009, USP 32-NF 27.
32. pH-indicator strips were purchased from EMD Chemicals Inc., pn: 9578-3.
33. U. S. Pharmacopeia <85>. Bacterial Endotoxins Test; 2009, USP 32-NF 27.
34. U. S. Pharmacopeia <71>. Sterility Tests 2009, USP 32-NF 27.
35. U. S. Pharmacopeia <61>. Microbial Enumeration Tests; 2009, USP 32-NF 27.

CHAPTER 28

SYNTHESIS OF 3-AMINO-4-[2-(*N*-METHYL-*N*-[^{11}C]METHYL-AMINO-METHYL)PHENYLSULFANYL]-BENZONITRILE ([^{11}C]DASB)

DANIELA HAEUSLER AND MARKUS MITTERHAUSER

Department of Nuclear Medicine, Medical University of Vienna, Vienna, Austria
Department of Pharmaceutical Technology and Biopharmaceutics, University of Vienna, Vienna, Austria

WOLFGANG WADSAK

Department of Nuclear Medicine, Medical University of Vienna, Vienna, Austria
Department of Inorganic Chemistry, University of Vienna, Vienna, Austria

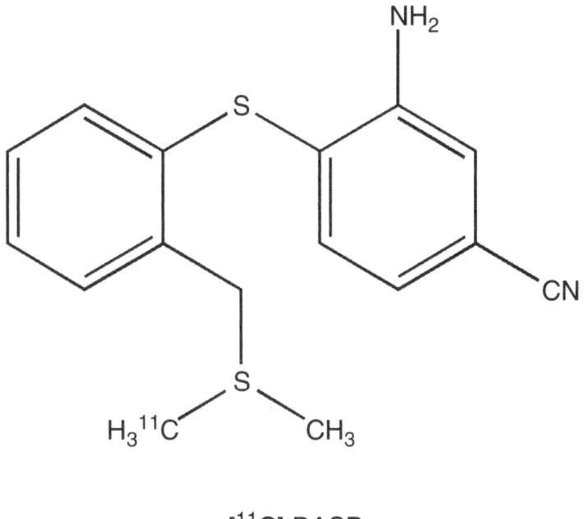

[^{11}C] DASB

1 INTRODUCTION

The serotonin transporter (SERT) influences and controls extracellular serotonin levels in the synaptic cleft by decreasing the serotonin level through reuptake

Radiochemical Syntheses, Volume 1: Radiopharmaceuticals for Positron Emission Tomography, First Edition. Edited by Peter J. H. Scott, Brian G. Hockley.

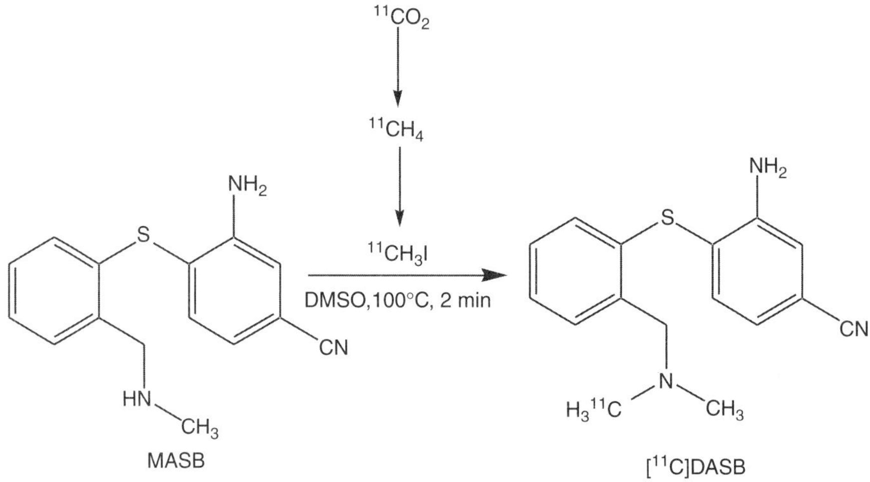

FIGURE 1 Radiosynthesis of [^{11}C]DASB.

and is involved in the pathophysiology of psychiatric disorders, for example, schizophrenia, mood disorders, depression, and anxiety [1–4]. Hence, it is the target for an important class of antidepressant drugs, the so-called SSRIs (selective serotonin reuptake inhibitors). SERT imaging, therefore, is an important tool in neuropsychiatric research and therapy monitoring [5–7]. For the visualization and quantification of the SERT, many highly selective carbon-11 and fluorine-18 labeled positron emission tomographic (PET) tracers have been developed and synthesized. The most extensively used tracers are [^{11}C]McN5256 [8], [^{11}C]DASB [9–12] (for reaction scheme see Fig. 1), and [^{11}C]MADAM [13].

[^{11}C]DASB combines all major prerequisites for a successful SERT-ligand, providing excellent biological properties and in vivo behavior [14,15], which makes it an excellent SERT-tracer for clinical use: high affinity (1.1 nM); excellent selectivity (NET/SERT = 1230 and DAT/SERT = 1300) NET, norepinephrine transporter; DAT, dopamine transporter; high specific binding; reversible, high brain uptake; and binding equilibrium within a reasonable timeframe [9,10].

2 SYNTHESIS PROCEDURES

CAUTION: All radiochemical syntheses must be carried out using appropriate equipment in a facility authorized for the use of radioactive materials. Personal protective equipment must be worn and all local radiation safety laws followed.

2.1 Preparation of the Synthesis Module

All parts used before high performance liquid chromatography (HPLC) purification (reactor, HPLC injector, tubing) were rinsed with dimethyl sulfoxide (DMSO) [16] and/or acetone [17] and dried in a stream of helium. In detail, Vial 2, tubing toward the reactor, transferring needle, tubing toward the HPLC injector, and the injector itself were rinsed with 5 mL of acetone and dried in a stream of helium for 3 min (100 mL/min). The reactor, transferring needle, tubing toward the HPLC injector, and the injector itself were subsequently rinsed with 5 mL of water-free DMSO and dried in a stream of helium for another 3 min (100 mL/min). Parts used after preparative HPLC [18] (i.e., solid phase extraction (SPE) tubing, dilution flask, product collection vial, product outlet tubing) were cleaned using ethanol [19] (5 mL) and physiological saline (0.9%) [20] (15 mL). A scheme of the synthesis module [21] is presented in Fig. 2.

Vessels were filled as follows:

Vial 2: 1 mL of HPLC eluent (i.e., acetonitrile [22] and 0.1 M ammonium acetate [23] (40/60; v/v)) for quenching of the reaction and dilution before HPLC injection;

Flask 3: 100 mL of sterile water for dilution of the lipophilic HPLC fraction after peak cutting;

Vial 4: 5 mL of physiological saline (0.9%) [20] for final rinsing of the SPE cartridge [24] and dilution of the eluate;

Vial 5: 1.5 mL ethanol [19] for elution of the SPE cartridge [24];

Vial 6: 10 mL sterile water for washing the SPE cartridge [24].

A C18 SPE cartridge [24] was preconditioned using 5 mL of ethanol [19] and 10 mL of sterile water, dried, and connected to its designated position. The product collection vial was filled with 4 mL of physiological saline (0.9%) [20],

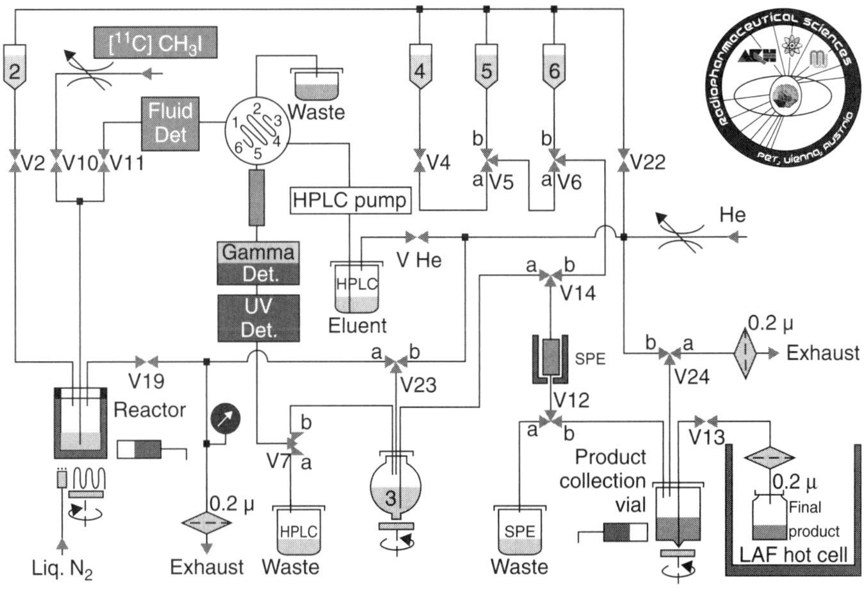

FIGURE 2 Illustration of the automated procedure for the radiosynthesis and purification of [^{11}C]DASB with a commercial C11-synthesizer.

1 mL of 3% saline [25], and 1 mL of phosphate buffer (125 mM) [26]. The reactor was filled with 1 mg MASB (*N*-methyl-2-(2-amino-4-cyanophenylthio)-benzylamine) [27] dissolved in 0.5 mL DMSO [16] and subsequently placed into the heating block.

2.2 Production of [^{11}C]CH_3I

[^{11}C]CO_2 was produced at the GE PETtrace cyclotron [28] by irradiation of a gas target containing N_2 and 0.5% O_2 using the $^{14}N(p,\alpha)^{11}C$ nuclear reaction with 16.5 MeV protons. Typical beam currents were 50–60 μA and irradiation was stopped as soon as the desired activity level was reached (i.e., 1.9–2.4 Ci (70–90 GBq)). [^{11}C]CH_3I was produced using the gas phase conversion described by Larsen et al. in 1997 [29] in an automated synthesis module [30]. Briefly, [^{11}C]CO_2 was trapped on a molecular sieve (4 Å) and converted to [^{11}C]CH_4 within 120 s in the presence of Ni catalyst [31] and hydrogen at 360°C. Subsequently, [^{11}C]CH_4 was reacted with elemental iodine [32] at 720°C to [^{11}C]CH_3I and recirculated for 300 s. [^{11}C]CH_3I was trapped online on Porapak N [33] in a small glass column and finally released by heating to 190°C in a stream of helium (45 mL/min) within 160 s.

2.3 Synthesis of [^{11}C]DASB

Gaseous [^{11}C]methyl iodide produced by the MeI-Microlab module [30] was delivered through V10 into the reactor of the previously prepared synthesizer system and trapped online in the reaction mixture containing 1 mg MASB [27] in 500 μL of DMSO [16] at 22°C. The reactor was heated to 100°C and kept at this temperature for 2 min to achieve conversion of MASB to [^{11}C]DASB. After cooling down to room temperature (22–25°C; 0.8 min), the reaction mixture was quenched by the addition of 1 mL of HPLC eluent (through V2) and subsequently transferred to the 5-mL injection loop of the HPLC system through V11 passing the fluid detector.

2.4 Purification and Formulation of [^{11}C]DASB

Injection of the crude reaction mixture to the semipreparative reverse phase high performance liquid chromatographic (RP-HPLC) column [34] was controlled by an automated fluid detector. Semipreparative HPLC was performed with acetonitrile [22] and 0.1 M ammonium acetate [23] (40/60; v/v) as the mobile phase with a flow of 8 mL/min (pressure: 150–160 bar). Chromatograms were registered using a UV detector (254 nm) and a NaI radioactivity detector in series. An example is given in Fig. 3.

The [^{11}C]DASB fraction was cut into flask 3 through V7 and diluted with 100 mL sterile water. The resulting solution was then pushed over a C18 SPE cartridge [24]. After washing with 10 mL of water (Vial 6), the pure product was completely eluted with 1.5 mL of ethanol [19] (Vial 5) and 5 mL of 0.9% saline [20] (Vial 4). Formulation was done with a further 4 mL of physiological saline (0.9%) [20], 1 mL of saline solution (3%) [25], and 1 mL of physiological phosphate buffer (125 mM) [26] in the product collection vial. This final solution was

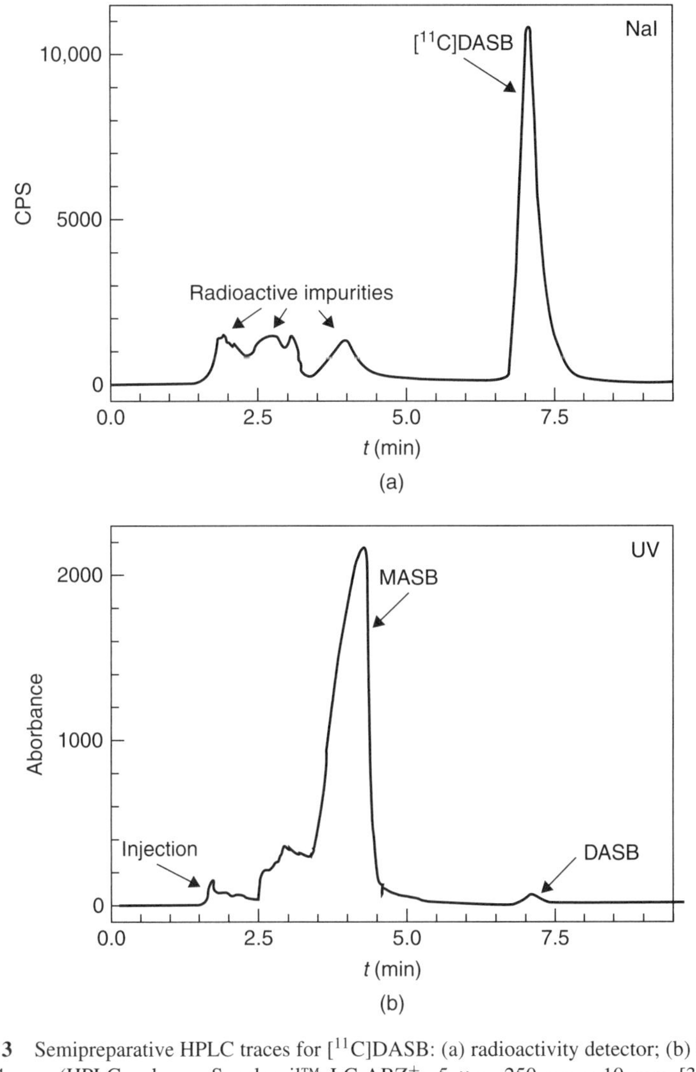

FIGURE 3 Semipreparative HPLC traces for [^{11}C]DASB: (a) radioactivity detector; (b) UV detector at 254 nm. (HPLC column: Supelcosil™ LC-ABZ$^+$; 5 μm, 250 mm × 10 mm [34]; HPLC conditions: mobile phase: acetonitrile [22] and 0.1 M ammonium acetate [23] (40/60; v/v); flow rate: 8 mL/min; pressure: 150–160 bar).

transferred to a shielded laminar-air-flow hot cell and sterile-filtered (0.22 μm) [35] into a sterile 25-mL vial [36] containing another 5 mL of physiological saline (0.9%) [20].

3 QUALITY CONTROL PROCEDURES

CAUTION: All radiopharmaceuticals produced for clinical use must have local regulatory approval before use in humans. Trained personnel must carry out

TABLE 1 QC Data for Three Repeat Runs of [^{11}C]DASB

QC Test	Release Criteria	Run 1	Run 2	Run 3
Yield mCi (GBq)	NA	86 (3.2)	111 (4.1)	121 (4.5)
Visual inspection	Clear and colorless	Pass	Pass	Pass
Radiochemical identity (min)	HPLC RT 6.5–7.2	6.72	6.80	6.75
Radiochemical purity (%)	≥95	99.9	99.8	99.9
Specific activity (GBq/μmol)	NA	81.5	132.8	53.6
Precursor amount [^{12}C]MASB (μg/mL)	NA	0.053	0.003	0.012
Residual solvent analysis	DMSO <5000 ppm	nd	nd	nd
	Acetonitrile <410 ppm	2.3 ppm	5 ppm	3 ppm
	Ethanol ≤ 10%	8.5%	8.5%	8.5%
Dose pH	5.0–8.5	7.5	7.6	7.3
Dose osmolality (mOsm/kg)	190–370	280	320	292
Sterile filter integrity test (psi)	>50	Pass	Pass	Pass
Radionuclidic identity ($t_{1/2}$) (min)	20.3 ± 2	19.8	20.1	20.5
Endotoxin analysis (IU/mL)	<17.5	<1	<1	<1
Sterility testing	No colony growth out to 14 d	Pass	Pass	Pass

NA, not applicable; RT, retention time; nd, not detected.

quality control (QC) procedures, and each dose must meet all established QC criteria before release to the clinic.

QC procedures for [^{11}C]DASB, based on the current requirements for radiopharmaceuticals laid out both in the European Pharmacopoeia [37] and in the US Pharmacopeia [38], are summarized below and can be performed within 9 min (except microbial and endotoxin testing). Complete QC data for three repeat batches of [^{11}C]DASB produced using the methods described herein are summarized in Table 1.

3.1 Visual Inspection

The [^{11}C]DASB dose is examined behind a 511-keV shielding lead glass window: the solution must be clear, colorless, and free of particles.

3.2 Radiochemical Identity and Radiochemical Purity

Chemical and radiochemical impurities were detected and quantified using analytical radio-HPLC. Acetonitrile [22] and 0.1 M ammonium formate [39] (55/45; v/v) were used as the mobile phase with a flow rate of 2 mL/min (pressure: 55–65 bar) on a reversed phase column [40]. The HPLC analysis was completed within 8 min, the retention time of the precursor (MASB [27]) was 4.4–4.8 min ($k' = 2.2$–2.7), and the product [^{11}C]DASB was eluted with a retention time of 6.5–7.2 min ($k' = 3.7$–4.6). The chemical identity of [^{11}C]DASB was determined

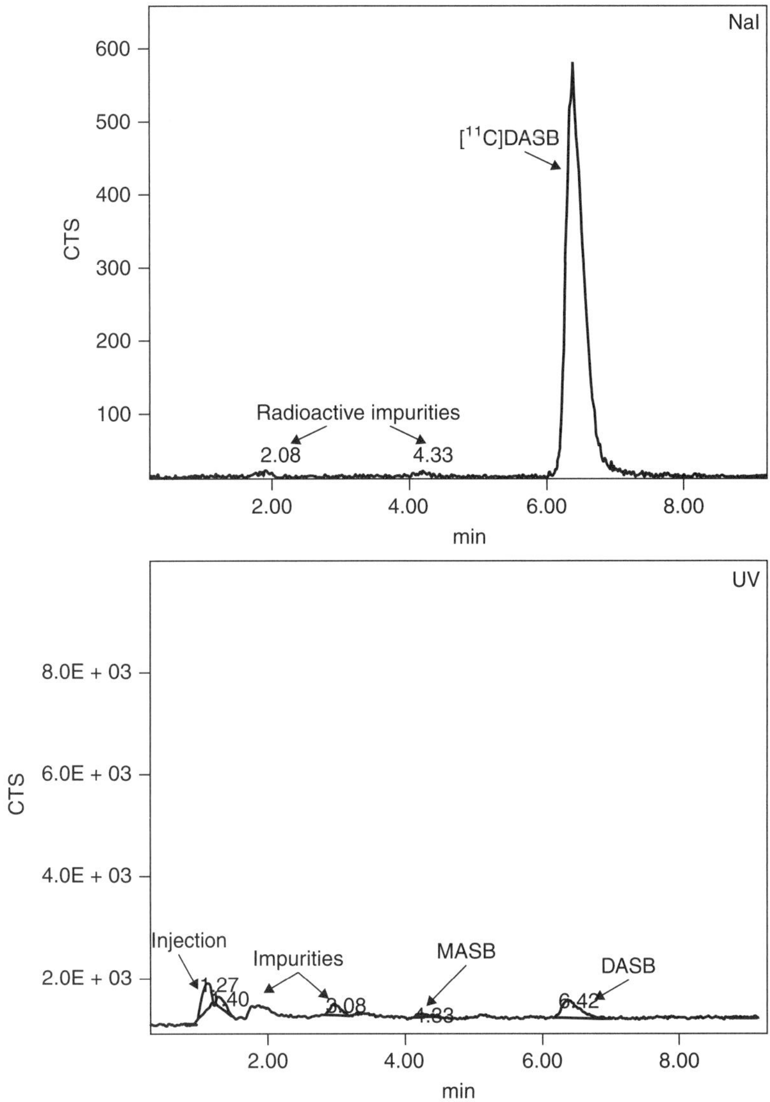

FIGURE 4 Analytical HPLC traces for [^{11}C]DASB: (a) radioactivity detector; (b) UV detector at 254 nm. (HPLC column: Prodigy 5 μm Phenyl-3 (PH-3); 250 × 4.6 mm [40]; HPLC conditions: mobile phase: acetonitrile [22] and 0.1 M ammonium formate [39] (55/45; v/v); flow rate: 2 mL/min; pressure: 55–65 bar).

by co-injection of the unlabeled reference compound, [^{12}C]DASB [41]. To check for contaminations with higher retention factors, we investigated the RP-HPLC chromatograms for 20 min, but no substance (both in the UV and the radioactivity trace) was observed eluting from the HPLC after 8 min. A sample chromatogram is given in Fig. 4.

In this example, [^{11}C]DASB accounts for 98.53% and the two radioactive impurities for 0.95% and 0.52%, respectively (in the radioactive trace). The peaks in the UV channel represent 0.050 μg/mL for MASB and 0.461 μg/mL for DASB.

3.3 Residual Solvents Analysis

Residual solvents were analyzed by gas chromatography (GC) with flame ionization detector (FID) [42] (carrier gas: He, flow: 2.7 mL/min, 45°C (2.5 min); 20°C/min to 110°C; 30°C/min to 200°C; 200°C (10 min); FID: 270°C).

The limits were <5000 ppm for DMSO, ≤10% for ethanol, and <410 ppm for acetonitrile [43].

3.4 Specific Activity

The specific activity (a_S) was defined as the result of the division of measured radioactivity of [^{11}C]DASB (gigabecquerel) by the amount of [^{12}C]DASB (micromoles). The amount of [^{12}C]DASB was determined using HPLC. (A calibration curve is determined by measuring known concentrations of an authentic DASB sample [41], recording the corresponding peak areas and fitting the data by linear regression). Thus, the specific activity is calculated by the following Equation 1:

$$a_S[\text{GBq}/\mu\text{mol}] = A[\text{GBq}]/(c_{\text{DASB}}\ [\mu\text{g/mL}] \times V[\text{mL}]/283.39\ [\text{g/mol}]) \qquad (1)$$

A is the measured radioactivity; c_{DASB} is the calculated concentration of unlabeled DASB through HPLC; V is the volume of the product solution.

3.5 Dose pH

The pH of the [^{11}C]DASB dose was analyzed by a WTW pH 537 Microprocessor pH Meter [44]. The limits of the pH values for this parenteral preparation were set to be 5.0–8.5.

3.6 Dose Osmolality

The osmolality of the [^{11}C]DASB dose was analyzed by a Wescor osmometer 5520 [45]: a 10-μL aliquot of the dose was applied onto a paper filter inside the osmometer chamber. The limits of the osmolality for this parenteral preparation were set to be 190–370 mosmol/kg.

3.7 Sterile Filter Integrity Test

The sterile filter [35] from the aseptic filling of the final [^{11}C]DASB solution (with needle still attached) was connected to a nitrogen supply via a regulator. The needle was submerged in water, and the nitrogen pressure was gradually increased. If the pressure was raised above the filter acceptance pressure (50 psi) without seeing a stream of bubbles, the filter was considered intact.

3.8 Radionuclidic Identity

γ-Spectra were recorded within a Packard Cobra II Auto-Gamma counter [46] and peaks analyzed. Only at 511 and 1022 keV, peaks were allowed. Since all

PET radioisotopes yield the same γ-trace, additionally, activities were measured at different time points using a Veenstra VDC404 Radioisotope Dose Calibrator [47], and the half-life was calculated using Equation 2. The calculated half-life must be 20.3 ± 2 min.

$$t_{1/2} = -\ln 2 \times (t_2 - t_1)/\ln\left(\frac{A_2}{A_1}\right) \tag{2}$$

In Equation 2, A_1 is the starting activity at t_1 and A_2 is the end activity at t_2.

3.9 Endotoxin Analysis

Bacterial endotoxin content in doses of [^{11}C]DASB was analyzed (when the probes were physically decayed) by an external laboratory [48] where the analyses were performed according to the current Pharmacopoeia Europaea [49] and US Pharmacopeia [50] guidelines using the limulus amoebocyte lysate (LAL) test [51]: For this, an aliquot of the decayed [^{11}C]DASB dose was added to the LAL and incubated at 37°C for 60 min; then, the extent of clotting was compared against a negative control and a positive standard. The limit is <17.5 IU/mL. The determination of this parameter was always done after injection to the patient when radioactivity was decayed, in accordance with the regulations laid down in the pharmacopeias [37,38].

3.10 Sterility Testing

Sterility testing was performed (when the probes were physically decayed) according to the current Pharmacopoeia Europaea [52] standard procedures by an external laboratory [48], whereby fluid thioglycolate media (FTM, sensitive for anaerobes and fungi) and soybean casein digest media (SCDM, sensitive for aerobes) [52,53] were used and aliquots of the decayed [^{11}C]DASB dose were added. The FTM samples were incubated at 30–35°C for 14 days, and the SCDM samples were incubated at 20–25°C and were visually compared against a negative control and a positive standard after a 14-day incubation period. The determination of this parameter was always done after injection to the patient when radioactivity was decayed, in accordance with the regulations laid down in the pharmacopeias [37,38].

CHEMICAL ABSTRACTS NOMENCLATURE (REGISTRY NUMBER)

Acetone (67-64-1)

Acetonitrile (75-05-8)

Ammonium acetate (631-61-8)

Ammonium formate (540-69-2)

Benzonitrile, 3-amino-4-[[2-[(dimethylamino)methyl]phenyl]thio] (=3-amino-4-[(2-(*N,N*-dimethylamino-methyl)phenylsulfanyl]-benzonitrile;= *N,N*-dimethyl-2-(2-amino-4-cyanophenylthio)benzylamine;=DASB) (627490-01-1)

Benzonitrile, 3-amino-4-[[2-[(methylamino)methyl]phenyl]thio] (=*N*-methyl-2-(2-amino-4-cyanophenylthio)-benzylamine;=*N*-desmethyl-DASB;= MASB) (296774-10-2)

Dimethylsulfoxide (67-68-5)

Disodiumhydrogenphosphate dihydrate (10028-24-7)

Ethanol (64-17-5)

Iodine (7553-56-2)

Ni (Shimalite 80/100 reduced) (7440-02-0)

Porapak N, Porapak™ Porous Polymer Adsorbent matrix, 80–100 mesh (no CAS no. available)

Sodium chloride (7647-14-5)

Sodium dihydrogenphosphate monohydrate (10049-21-5)

REFERENCES AND NOTES

1. Frankle WG, Narendran R, Huang Y, Hwang DR, Lombardo I, Cangiano C, Gil R, Laruelle M, Abi-Dargham A. Biol Psychiatry 2005;57:1510–1516.
2. Nikolaus S, Antke C, Beu M, Müller HW. Rev Neurosci 2010;21:119–139.
3. Gerretsen P, Pollock BG. Expert Opin Drug Metab Toxicol 2008;4:1465–1478.
4. Hesse S, Barthel H, Schwarz J, Sabri O, Müller U. Neurosci Biobehav Rev 2004;28:547–563.
5. Kasper S, Sacher J, Klein N, Mossaheb N, Attarbaschi-Steiner T, Lanzenberger R, Spindelegger C, Asenbaum S, Holik A, Dudczak R. Int Clin Psychopharmacol 2009;24:119–125.
6. Hinz R, Selvaraj S, Murthy NV, Bhagwagar Z, Taylor M, Cowen PJ, Grasby PM. J Cereb Blood Flow Metab 2008;28:1478–1490.
7. Erlandsson K, Sivananthan T, Lui D, Spezzi A, Townsend CE, Mu S, Lucas R, Warrington S, Ell PJ. Eur J Nucl Med Mol Imaging 2005;32:1329–1336.
8. Suehiro M, Scheffel U, Dannals RF, Ravert HT, Ricaurte GA, Wagner HN Jr. J Nucl Med 1993;34:120–127.
9. Wilson AA, Garcia A, Jin L, Houle S. Nucl Med Biol 2000;27:529–532.
10. Wilson AA, Ginovart N, Schmidt M, Meyer JH, Threlkeld PG, Houle S. J Med Chem 2000;43:3103–3110.
11. Haeusler D, Mien LK, Nics L, Ungersboeck J, Philippe C, Lanzenberger RR, Kletter K, Dudczak R, Mitterhauser M, Wadsak W. Appl Radiat Isot 2009;67:1654–1660.
12. Solbach C, Reischl G, Machulla HJ. Radiochim Acta 2004;92:341–344.
13. Tarkiainen J, Vercouillie J, Emond P, Sandell J, Hiltunen J, Frangin Y, Guilloteau D, Halldin C. J Labelled Comp Radiopharm 2001;44:1013–1023.
14. Frankle WG, Huang Y, Hwang DR, Talbot PS, Slifstein M, Van Heertum R, Abi-Dargham A, Laruelle M. J Nucl Med 2004;45:682–694.
15. Meyer JH. J Psychiatry Neurosci 2007;32:86–102.
16. Dimethylsulfoxide (DMSO, >99.9%, anhydrous; part No. 276855) was obtained from Sigma Aldrich (Vienna, Austria).
17. Acetone (HPLC grade LiChrosolv; part No. 1.00020.1000) was purchased from Merck (Darmstadt, Germany).
18. The preparative HPLC system consisted of a Rheodyne injector, a Sykam S1021 pump, and a UV detector (254 nm) and a radioactivity detector in series.
19. Ethanol was purchased from Merck (Darmstadt, Germany; part.No. 1.00986.1000) and used as received.

20. Saline 0.9% was purchased from Noriderm Enterprises, Nicosia, Cyprus (part.No. 1-28379) and used as received.
21. The fully automated synthesizer was a C11-methylation module originally purchased from Nuclear Interface® (now General Electric Medical Systems, Uppsala, Sweden).
22. Acetonitrile in HPLC grade was purchased from Merck, Darmstadt, Germany (part.No. 1.00029.1000) and used as received.
23. Ammonium acetate solution, 0.1 M, was prepared by dissolving 0.1 mol of ammonium acetate (Fluka, Buchs, Switzerland; part No. 17836) in 1000 mL of bidistilled water.
24. C18 Plus Sep-Pak® cartridges (part No. WAT020515) were purchased from Waters, Milford, MA, USA.
25. Saline 3% was obtained from Landesapotheke Salzburg, Austria.
26. Phosphate buffer, 125 mM, was prepared by dissolving 0.224 g sodium dihydrogenphosphate monohydrate and 1.935 g disodiumhydrogenphosphate dihydrate in 100 mL bidistilled water.
27. GMP grade precursor (MASB) was obtained from ABX (Advanced Biochemical Compounds, Radeberg, Germany; product.no. 4301).
28. The GE PETtrace cyclotron was from General Electric Medical Systems, Uppsala, Sweden.
29. Larsen P, Ulin J, Dahlstrom K, Jensen M. Appl Radiat Isot 1997;48:153–157.
30. MeI-Microlab, General Electric Medical Systems, Uppsala, Sweden.
31. Ni catalyst (Shimalite Ni, reduced, 80/100 mesh) was purchased from Shimadzu (Kyoto, Japan).
32. Iodine (sublimated GR for analysis ACS,ISO,Reag. Ph Eur, product no. 1.04761.0100) was purchased from Merck (Darmstadt, Germany).
33. Porapak® N (80/100 mesh) was obtained from Alltech Associates, Inc. (Grace GmbH & Co. KG, Vienna, Austria).
34. The semipreparative column was a Supelcosil™ LC-ABZ$^+$; 5 μm, 250 mm × 10 mm; (No. 59179) from Supelco, Bellfonte, PA, USA.
35. Millex GV low protein binding 0.22-μm filters were purchased from Millipore, Bedford, MA, USA (part.No. SLGV 033RS) and used as received.
36. Sterile vials, 25 mL, were purchased from Mallinckrodt Medical, Petten, Holland.
37. Pharmacopeia Europea <6.0/0125>. Radioactive pharmaceuticals. Ph. Eur. 6; 2008.
38. U.S. Pharmacopeia <823>. Radiopharmaceuticals for Positron Emission Tomography Compounding. USP 32-NF 27, 2009.
39. Ammonium formate, 0.1 M, was prepared by dissolving 0.1 mol of ammonium formate (Fluka, Buchs, Switzerland; part No. 17843) in 1000 mL of bidistilled water.
40. The analytical HPLC column (Prodigy 5 μm Phenyl-3 (PH-3); 250 × 4.6 mm) was obtained from Phenomenex (Aschaffenburg, Germany).
41. The reference standard (DASB) was obtained from ABX (Advanced Biochemical Compounds, Radeberg, Germany; product no. 4310).
42. Gas chromatography was performed using an HP 6890 series system (Agilent, Santa Clara, CA, USA) with flame ionization detector (FID) on a HP innowax capillary column.
43. European Pharmacopoeia 6th edition (2.4.24). Identification and Control of Residual Solvents. Ph.Eur. 6, Strasbourg, France; 2008.
44. The WTW pH 537 pH meter was purchased from WTW (Weilheim, Germany).
45. The Wescor osmometer 5520 was purchased from Sanova Medical Systems, Vienna, Austria.
46. The Cobra II Auto-Gamma counter was purchased from Packard, Meriden, CT, USA (now: Perkin Elmer).
47. The activimeter was a Veenstra VDC-404 and purchased from Veenstra (Joure, The Netherlands).
48. The external laboratory that performed both endotoxin analyses and sterility testings was the research laboratory of the Department for Infectious Diseases and Tropical Medicine of the Medical University of Vienna, Vienna, Austria.
49. European Pharmacopoeia 6th edition (2.06.14). Determination of Bacterial Endotoxins. Ph.Eur. 6, Strasbourg, France; 2008.
50. U.S. Pharmacopeia <85>. Bacterial Endotoxins Test. USP 32-NF 27; 2009.

51. The Pyrochrome® Limulus Amoebocyte Lysate (LAL)-Test was purchased from Cape Cod, East Falmouth, MA, USA.
52. European Pharmacopoeia 6th edition (2.06.01). Determination of Sterility. Ph.Eur. 6, Strasbourg, France; 2008.
53. U.S. Pharmacopeia <71>. Sterility Tests. USP 32-NF 27, 2009.

CHAPTER 29

SYNTHESIS OF [^{11}C]ACETATE

FILIPPO LODI, CLAUDIO MALIZIA, AND STEFANO BOSCHI

PET Radiopharmacy, Nuclear Medicine Unit, Policlinico S.Orsola-Malpighi, Bologna, Italy

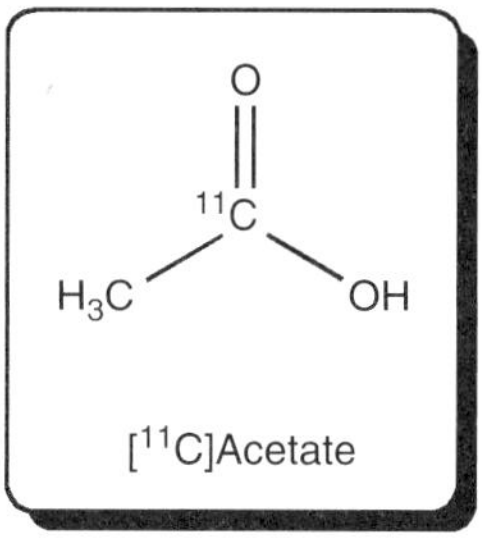

1 INTRODUCTION

Acetate is a metabolic substrate of β-oxidation for synthesis of cholesterol and lipids. This metabolite was labeled with ^{11}C and used as positron emission tomographic (PET) radiopharmaceutical for the study of myocardial oxidative metabolism [1,2] and recently as an oncological tracer [3]. [^{11}C]Acetate was employed in the imaging of prostate and kidney tumors [4–7] because is not eliminated through urination, allowing a better imaging of this kind of tumors than 2-[^{18}F]fluoro-2-deoxy-D-glucose ([^{18}F]FDG).

2 SYNTHESIS PROCEDURES

CAUTION: All radiochemical syntheses must be carried out using appropriate equipment in a facility authorized for the use of radioactive materials. Personal protective equipment must be worn and all local radiation safety laws followed.

2.1 Production of [^{11}C]CO_2

[^{11}C]CO_2 was produced by nuclear reaction $^{14}N(p,\alpha)^{11}C$ on a mixture of nitrogen (N60 purity grade) and 1% oxygen using a high performance ^{11}C target in a

Radiochemical Syntheses, Volume 1: Radiopharmaceuticals for Positron Emission Tomography, First Edition. Edited by Peter J. H. Scott, Brian G. Hockley.

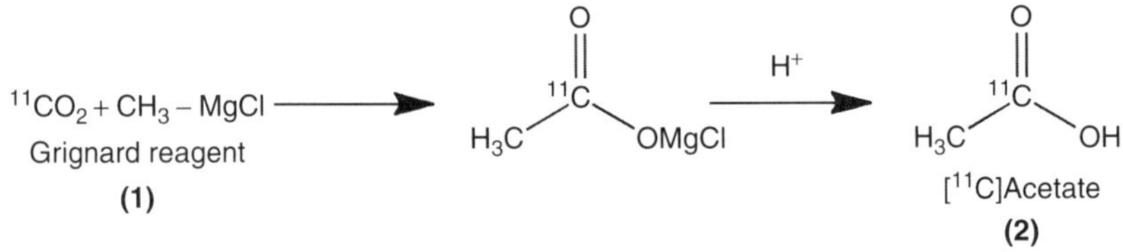

FIGURE 1 Synthesis of [^{11}C]acetate.

PETtrace cyclotron (16.5 MeV protons) [8]. The target was irradiated (70 μA current for 12 min) to produce ~3 Ci (111 GBq) of [^{11}C]CO_2.

2.2 Trapping and Purification of [^{11}C]CO_2

[^{11}C]CO_2 was delivered into TRACERlab FX_C ^{11}C methylation module [8] modified for the [^{11}C]acetate synthesis [9] and trapped in a coil cooled at −160°C with liquid nitrogen. The coil was then flushed with helium (20 mL/min for 1 min) to remove the gas target impurities.

2.3 [^{11}C]Carboxylation of Grignard Reagent by [^{11}C]CO_2

The coil was warmed at 90°C, and [^{11}C]CO_2 was delivered under a stream of helium (10 mL/min) into the reactor and bubbled into a solution of Grignard reagent methylmagnesium chloride (**1;** 0.1 M in tetrahydrofuran, 1 mL) [10] at room temperature within 3 min. After the reaction, 4 mL of CH_3COOH 1 mM [11] was added to hydrolyze the product (Fig. 1).

2.4 Purification of [^{11}C]Acetate

Disposable columns were conditioned before use as follows: PS–H^+, PS–Ag^+ with 10 mL of water, Maxiclean SAX was previously activated with 10 mL of ethanol 70% [12], 5 mL of NaCl 0.9% [13], and 10 mL of sterile water.

The reactor solution was delivered, under helium flow, through the solid phase extraction chromafix columns PS–H^+, PS–Ag^+ [14] and the anion-exchange column Maxiclean SAX [15]. The reactor was then rinsed with 4 mL of CH_3COOH 1 mM and the solution delivered though the columns. Finally, the Maxiclean SAX column was washed with 10 mL of sterile water and eluted with 10 mL of NaCl 0.9% solution [16,17]. The final solution was bubbled with helium for 2 min to remove the eventual residual of [^{11}C]CO_2 and was sterilized with 0.22-μm filter [18] in a sterile vial [19]. The module control panel is displayed in Fig. 2.

3 QUALITY CONTROL PROCEDURES

CAUTION: All radiopharmaceuticals produced for clinical use must have local regulatory approval before use in humans. Trained personnel must carry out quality control (QC) procedures, and each dose must meet all established QC criteria before release to the clinic.

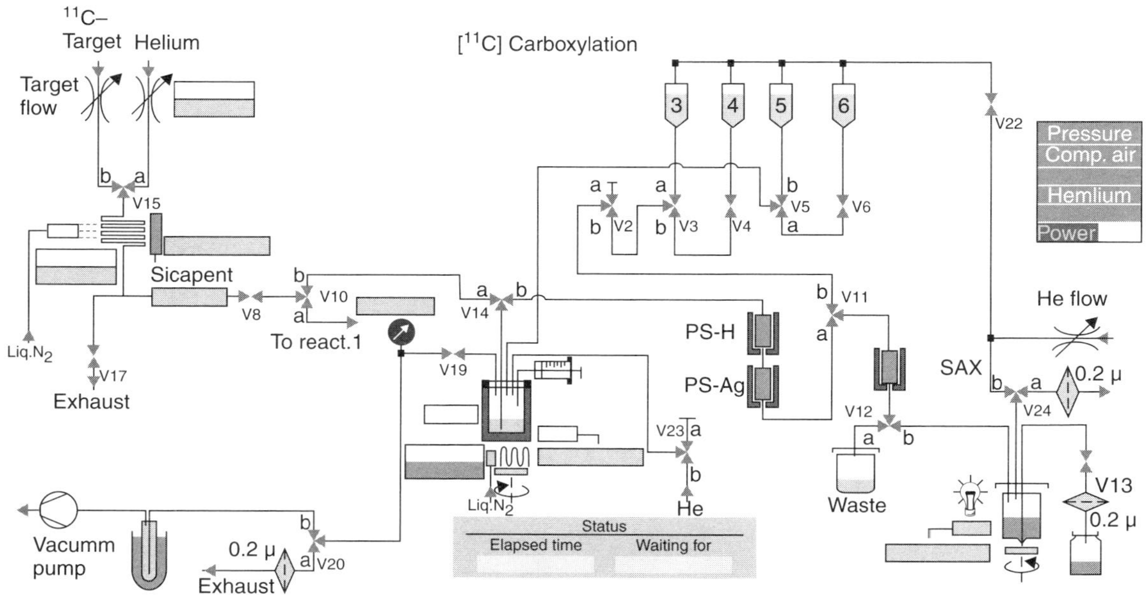

FIGURE 2 Module control panel for the synthesis of [^{11}C]acetate.

TABLE 1 QC Data for Three Repeat Runs of [^{11}C]Acetate

QC Test	Release Criteria	Run 1	Run 2	Run 3
Yield mCi (GBq) (DC, EOB)	N/A	2009 (74.3) 66%	1674 (61.9) 56%	2035 (75.3) 67%
Visual inspection	Clear, colorless	Clear, colorless	Clear, colorless	Clear, colorless
Radiochemical identity (min)	Method 1 (Eu. Ph.): RT = 6.3–7.7	RT = 7	RT = 7.1	RT = 7
	Method 2: RT = 2.7–3.3	RT = 3	RT = 3	RT = 2.9
Radiochemical purity (%)	Method 1 (Eu. Ph.): ≥95	>99	>99	>99
	Method 2: ≥95	>99	98.5	99
Chemical purity	<Standard area (28 mg/V)	<Standard area	<Standard area	<Standard area
pH	4.5–8.5	6.0	6.0	6.5
Residual solvents analysis	Ethanol <50 mg/V	0.04 mg/mL	0.05 mg/mL	0.05 mg/mL
	Acetone <50 mg/V	0.5 mg/mL	0.4 mg/mL	0.1 mg/mL
	Tetrahydrofuran (THF) <7.2 mg/V	0.04 mg/mL	0.1 mg/mL	0.09 mg/mL
Radionuclidic purity (%)	≥99	>99.99	>99.99	>99.99
Radionuclidic identity ($t_{1/2}$) (min)	19.9–20.9	20.3	20.4	20.4
Endotoxin analysis	<175/V IU/mL.	Pass	Pass	Pass
Sterility testing	According to Eu. Ph.	Pass	Pass	Pass

EOB, end of bombardment; N/A, not applicable; RT, retention time; DC, decay corrected.

QC of [^{11}C]acetate was based on the sodium acetate [1-^{11}C] injection monograph reported in the European Pharmacopoeia (Eu. Ph.) [20]. Complete QC results for three batches of [^{11}C]acetate produced with the method previously described are reported in Table 1. Each of three batches met the QC criteria.

3.1 Visual Inspection

[^{11}C]Acetate injectable solution must be clear and colorless.

3.2 Radiochemical Identity

Radiochemical identity was determined with two different high performance liquid chromatographic (HPLC) methods: method 1 (Eu. Ph. method) using a strong anion-exchange column CarboPac PA10 column (4 × 250 mm) [21] with NaOH 0.1 M [22] as mobile phase at a flow rate of 1 mL/min; method 2 [17] with PLRP-S 100A, 5 mm, 150 × 4.6 mm column [23] with $KH_2PO_4$0.4 M/acetonitrile 99.5/0.5 [24,25] as mobile phase at a flow rate of 1 mL/min. HPLCs were equipped with radio and UV detectors ($\lambda = 220$ nm). The retention time of [^{11}C]acetate is compared to that of sodium acetate reference standard [26] and must be ±10%.

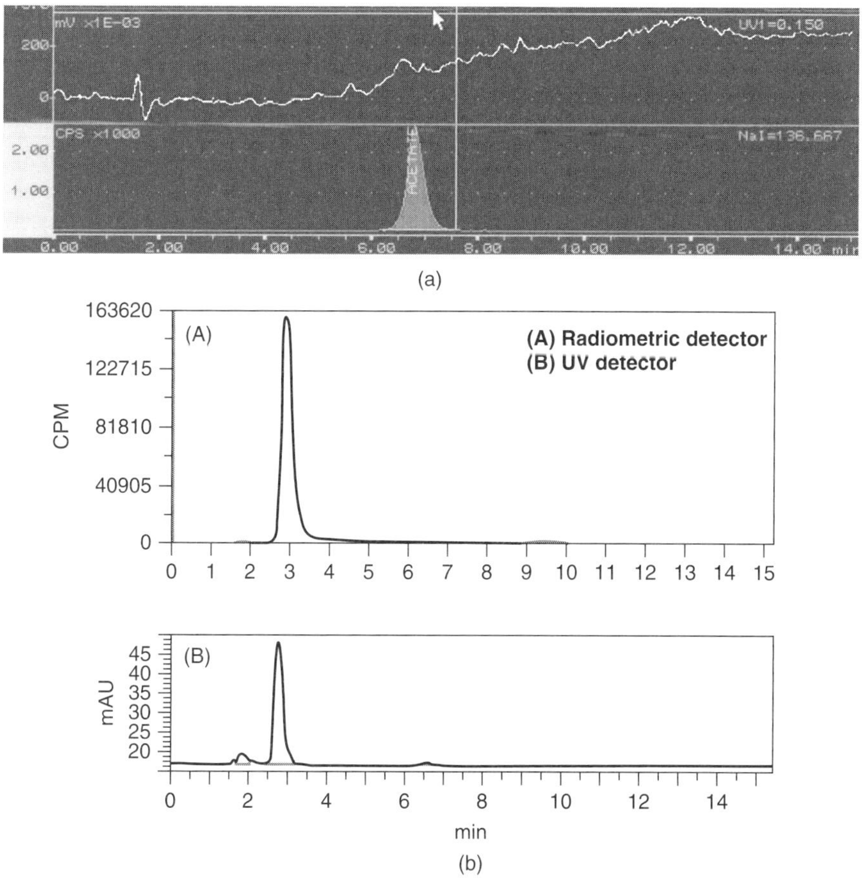

FIGURE 3 Analytical UV and radioactive HPLC traces for [^{11}C]acetate: (a) method 1 (Eu. Ph. method), (b) method 2.

3.3 Radiochemical Purity

Radiochemical purity was determined with two HPLC methods described above. Radiochemical purity must be a minimum of 95%. Representative analytical radioactive HPLC traces are displayed in Fig. 3.

3.4 Chemical Purity

Chemical purity was assessed with two HPLC methods described to determine radiochemical identity and radiochemical purity using a UV detector ($\lambda = 220$ nm) (Fig. 3). The area of acetate mass peak obtained by injecting the final solution of [^{11}C]acetate must be lower than the corresponding area of a sodium acetate reference solution: 28 mg of sodium acetate [26] dissolved in water and diluted to maximum recommended dose in milliliters (V).

3.5 Residual Solvents Analysis

Residual organic solvents were measured by gas chromatograph [27] on a capillary column RTX-200 (30 m × 0.3 mm) [28]. The limits for the residual solvents reported in the residual solvents monograph of the Eu. Ph. are 50 mg/V for acetone and ethanol (class 3) (from TRACERlab FX_C ^{11}C methylation module cleaning procedure) and 7.2 mg/V for tetrahydrofuran (class 2).

3.6 pH

pH was measured by a standard pH meter [29]. pH value of the injectable solution must be between 4.5 and 8.5.

3.7 Radionuclidic Identity

The half-life (19.9–20.9 min) was calculated by measuring activities using Capintec CRC® -15R radioisotope dose calibrator [30] and using the following equation:

$$t_{1/2} = -\ln 2(\text{time difference}/(\ln(\text{ending activity}/\text{starting activity})))$$

3.8 Radionuclidic Purity

Radionuclidic purity was determined by high resolution γ-ray spectrometry [31]. The spectrum obtained with [^{11}C]acetate solution does not differ significantly from that obtained with a standardized fluorine-18 solution. Radiochemical purity must be a minimum of 99% of the total radioactivity.

3.9 Endotoxin Analysis

Endotoxin contents were measured by the conventional limulus amebocyte lysate (LAL) test, using a gel clotting assay [32]. The limit is less than 175/V IU (international units)/mL.

3.10 Sterility Testing

Tests for sterility were done according to the Eu. Ph. [33].

Samples of [^{11}C]acetate were placed onto fluid thioglycolate media (FTM) plates and soybean casein digest media (SCDM) plates. [^{11}C]Acetate plates were incubated along with positive and negative controls for 14 days. FTM plates were incubated at 32°C, and SCDM plates were incubated at 22°C according to current Eu. Ph. guidelines.

CHEMICAL ABSTRACTS NOMENCLATURE (REGISTRY NUMBER)

Acetic acid (64-19-7)

Acetonitrile (75-05-8)

Ethanol (64-17-5)

Furan, tetrahydro- (109-99-9)

Magnesium, chloromethyl- (676-58-4)

REFERENCES AND NOTES

1. Brown MA, Myears DW, Bergmann SR. J Nucl Med 1989;30:187–193.
2. Buxton DB, Schwaiger M, Nguyen A, Phelps ME, Schelbert HR. Circ Res 1988;63:628–634.
3. Menendez JA, Lupu R. Nat Rev Cancer 2007;7:763–777.
4. Oyama N, Akino H, Kanamaru H, Suzuki Y, Muramoto S, Yonekura Y, Sadato N, Yamamoto K, Okada K. J Nucl Med 2002;43:181–186.
5. Albrecht S, Buchegger F, Soloviev D, Zaidi H, Vees H, Khan HG, Keller A, Bischof Delaloye A, Ratib O, Miralbell R. Eur J Nucl Med Mol Imaging 2007;34(2):185–196.
6. Morris MJ, Scher HI. Eur J Nucl Med Mol Imaging 2007;34(2):181–184.
7. Shreve PD, Chiao PC, Humes HD, Schwaiger M, Gross MD. J Nucl Med 1995;36:1595–1601.
8. G.E. Healthcare, USA.
9. Lodi F, Trespidi S, Di Pierro D, Marengo M, Farsad M, Fanti S, Franchi R, Boschi S. Appl Radiat Isot 2007;65(6):691–695.
10. Methylmagnesium chloride (0.1 M in tetrahydrofuran, 1 mL) was purchased from ABX, Germany (product no. 848.0001).
11. Glacial acetic acid (99–100%) was purchased from Sigma Aldrich and diluted to 1 mM concentration (product no. 338826-25).
12. Absolute ethanol was purchased from Carlo Erba reagenti and diluted with water to 70% (product no. 4146052).
13. Sodium chloride 0.9% was purchased by B.Braun (product no. 3631478).
14. Chromafix columns PS–H^+, PS–Ag^+ were obtained from Macherey-Nagel (product no. 731867, 731866).
15. Maxiclean SAX columns were obtained from Altech (product no. 21907).
16. Roeda D, Dolle' F, Crouzel C. Appl Radiat Isot 2002;57:857–860.
17. Soloviev D, Tamburella C. Appl Radiat Isot 2006;64:995–1000.
18. Millex-GS sterile 0.22-μm filters were purchased from Millipore (part no. SLGSV255F).
19. Sterile and apyrogenic Pyrovac vials were purchased from Acila (1960500).
20. European Pharmacopoeia 6.6 Edition.
21. Strong anion exchange column CarboPac PA10 column was purchased by Dionex (product no. 046110).
22. NaOH 1 M in water was obtained from Merck (Titrisol) and diluted to 0.1 M with distilled water (product no. 1.09959.0001).
23. PLRP-S column was obtained from Polymer Laboratories, UK (product no. 1111-3500).
24. KH_2PO_4 (98–100.5%) was obtained from Carlo Erba reagenti (product no. 361507).
25. Acetonitrile for HPLC was obtained from Carlo Erba reagenti (product no. 412392000).
26. Sodium acetate trihydrate was obtained from Sigma Aldrich (product no. 71188).
27. Varian, CP3380.
28. GC capillary column RTX-200 was purchased from Restek (product no. 15054).
29. pH meter was obtained from Mettler Toledo (MP225).
30. Capintec, Inc., USA.
31. Camberra, distributed in Italy by Technology Nuclear Electronics, Milan, Italy.
32. Lonza, Verviers, Belgium.
33. European Pharmacopoeia: Sterility chapter (2.6.1).

CHAPTER 30

SYNTHESIS OF *N*-(2-[^{11}C]METHOXYBENZYL)-*N*-(4-PHENOXYPYRIDIN-3-YL)ACETAMIDE ([^{11}C]PBR28)

QI-HUANG ZHENG, MIN WANG, AND BRUCE H. MOCK

Department of Radiology and Imaging Sciences, Indiana University School of Medicine, Indianapolis, Indiana

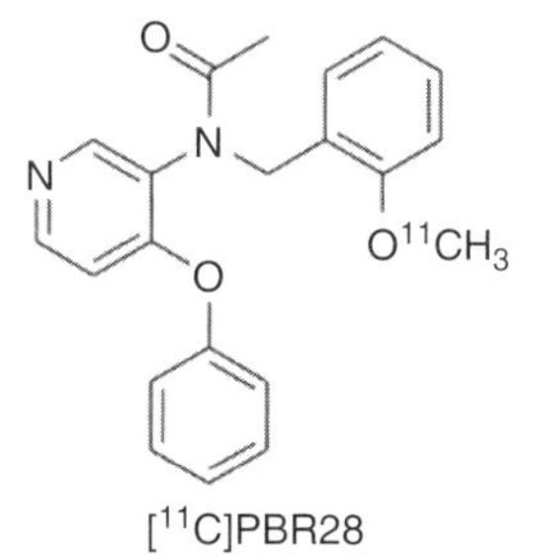

1 INTRODUCTION

Translocator protein (TSPO) 18 kDa, formerly known as the peripheral benzodiazepine receptor (PBR) [1], is a protein found in the lung, liver, heart, spleen, kidney, adrenals, brain, glial cells, masts cell, and macrophages and is implicated in numerous nervous system disorders such as epilepsy, cerebral ischemia, nerve injury, neurodegenerative diseases, and immune system diseases such as cancer [2]. Brain TSPO density increases in several neuropathological conditions and after experimental injuries to the central nervous system [3]. As a result, TSPO is an attractive target for molecular imaging of neuroinflammation such as Alzheimer's disease and tumor progression using positron emission tomography (PET) [4]. The prototypical TSPO-selective PET radioligand is [^{11}C]PK11195; however, it is reported to have many limitations such as low brain uptake and low sensitivity [5]. These limitations have motivated investigators to search for new TSPO PET radioligands. Promising candidates progressing to human PET studies include [^{11}C]DAA1106, [^{18}F]FEDAA1106, and [^{11}C]PBR28 [6–8]. The new

Radiochemical Syntheses, Volume 1: Radiopharmaceuticals for Positron Emission Tomography, First Edition. Edited by Peter J. H. Scott, Brian G. Hockley.

PET radioligand [^{11}C]PBR28 (*N*-(2-[^{11}C]methoxybenzyl)-*N*-(4-phenoxypyridin-3-yl)acetamide, IC_{50} 0.658 nM) was originally developed by Innis and Pike et al. at the National Institute of Mental Health (NIMH) [8–11]. Wishing to study this compound in our laboratory, we investigated a fully automated synthesis of [^{11}C]PBR28 using [^{11}C]methyl triflate ([^{11}C]CH_3OTf) [12,13].

2 SYNTHESIS PROCEDURES

CAUTION: All radiochemical syntheses must be carried out using appropriate equipment in a facility authorized for the use of radioactive materials. Personal protective equipment must be worn and all local radiation safety laws followed.

2.1 Synthesis of Precursor Desmethyl-PBR28 and Reference Standard PBR28

Precursor desmethyl-PBR28 (*N*-(2-hydroxybenzyl)-*N*-(4-phenoxypyridin-3-yl)acetamide) and reference standard PBR28 (*N*-(2-methoxybenzyl)-*N*-(4-phenoxypyridin-3-yl)acetamide) were synthesized in our laboratory as previously reported [14].

2.2 Production of [^{11}C]CH_3OTF

[^{11}C]CO_2 was produced by the ^{14}N(p,α)^{11}C nuclear reaction in a specialty gas consisted of 1% oxygen in nitrogen purchased from Praxair (Indianapolis, IN, USA) as the target gas. Using the Siemens RDS-111 Eclipse cyclotron [15]. High specific activity [^{11}C]CH_3OTf was produced via methyl bromide (CH_3Br) gas-phase production method [13] using the Eckert & Ziegler [^{11}C]methyl iodide/triflate Modular-Lab system [16].

2.3 Synthesis of [^{11}C]PBR28

As indicated in Fig. 1, the radiosynthesis was performed in an in-house automated multipurpose ^{11}C-radiosynthesis module that allows for measurement of specific radioactivity during synthesis [17,18]. In a small 5-mL reaction vial, the precursor desmethyl-PBR28 (0.5–1.0 mg) was dissolved in 500 μL of acetonitrile. To this solution was added ~1 mg of dry sodium hydride (NaH). [^{11}C]CH_3OTf was delivered to the reaction vial at room temperature until the trapped radioactivity

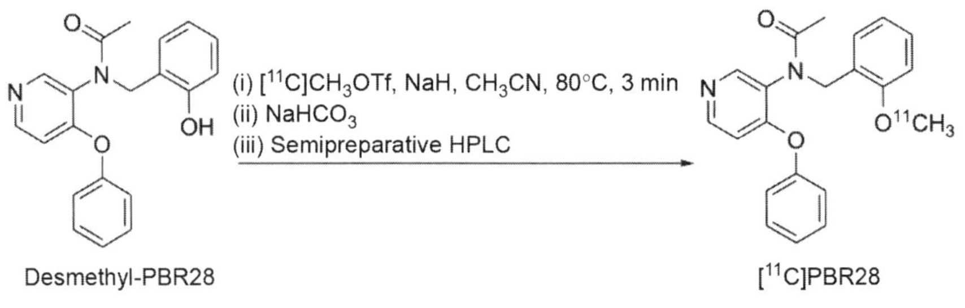

FIGURE 1 Synthesis of [^{11}C]PBR28.

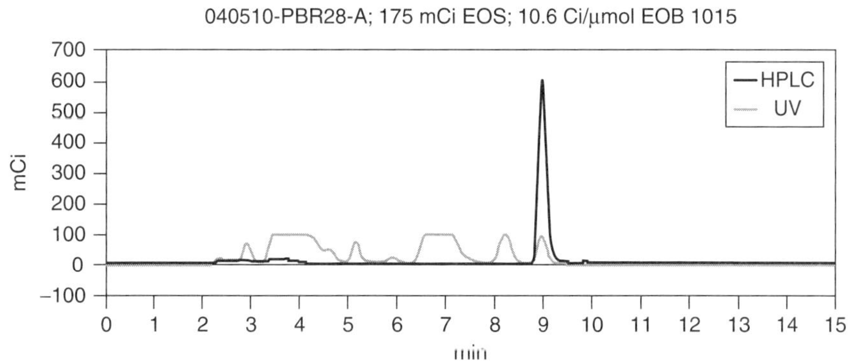

FIGURE 2 Semipreparative radioactive HPLC and UV traces for [^{11}C]PBR28.

reached a maximum (~2 min), and the reaction vial was isolated and reacted at 80°C for 3 min.

2.4 Purification and Formulation of [^{11}C]PBR28

After the reaction was complete, the reaction vial contents were diluted with aqueous $NaHCO_3$ solution (2 mL, 0.1 M), transferred to a 3-mL injection loop, and injected onto the semipreparative high performance liquid chromatographic (HPLC) column for purification (Phenomenex Prodigy, S-5 μm, 12 nm, 10 × 250 mm C-18 column, P/No. 00G-4097-N0 [19]; 52% CH_3CN/H_2O mobile phase; flow rate 5.0 mL/min; and UV (254 nm) and γ-ray (PIN diode) detectors). A representative semipreparative chromatogram from the radiosynthesis system is shown in Fig. 2. The product fraction was collected at ~9 min, diluted with 30 mL of Milli-Q water, passed through a Sep-Pak Plus C-18 cartridge (Waters, Milford, MA, P/No. WAT020515) [20], and washed with 5 × 4 mL of Milli-Q water. The trapped product was then eluted with ethanol (2 × 0.5 mL) and sterile saline (10 mL) and sterile-filtered through a Millex-FG membrane filter (0.2 μm, 25 mm) (Millipore, Bedford, MA, Cat. No. SLFG025LS) [21] into a sterile product vial.

3 QUALITY CONTROL PROCEDURES

CAUTION: All radiopharmaceuticals produced for clinical use must have local regulatory approval before use in humans. Trained personnel must carry out quality control (QC) procedures, and each dose must meet all established QC criteria before release to the clinic.

QC procedures for [^{11}C]PBR28, based on the current requirements for radiopharmaceuticals laid out in the United States Pharmacopeia (USP) [22], are summarized below. Complete QC data for three repeat batches of [^{11}C]PBR28 produced using the method disclosed herein are summarized in Table 1. Each of the three doses met all of the established QC criteria.

TABLE 1 QC Data for Three Repeat Runs of [^{11}C]PBR28

QC Test	Release Criteria	Run 1	Run 2	Run 3
Yield (%)	NA	85	83	80
Visual inspection	Clear, colorless	Clear, colorless	Clear, colorless	Clear, colorless
Radiochemical identity	RRT = 0.9–1.1	1.0	1.0	1.0
Radiochemical purity (%)	≥90	100	99	100
Chemical purity (precursor limit) (μg/dose)	<1	0.2	0.3	0.3
Specific activity at TOHA (Ci/μmol)	0.5	1.5	2.0	2.0
Residual solvent analysis (mg/mL)	$CH_3CN \leq 0.41$	0.11	0.20	0.17
Dose pH	4.5–7.5	6.0	6.5	6.0
Sterile filter integrity test (psi)	≥13	≥13	≥13	≥13
Radionuclidic identity ($t_{1/2}$) (min)	18.4–22.4	20	20	20
Endotoxin analysis (EU/mL)	≤17.5	≤2	≤2	≤2
Sterility testing	No colony growth out to 14 d	Pass	Pass	Pass

NA, not applicable.

3.1 Visual Inspection

[^{11}C]PBR28 should be clear and colorless with no visual evidence of cloudiness or particulate matter. [^{11}C]PBR28 should be stored in a sterile glass product vial (prelabeled with the appropriate batch number) at room temperature and in a lead-shielded container.

3.2 Radiochemical Identity

HPLC analysis of radiochemical identity was conducted using a Waters Breeze HPLC System [20] fitted with a variable wavelength UV detector and a Carroll-Ramsey γ-detector [23] (column: Phenomenex Prodigy S-5 μm ODS-3v, 100A, C18, 4.6 × 250 mm, P/No. 00G-3300-E0 [19]; mobile phase: 40% acetonitrile, 60% 0.1 M aqueous $NaHCO_3$, pH 8.5; flow rate 2.0 mL/min; $\lambda = 254$ nm). The retention time of [^{11}C]PBR28 is compared to that of the [^{12}C]PBR28 reference standard [14] and must be within ±10% (the relative retention time (RRT) must be 0.9–1.1). In addition, the retention time of [^{11}C]PBR28 on the semipreparative HPLC column used for postreaction purification shall correspond (±15%) to that of a standard solution of [^{12}C]PBR28.

3.3 Radiochemical Purity

An aliquot of the final [^{11}C]PBR28 formulation is subjected to analytical HPLC using a radioactivity detector. The area of the radioactivity peak corresponding

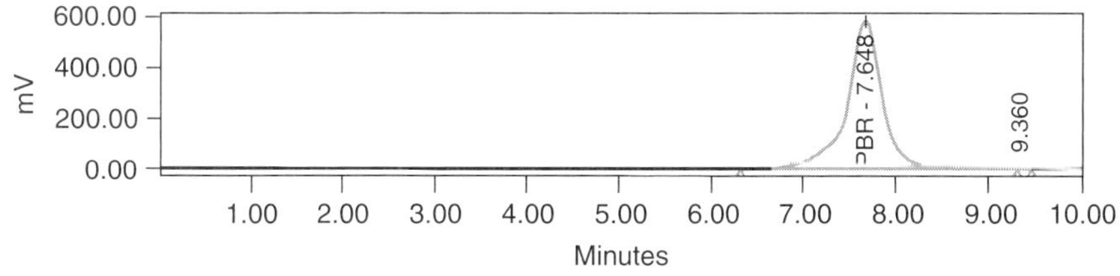

FIGURE 3 Analytical radioactive HPLC trace for [^{11}C]PBR28.

to [^{11}C]PBR28 shall represent ≥90% of the sum of the areas of all radioactive peaks as shown in Fig. 3.

3.4 Chemical Purity

An aliquot of the final [^{11}C]PBR28 formulation is subjected to analytical HPLC using a UV detector operated at 254 nm wavelength. Such analysis will be conducted simultaneously with the analysis mentioned in Section 3.3. The summed areas of all non-PBR28 UV peaks is converted to microgram per milliliter using the reference PBR28 calibration curve rather than expressed as a fixed percentage as shown in Fig. 4. The limit for these impurities has been established at 1 μg per administered dose or 10% of the permitted PBR-28 mass (see Section 3.5).

3.5 Specific Activity

The specific activity of [^{11}C]PBR28 shall be not less than 0.5 Ci/μmol at the time of human administration (TOHA). Specific activity of [^{11}C]PBR28 (as curie per micromole at end of bombardment (EOB)) is automatically determined immediately after the product peak elutes from the semipreparative HPLC column within the [^{11}C]radiosynthesis module. By integrating the area under the decay-corrected radioactivity peak and the UV peak from the semipreparative chromatogram, and by comparing the peak areas with previously stored calibration curves for each detector, the specific activity at EOB (as curie per micromole) is calculated by dividing the radioactivity measurement (millicurie at EOB) by the mass measurement (nanomoles). The expiration period (or the time required for the measured

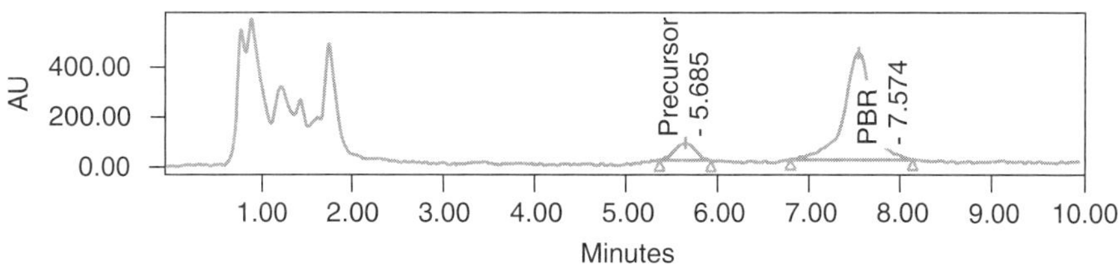

FIGURE 4 Analytical UV HPLC trace for [^{11}C]PBR28.

specific activity to fall below 0.5 Ci/μmol) can be calculated using the following equation:

$$\text{Expiration period} = \ln[(0.5\ \text{Ci}/\mu\text{mol})/(\text{calculated SpAct})]/(-0.0342)\ \text{in minutes}$$

Adding the expiration period (in minutes) to the EOB time shall determine the exact time of the day at which the [^{11}C]PBR28 batch will expire and can no longer be administered to a human subject. Such a specific activity limit ensures that no more than 10.4 μg of PBR-28 is administered per 15-mCi dose of ^{11}C-radioactivity.

3.6 Residual Solvent Analysis

Per USP Chapter 467 limits on volatile organic impurities, there shall be $\leq$0.41 mg/mL of acetonitrile in the final formulation of [^{11}C]PBR28. Such a determination is made by injecting 0.5 μL of the final formulation into a gas chromatograph equipped with a capillary column and flame ionization detector (FID) and comparing the peak area corresponding to acetonitrile to the peak area measured by injecting 0.5 μL of a known 0.41 mg/mL acetonitrile standard solution into that same gas chromatography (GC) system on the same day.

3.7 Dose pH

The pH of the [^{11}C]PBR28 product solution is analyzed by applying a small amount of the product solution to colorpHast® pH 2.0–9.0 nonbleeding pH-indicator strips [24]. The pH of [^{11}C]PBR28 should be between 4.5 and 7.5.

3.8 Sterile Filter Integrity Test

The integrity of the 0.2-μm Millex-FG filter is confirmed by a bubble point pressure test. After the synthesis and formulation of [^{11}C]PBR28 is completed, and before the release of product for human administration, the filter assembly shall be rinsed with methanol and tested to meet a bubble point specification of $\geq$13 psi. Any bubble observed through the outlet needle at a pressure of $<$13 psi will indicate a ruptured or defective membrane filter, and would require that the product formulation be resterilized with a new membrane filter assembly into a new sterile glass product vial before release for human administration.

3.9 Radionuclidic Identity

Repeat assay of an aliquot of the final product, performed using a Capintec CRC®-15R radioisotope dose calibrator [25] of established accuracy, shall result in a calculated half-life of between 18.4 and 22.4 min ($t_{1/2}$ of the carbon-11 radionuclide = 20.4 min). The half-life was calculated using the following equation.

$$t_{1/2} = -\ln 2(\text{time difference}/(\ln(\text{ending activity}/\text{starting activity})))$$

3.10 Endotoxin Analysis

Endotoxin content of the [^{11}C]PBR28 product solution was analyzed by a Charles River Laboratories EndoSafe Portable Testing System [26]. According to the USP, the final [^{11}C]PBR28 formulation shall contain no more than 175 EU/*V* (USP endotoxin units), in which *V* is the maximum volume of human subject administration in milliliters.

3.11 Sterility Testing

Aliquots of [^{11}C]PBR28 product solution were placed into fluid thioglycolate medium (FTM) tubes and soybean casein digest agar medium (SCDAM) tubes. FTM is used to test for anaerobes, aerobes, and microaerophiles, while SCDAM is used to test for nonfastidious and fastidious microorganisms [27]. [^{11}C]PBR28 tubes were incubated along with positive and negative controls for 14 days. FTM tubes were incubated at 32°C, and SCDAM tubes were incubated at 22°C according to current USP guidelines [28]. Tubes were visually inspected on the third, eighth, and fourteenth days of the test period and compared to the positive and negative standards. Positive standards must show growth (turbidity) in the tubes, and [^{11}C]PBR28/negative controls must have no culture growth after 14 days to be indicative of sterility.

CHEMICAL ABSTRACTS NOMENCLATURE (REGISTRY NUMBER)

Acetamide, *N*-[(2-hydroxyphenyl)methyl]-*N*-(4-phenoxy-3-pyridinyl)- (desmethyl-PBR28) (1005325-42-7)

Acetamide, *N*-[(2-methoxyphenyl)methyl]-*N*-(4-phenoxy-3-pyridinyl)- (PBR28) (253307-65-2)

Acetamide, *N*-[[2-(*methoxy*-^{11}C)phenyl]methyl]-*N*-(4-phenoxy-3-pyridinyl)- ([^{11}C]PBR28) (1005325-43-8)

Acetonitrile (75-05-8)

Ethanol (64-17-5)

Methyl triflate (333-27-7)

Sodium bicarbonate (144-55-8)

Sodium hydride (7646-69-7)

REFERENCES AND NOTES

1. Papadopoulos V, Baraldi M, Guilarte TR, Knudsen TB, Lacapere J-J, Lindemann P, Norenberg MD, Nutt D, Weizman A, Zhang M-R, Gavish M. Trends Pharmacol Sci 2006;27:402–409.
2. Wang M, Gao M, Hutchins GD, Zheng Q-H. Eur J Med Chem 2009;44:2748–2753.
3. Benavides J, Fage D, Carter C, Scatton B. Brain Res 1987;421:167–172.
4. Probst KC, Izquierdo D, Bird JLE, Brichard L, Franck D, Davis JR, Fryer TD, Richards HK, Clark JC, Davenport AP, Weissberg PL, Warburton EA. Nucl Med Biol 2007;34:439–446.
5. Venneti S, Lopresti BJ, Wiley CA. Prog Neurobiol 2006;80:308–322.

6. Yasuno F, Ota M, Kosaka J, Ito H, Higuchi M, Doronbekov TK, Nozaki S, Fujimura Y, Koeda M, Asada T, Suhara T. Biol Psychiatry 2008;64:835–841.
7. Fujimura Y, Ikoma Y, Yasuno F, Suhara T, Ota M, Matsumoto R, Nozaki S, Takano A, Kosaka J, Zhang MR, Nakao R, Suzuki K, Kato N, Ito H. J Nucl Med 2006;47:43–50.
8. Brown AK, Fujita M, Fujimura Y, Liow JS, Stabin M, Ryu YH, Imaizumi M, Hong J, Pike VW, Innis RB. J Nucl Med 2007;48:2072–2079.
9. Imaizumi M, Kim HJ, Zoghbi SS, Briard E, Hong J, Musachio JL, Ruetzler C, Chuang DM, Pike VW, Innis RB, Fujita M. Neurosci Lett 2007;411:200–205.
10. Imaizumi M, Briard E, Zoghbi SS, Gourley JP, Hong J, Fujimura Y, Pike VW, Innis RB, Fujita M. Neuroimage 2008;39:1289–1298.
11. Briard E, Zoghbi SS, Imaizumi M, Gourley JP, Shetty HU, Hong J, Cropley V, Fujita M, Innis RB, Pike VW. J Med Chem 2008;51:17–30.
12. Jewett DM. Int J Radiat Appl Instrum A 1992;43:1383–1385.
13. Mock BH, Mulholland GK, Vavrek MJ. Nucl Med Biol 1999;26:467–471.
14. Wang M, Yoder KK, Gao M, Mock BH, Xu X-M, Saykin AJ, Hutchins GD, Zheng Q-H. Bioorg Med Chem Lett 2009;19:5636–5639.
15. Siemens, USA.
16. Eckert & Ziegler Isotope Products GmbH, Berlin, Germany.
17. Mock BH, Zheng Q-H, DeGrado TR. J Labelled Comp Radiopharm 2005;48:S225.
18. Mock BH, Glick-Wilson BE, Zheng Q-H, DeGrado TR. J Labelled Comp Radiopharm 2005;48:S224.
19. Phenomenex, Inc., Torrance, CA, USA.
20. Waters Technologies Corporation, Milford, MA, USA.
21. Millipore Corporate Headquarters, Billerica, MA, USA.
22. U.S. Pharmacopeia <823>. Radiopharmaceuticals for Positron Emission Tomography Compounding. USP 32-NF 27; 2009.
23. Carroll-Ramsey Associates, Berkeley, CA, USA.
24. EMD Chemicals, Inc., USA.
25. Capintec, Inc., USA.
26. Charles River Laboratories, USA.
27. Becton, Dickinson and Company, USA.
28. U.S. Pharmacopeia <71>. Sterility Tests. USP 32-NF 27; 2009.

PART III

OTHER RADIOPHARMACEUTICALS

CHAPTER 31

SYNTHESIS OF [^{13}N]AMMONIA ([^{13}N]NH_3)

PETER J. H. SCOTT

Department of Radiology, University of Michigan School of Medicine, Ann Arbor, Michigan

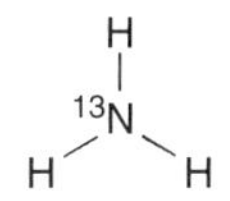

[^{13}N]NH_3

1 INTRODUCTION

[^{13}N]Ammonia has been used routinely as a radiotracer in cardiac positron emission tomographic (PET) imaging for decades [1–4] and has a well-established role in the diagnosis and care of adult patients with coronary artery disease. Moreover, [^{13}N]ammonia is considered a safe and effective radiopharmaceutical and was approved by the US Food and Drug Administration (FDA) for PET imaging of the myocardium under rest and pharmacological stress conditions to evaluate myocardial perfusion in patients with suspected or existing coronary artery disease [5].

2 SYNTHESIS PROCEDURES

CAUTION: All radiochemical syntheses must be carried out using appropriate equipment in a facility authorized for the use of radioactive materials. Personal protective equipment must be worn and all local radiation safety laws followed.

2.1 Production of [^{13}N]Nitrites

Milli-Q [^{16}O]H_2O (1.5 mL) [6] was loaded into the [^{13}N]nitrogen target [7] of a General Electric Medical Systems (GEMS) PETTrace cyclotron [8]. The target

Radiochemical Syntheses, Volume 1: Radiopharmaceuticals for Positron Emission Tomography, First Edition. Edited by Peter J. H. Scott, Brian G. Hockley.

FIGURE 1 Synthesis of [^{13}N]ammonia.

was bombarded (30 μA beam for 6 min) to generate [^{13}N]NO_3^- and [^{13}N]NO_2^- enriched water by the $^{16}O(p,\alpha)^{13}N$ nuclear reaction.

2.2 Synthesis of [^{13}N]Ammonia

Production of [^{13}N]ammonia (Fig. 1) is carried out in a GE PETtrace [^{13}N]ammonia process module. Sodium hydroxide (3 × 100 mg pellets) [9] and DeVarda's alloy [10] (100 mg) are placed into the reaction vessel of the ammonia cassette [11], and 0.9% sodium chloride for injection [12] (4 mL) is added to the trapping vessel. [^{13}N]NO_3^-- and [^{13}N]NO_2^--enriched water is delivered into the reaction vessel from the GEMS PETtrace cyclotron, and both anions get reduced to [^{13}N]NH_3 in the presence of aqueous sodium hydroxide and DeVarda's alloy. Concomitantly with this reduction, the heat generated during the exothermic dissolution of sodium hydroxide in water is sufficient to distill [^{13}N]NH_3 over to the trapping vessel where it is dissolved in 0.9% sodium chloride for injection to yield 4 mL of formulated drug product. The total manufacturing process takes approximately 20 min per batch and is completely automated by the GEMS PETtrace cyclotron. The final drug product was then passed through a sterile filter [13] into a sterile vial [14] to provide [^{13}N]ammonia (typically 50–70 mCi (1.85–2.59 GBq)) in an isotonic solution released for quality control (QC).

3 QUALITY CONTROL PROCEDURES

CAUTION: All radiopharmaceuticals produced for clinical use must have local regulatory approval before use in humans. Trained personnel must carry out QC procedures, and each dose must meet all established QC criteria before release to the clinic.

QC procedures for [^{13}N]ammonia, based on the current requirements for radiopharmaceuticals laid out in the United States Pharmacopeia (USP) [15], are summarized below. In accordance with the USP, and reflecting the short half-life of nitrogen-13, one lot of [^{13}N]ammonia is prepared per day for analysis, and all pertinent QC is carried out on this lot. Subsequent doses are submitted directly to the clinic without QC. Complete QC data for three repeat batches of [^{13}N]ammonia produced using the method described herein are summarized in Table 1. Each of the three doses met all of the established QC criteria.

3.1 Visual Inspection

The [^{13}N]ammonia dose is examined behind a PET L-block and must be clear, colorless, and free of particulate matter.

TABLE 1 QC Data for Three Repeat Runs of [^{13}N]Ammonia

QC Test	Release Criteria	Run 1	Run 2	Run 3
Yield mCi (GBq)	NA	63 (2.3)	71 (2.6)	69 (2.6)
Visual inspection	Clear, colorless	Clear, colorless	Clear, colorless	Clear, colorless
Radiochemical identity	RRT = 0.9–1.1	1.0	1.0	1.0
Radiochemical purity (%)	≥95	99.9	100	99.9
Dose pH	4.5–7.5	5.0	5.0	5.0
Sterile filter integrity test (psi)	>50	>50	>50	>50
Radionuclidic identity ($t_{1/2}$) (min)	9.5–10.5	9.89	9.91	9.91
Endotoxin analysis (EU/mL)	≤43.75	≤2	≤2	≤2
Sterility testing	No colony growth out to 14 d	Pass	Pass	Pass
Residual alumina (μg/mL)	<10	<1.4	<2.0	<1.4
Radionuclidic purity (%)	>99.9	>99.9	>99.9	>99.9

NA, not applicable.

3.2 Radiochemical Identity

Radiochemical identity was determined using a Shimadzu 10AD LC system [16] fitted with a Shimadzu CDD-10AVP conductivity detector and a Bioscan SS420X γ-detector [17] (column: Phenomenex Luna SCX analytical column (4.6 × 50 mm) [18]; mobile phase: 40 mM aqueous potassium nitrate, pH 3.1 [19], flow rate: 0.5 mL/min). The retention time of [^{13}N]NH_3 is compared to that of the [^{14}N]NH_4Cl reference standard [20] and must be ±10% (relative retention time (RRT) must be 0.9–1.1).

3.3 Radiochemical Purity

Radiochemical identity was determined using a Shimadzu 10AD LC system [16] fitted with a Shimadzu CDD-10AVP conductivity detector and a Bioscan SS420X γ-detector [17] (column: Phenomonex Luna SCX analytical column (4.6 × 50 mm) [18]; mobile phase: 40 mM aqueous potassium nitrate, pH 3.1 [19], flow rate: 0.5 mL/min). Radiochemical purity must be >95%. Representative analytical high performance liquid chromatograhic (HPLC) traces are displayed in Fig. 2.

3.4 Dose pH

The pH of the [^{13}N]ammonia dose was analyzed by applying a small amount of the dose to colorpHast® pH 2.0–9.0 nonbleeding pH-indicator strips [20] and determined by visual comparison to the scale provided. Dose pH must be 4.5–7.5.

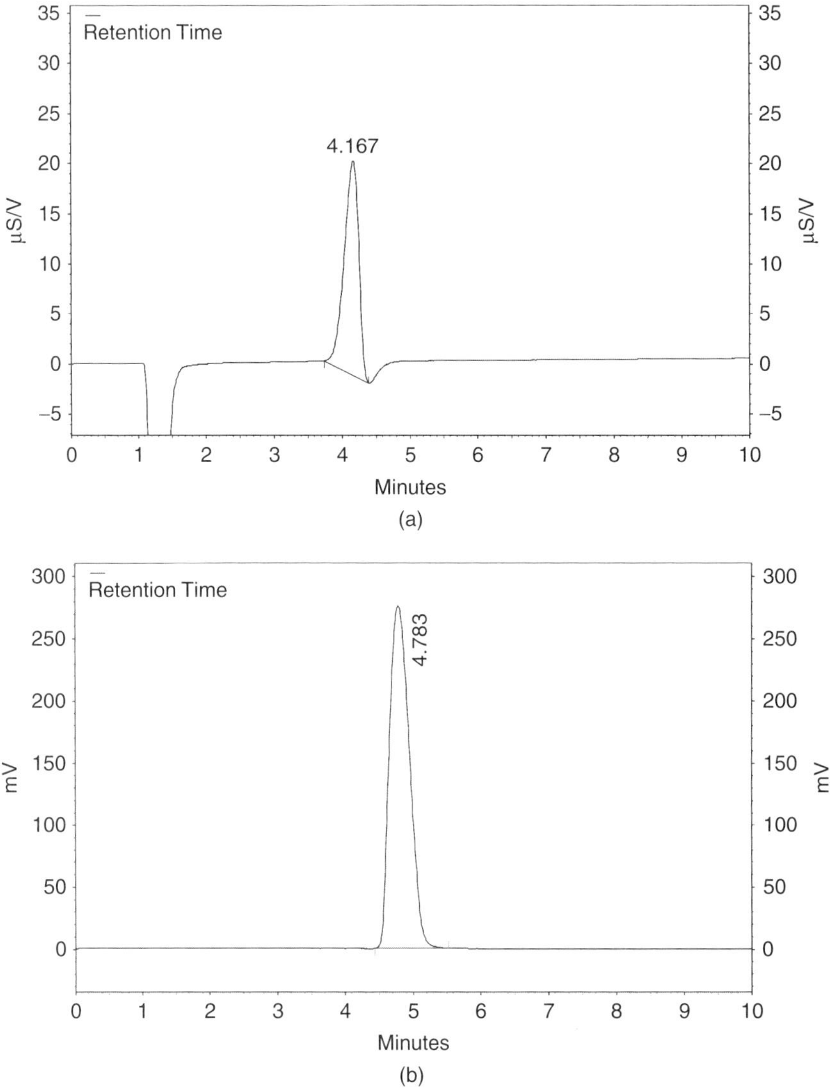

FIGURE 2 Typical analytical conductivity (a) and radioactivity (b) HPLC traces for [^{13}N]ammonia.

3.5 Sterile Filter Integrity Test

The sterile filter from the [^{13}N]ammonia (with needle still attached) was connected to a nitrogen supply via a regulator. The needle was submerged in water, and the nitrogen pressure was gradually increased. If the pressure was raised above the filter acceptance pressure (50 psi) without seeing a stream of bubbles, the filter was considered intact. If a stream of bubble occurs at <50 psi, the test fails.

3.6 Radionuclidic Identity

Activities were measured using a Capintec CRC®-15R Radioisotope Dose Calibrator [21], and half-life was calculated using Equation 1. Calculated half-life must be 9.5–10.5 min.

$$t_{1/2} = -\ln 2(\text{time difference}/(\ln(\text{ending activity}/\text{starting activity}))) \quad (1)$$

3.7 Endotoxin Analysis

Endotoxin content in doses of [^{13}N]ammonia was analyzed by a Charles River Laboratories EndoSafe® Portable Testing System [22] and according to the USP [23]. Doses must contain ≤175 endotoxin units (EU) per dose (≤43.75 EU/mL).

3.8 Sterility Testing

Samples of [^{13}N]ammonia were placed onto fluid thioglycolate media (FTM) plates and soybean casein digest media (SCDM) plates. FTM tubes are used to test for anaerobes, aerobes, and microaerophiles, whereas SCDM tubes are used to test for nonfastidious and fastidious microorganisms [24]. [^{13}N]ammonia tubes were incubated along with positive and negative controls for 14 days. FTM tubes were incubated at 32°C, and SCDM tubes were incubated at 22°C according to current USP guidelines [25]. Tubes were visually inspected on the third, eighth, and fourteenth days of the test period and compared to the positive and negative standards. Positive standards must show growth (turbidity) on the tubes and [^{13}N]ammonia/negative controls must have no culture growth after 14 days to be indicative of sterility.

3.9 Residual Alumina Analysis

DeVarda's alloy is an alloy of aluminum (44–46%), copper (49–51%), and zinc (4–6%). To determine if any of this alloy is contaminating the final formulated drug, representative lots 1–3 were submitted to EMSL Analytical for determination of residual aluminum [26]. Levels of aluminum were analyzed using inductively coupled plasma mass spectrometry (ICP-MS) and must be <10 μg/mL.

3.10 Radionuclidic Purity Testing

Radionuclidic purity is analyzed by γ-ray spectrometry, and doses of [^{13}N]ammonia from our facility are allowed to decay and then sent out to a contract research organization (CRO) (Dade Moeller and Associates [27]) to test for the presence of long-lived radioactive contaminants. Doses must have radionuclidic purity of >99.9%.

CHEMICAL ABSTRACTS NOMENCLATURE (REGISTRY NUMBER)

Ammonia (7664-41-7)
Ammonium chloride (12125-02-9)

Sodium hydroxide (1310-73-2)
DeVarda's alloy (8049-11-4)

For detailed supplier information, see Appendix 1 (page 335).

REFERENCES AND NOTES

1. Clark JC, Aigbirdhio FI. Chemistry of nitrogen-13 and oxygen-15. In: Welch MJ, Redvanly CS, editors. Handbook of radiopharmaceuticals. Chichester, West Sussex, England: John Wiley and Sons, Inc.; 2003. pp.119–140.
2. Phelps ME, Hoffman EJ, Raybaud C. Stroke 1977;8:694–702.
3. Harper PV, Schwartz J, Beck RN, Lathrop KA, Lembares N, Krizek H, Gloria I, Dinwoodie R, McLaughlin A, Stark VJ, Bekerman C, Hoffer PB, Gottschalk A, Resnekov L, al-Sadir J, Mayorga A, Brooks HL. Radiology 1973;108:613–617.
4. Walsh WF, Fill HR, Harper PV. Semin Nucl Med 1977;7:59–66.
5. vom Dahl J. Z Kardiol 2001;90:835–847.
6. Milli-Q $[^{16}O]H_2O$ was obtained from a Millipore Milli-Q Gradient A10 Water Purification System.
7. Nitrogen-13 target system and ammonia processing system for the production of $[^{13}N]NH_3$ (P5230JB).
8. GE Healthcare, USA.
9. Sodium hydroxide pellets were purchased from Sigma Aldrich (part no. 221465-500G).
10. DeVarda's alloy was purchased from Sigma Aldrich (part no. 269484-100G).
11. Ammonia cassettes were purchased from GE Healthcare (part no. E8680AD).
12. Sodium chloride, USP, 0.9%, was purchased from Hospira (part no. 0409-4888-50) and used as received.
13. Millex-GS sterile 0.22-μm filters were purchased from Millipore (part no. SLGSV255F) and used as received.
14. Ten-milliliter sterile vials were purchased from Hollister-Stier (part no. 7515ZA) and vented with a sterile Millex-FG vent filter (part no. SLFG025LS).
15. U.S. Pharmacopeia <823>. Radiopharmaceuticals for Positron Emission Tomography-Compounding. USP 32–NF 27; 2009.
16. Shimadzu Corporation, USA.
17. Bioscan, Inc., USA.
18. Analytical HPLC Column: Luna SCX column (4.6 × 50 mm), purchased from Phenomenex (part no. 00B-4398-E0); flow rate = 0.5 mL/min; analytical column was equilibrated with mobile phase for 30–60 min before use.
19. Analytical mobile phase: 40 mM aqueous potassium nitrate, pH adjusted to 3.1 with nitric acid.
20. Ammonium chloride reference standard ($[^{14}N]NH_4Cl$, 99.998%) was purchased from Sigma Aldrich (part no. 254134).
21. Capintec, Inc., USA.
22. Charles River Laboratories, USA.
23. U. S. Pharmacopeia <85>. Bacterial Endotoxins Test. USP 32–NF 27; 2009.
24. Becton, Dickinson and Company, USA.
25. U. S. Pharmacopeia <71>. Sterility Tests. USP 32–NF 27; 2009.
26. EMSL Analytical, Inc.
27. Dade Moeller and Associates, Gaithersburg, MD, USA.

CHAPTER 32

SYNTHESIS OF [^{68}Ga]GALLIUM DOTA-(TYR^{3})-OCTREOTIDE ACETATE ([^{68}Ga]-DOTATOC)

HARALD EIDHERR AND FRIEDRICH GIRSCHELE

Department of Nuclear Medicine, Medical University of Vienna, Vienna, Austria

MARKUS MITTERHAUSER

Department of Nuclear Medicine, Medical University of Vienna, Vienna, Austria
Department of Pharmaceutical Technology and Biopharmaceutics, University of Vienna, Vienna, Austria

WOLFGANG WADSAK

Department of Nuclear Medicine, Medical University of Vienna, Vienna, Austria
Department of Inorganic Chemistry, University of Vienna, Vienna, Austria

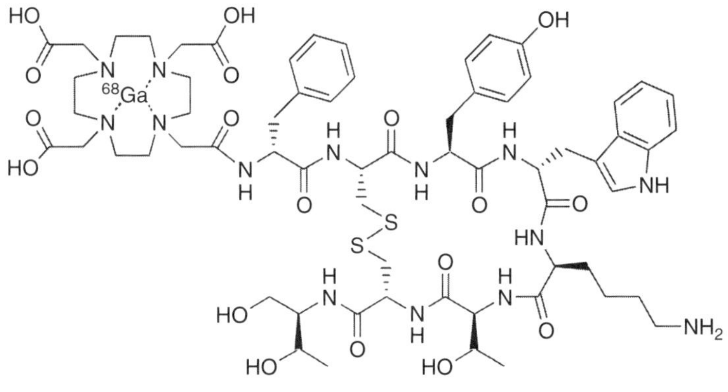

^{68}Ga-DOTATOC

Radiochemical Syntheses, Volume 1: Radiopharmaceuticals for Positron Emission Tomography, First Edition. Edited by Peter J. H. Scott, Brian G. Hockley.

1 INTRODUCTION

Positron emission tomography (PET) of somatostatin receptors (SSTRs) using Gallium-68 radiolabeled DOTA-(Tyr^3)-octreotide acetate (DOTATOC) is a sensitive and specific technique to visualize neuroendocrine tumors such as gastrinomas, insulinomas, and carcinoids that overexpress $SSTR_{1-5}$ whereby $SSTR_2$ and $SSTR_5$ are displayed predominantly [1–3]. In comparison to the formerly used [^{111}In]Indium DOTA-octreotide (DOTA-OCT), ^{68}Ga-DOTATOC has both better physical and pharmacokinetic properties and has therefore become a frequently asked PET-tracer in clinical routine [1–3].

Gallium-68 is an interesting short-lived positron emitter ($t_{1/2} = 67.6$ min, β^+ abundance = 89%) since it is available via a radionuclide generator system in which the long-lived germanium-68 ($t_{1/2} = 270.8$ days) decays directly to the ground state of its daughter nuclide, gallium-68, by pure electron capture. Even in 1961, the first ^{68}Ge/^{68}Ga generator [4] was developed in which germanium-68 was adsorbed on alumina and gallium-68 activity was eluted with ethylenediaminetetraacetic acid (EDTA) [4]. Later on, generators based on 1-deoxy-1-(methylamino)sorbitol-(*N*-methylglucamine) polymers were developed [5], which were eluted with citric acid. Other generators are based on tin(II) oxide [6] and are eluted with 0.6 M hydrochloric acid. Yet, the only successfully used and widely commercially available ^{68}Ge/^{68}Ga generators are based on titan(IV) oxide [7], which are eluted with normal 0.1 M hydrochloric acid in contrast to the other systems that require ultrapure solutions, milliQ-water, and so on, because of their high sensitivity against metallic impurities (Fe^{3+}, Ni^{3+}, Zn^{2+}, Sn^{4+}, Al^{3+}, etc.).

A typical titan(IV)-oxide-based ^{68}Ge/^{68}Ga generator [7] with a nominal ^{68}Ge activity of 1850 MBq (50 mCi) has a typical shelf-life of one year or 300 elutions and should be eluted daily for reproducible generator performance; in practice, it takes about 7 h after the previous elution until full ^{68}Ge/^{68}Ga equilibrium is reached and the maximum ^{68}Ga activity can be eluted (~70% of the available ^{68}Ge activity at a given time point).

There are three principle labeling modalities using ^{68}Ga: the first one is the so-called anion concentration method in which the eluted [^{68}Ga]Ga^{3+} ions are transformed into the anionic [[^{68}Ga]$GaCl_4^-$] complex by the addition of 5.5 M hydrochloric acid. Thus, this complex can be concentrated on a small anion exchange column and then eluted with a small amount of water (whereby the anionic complex disintegrates and the free [^{68}Ga]Ga^{3+} ions are released) [8]. The second method is the so-called cation concentration method in which the eluted [^{68}Ga]Ga^{3+} ions are directly retained on a small cation exchange column. Subsequent elution is performed with acetone/0.05 M hydrochloric acid (97.6/2.4, v/v) [9].

The two above-mentioned methods yield very pure and highly concentrated ^{68}Ga eluates but are rather time consuming. Owing to radioactive decay, the overall radiochemical yield is not better than that when using the third so-called fractionation method [10,11]. This fractionated elution is based on the fact that

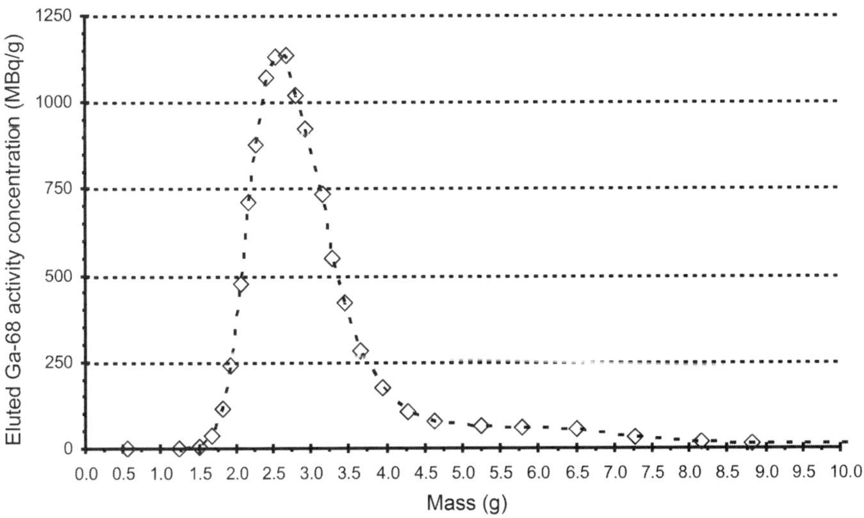

FIGURE 1 Elution profile of a titan(IV)-oxide-based 1850 MBq $^{68}Ge/^{68}Ga$ generator.

about 60–70% of the available ^{68}Ga activity at a given time point is eluted within a volume of 1.5–1.8 mL (Fig. 1). This method is easier to set up in a fully automated synthesizer (implementing GMP requirements), which is less time consuming and reliable [10,11].

2 SYNTHESIS PROCEDURES

CAUTION: All radiochemical syntheses must be carried out using appropriate equipment in a facility authorized for the use of radioactive materials. Personal protective equipment must be worn and all local radiation safety laws followed.

2.1 Production of [^{68}Ga]$GaCl_3$

[^{68}Ga]$GaCl_3$ was eluted from a $^{68}Ge/^{68}Ga$ generator [7] with 10 mL of 0.1 M hydrochloric acid [12]. For reproducible radiolabeling results, the generator should be eluted daily with pre-elution after extended pauses (e.g., weekends) to minimize trace ion contamination (e.g., Fe^{3+}, Zn^{2+}, Ti^{4+}) of the eluate, which would hamper the ^{68}Ga-complexation competitively.

For fractionated elution, an individual elution profile of each generator in the local labeling setup (depending on the type of radiosynthesis module used, length of tubing, etc.) has to be determined (Fig. 1). In the presented setup, the generator was linked to a Raytest SynChrom synthesizer [13] whereby typically a void volume of 0.4 mL was observed. The following 1.6 mL fraction contained only <1% of the elutable activity, whereas the next 1.8 mL fraction yielded about 60% of the achievable total ^{68}Ga activity. The final 6.2-mL fraction (~40% of ^{68}Ga activity) was discarded.

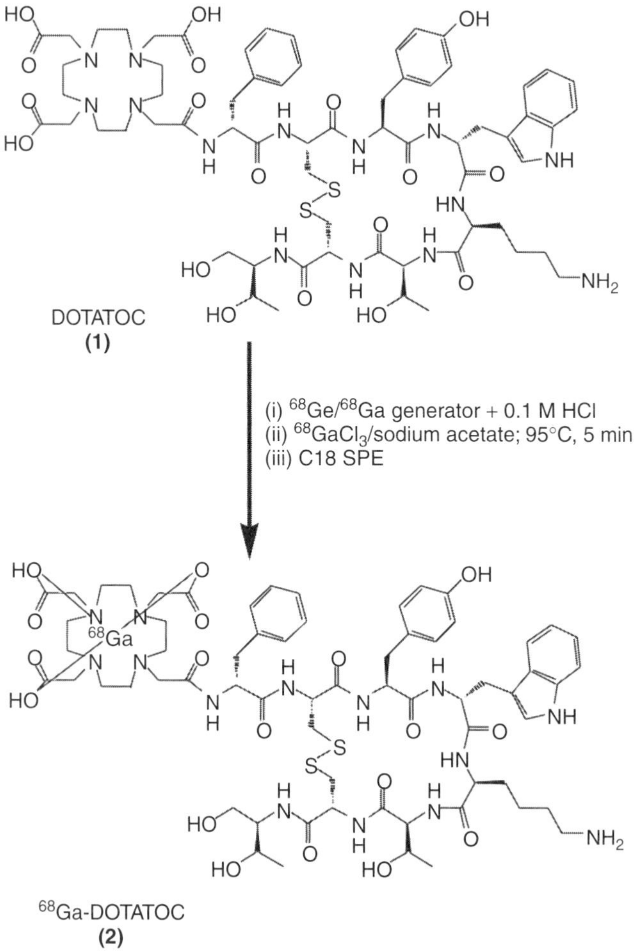

FIGURE 2 Synthesis of [^{68}Ga]gallium DOTA-(Tyr3)-octreotide acetate (^{68}Ga-DOTATOC).

2.2 Synthesis of ^{68}Ga-DOTATOC

The reaction scheme for the preparation of ^{68}Ga-DOTATOC is given in Fig. 2. A schematic illustration of the synthesizer setup is given in Fig. 3. All tubing of the module had to be rinsed first with ethanol [14] and subsequently with bidistilled water [15] before radiosynthesis. The radiochemical procedure was based on a previously published method [10,11] and performed on the Raytest SynChrom synthesizer [13]. The main ^{68}Ga activity in a volume of 1.8 mL of 0.1 M hydrochloric acid was eluted into the glass reaction vessel containing 155 μL of 1 M sodium acetate [16] and 50 μg of DOTA-(Tyr^3)-octreotide acetate (DOTATOC) [17] dissolved in 50 μL of 0.05 M hydrochloric acid [18]. For satisfactory formation of ^{68}Ga-DOTATOC, the total reaction volume was kept

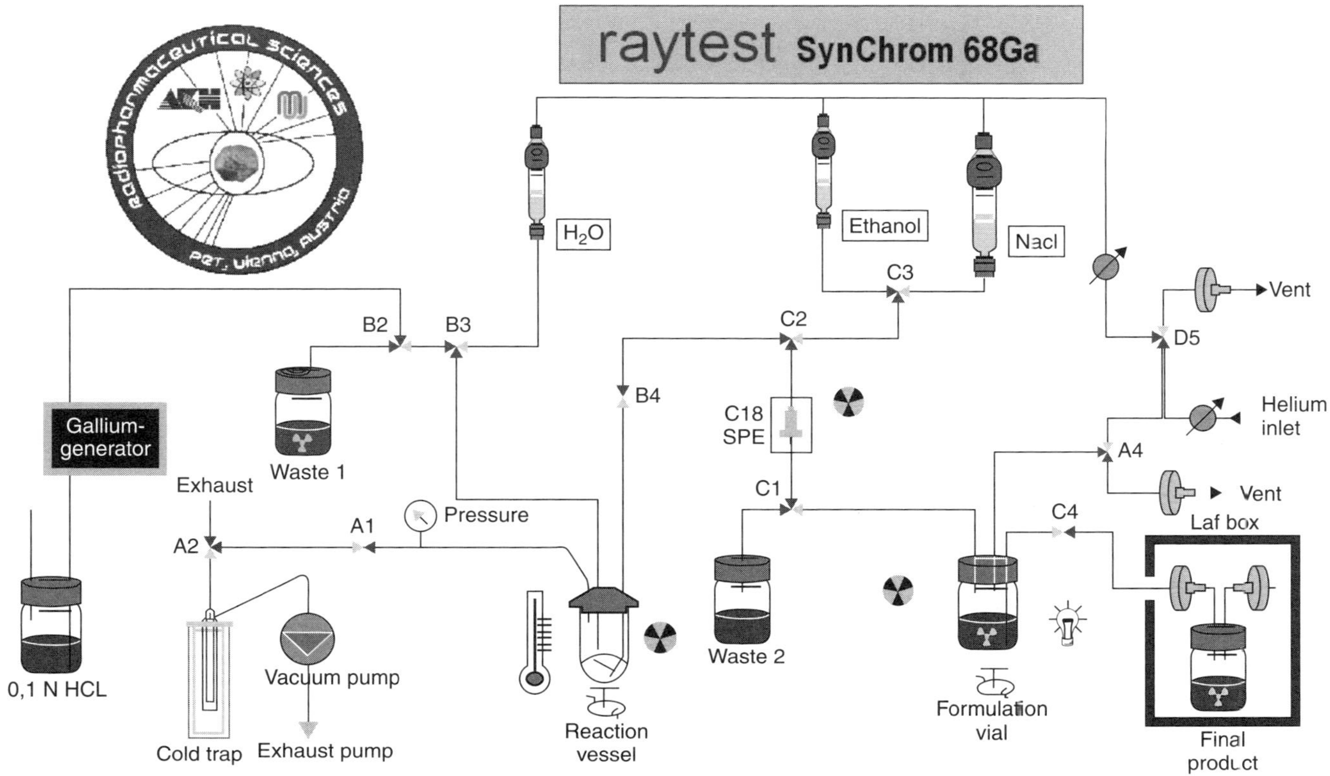

FIGURE 3 Reaction scheme for ^{68}Ga-DOTATOC (fractionation method).

below 2.0 mL at pH 2–3. The reaction solution was heated to 95°C for 5 min under stirring.

2.3 Purification and Formulation of ^{68}Ga-DOTATOC

For purification of the crude reaction mixture, the raw product was transferred to a RP18 solid phase extraction (SPE) cartridge [19] where ^{68}Ga-DOTATOC was retained and ^{68}Ga impurities passed through (into a waste vial). After washing with 2 mL of bidistilled water [15], the purified ^{68}Ga-DOTATOC was eluted from the cartridge with 0.6 mL of ethanol [14] followed by 2 mL of 0.9% saline [20] and was transferred to a formulation vial containing 6 mL of 0.9% saline [20] and 1 mL of 125 mM sodium phosphate buffer, pH 7.4. (This buffer was prepared by dissolving 0.224 g of sodium dihydrogenphosphate monohydrate [21] and 1.935 g of disodiumhydrogenphosphate dihydrate [22] in 100 mL of bidistilled water [15].) Finally, the resulting formulation (9.6 mL) was sterile filtered to avoid possible microbial contamination under aseptic conditions. A 0.22-µm filter [23] was used, and the product solution was collected in a 11-mL sterile vial [24] equipped with a vented air needle filter [25] to prevent liquid spillage. The whole automated synthesis procedure was completed within 18 ± 1 min.

3 QUALITY CONTROL PROCEDURES

CAUTION: All radiopharmaceuticals produced for clinical use must have local regulatory approval before use in humans. Trained personnel must carry out quality control (QC) procedures, and each dose must meet all established QC criteria before release to the clinic.

Since in Europe, no dedicated monograph on the preparation of ^{68}Ga-DOTATOC is in effect so far (but a draft has already been prepared), QC procedures for ^{68}Ga-DOTATOC are based on the current general requirements for radiopharmaceuticals laid out both in the European Pharmacopoeia [26] and in the US Pharmacopeia [27]. These procedures are summarized below and can be performed within 6 ± 2 min (except high performance liquid chromatography (HPLC), microbial, and endotoxin testing). Complete QC data for three batches of ^{68}Ga-DOTATOC produced using the method described herein are summarized in Table 1.

3.1 Visual Inspection

The ^{68}Ga-DOTATOC dose is examined behind a 511-keV shielding lead glass window: the solution must be clear, colorless, and free of particles.

3.2 Radiochemical Identity and Radiochemical Purity by HPLC Method

HPLC analysis of both radiochemical identity and radiochemical purity was conducted using a Merck Hitachi Lichrograph L-6200A [28] equipped with both an L-4000 UV-detector [28] and a Packard Radiomatic 150TR γ-detector in which

TABLE 1 QC Data for Three Back-to-Back Runs of ^{68}Ga-DOTATOC

QC Test	Release Criteria	Run 1	Run 2	Run 3
Yield mCi (MBq)	N/A	11.2 (413.0)	11.4 (421.0)	11.1 (409.0)
Visual inspection	Clear, colorless	Pass	Pass	Pass
Radiochemical identity, HPLC (min)	RT = 13.0 ± 10%	13.00	12.98	13.13
Radiochemical identity, ITLC	ITLC-1: R_f0.05–0.3	0.09	0.13	0.23
	ITLC-2: R_f0.7–1.0	0.86	0.76	0.88
Radiochemical purity (%)				
HPLC	>90	99.93	99.87	99.69
ITLC-1	>90	99.96	99.86	99.33
ITLC-2	>90	99.63	99.92	99.26
Specific activity (MBq/μmol)	>1990	11,743	11,971	11,630
Dose pH	5.0–8.5	7.39	7.41	7.40
Dose osmolality (mOsm/kg)	190–370	261	262	261
Sterile filter integrity test (psi)	>50	>50	>50	>50
Radionuclidic identity ($t_{1/2}$) (min)	67.9 ± 5	68.6	68.4	69.0
[^{68}Ge]Germanium content (%)	<0.06	$<4 \times 10^{-5}$	$<4 \times 10^{-5}$	$<4 \times 10^{-5}$
Endotoxin analysis (EU/mL)	<17.5	<1	<1	<1
Sterility testing	No colony growth out to 14 d	Pass	Pass	Pass

a Biotech (BT)—cell [29] was inserted. For the stationary phase, a Macherey-Nagel Nucleosil 100-5 C18 HPLC column was used [30]. The mobile phase was as follows: 0–2 min: 100% A; 2–9 min: linear gradient to 33% A and 67% B; 9–15 min: 33% A and 67% B; 15–16 min: linear gradient to 100% A; 16–30 min: 100% A. Flow rate: 1 mL/min, $\lambda = 220$ nm. Eluent A consisted of 0.1% trifluoroacetic acid [31] in bidistilled water [15]. Eluent B consisted of 60% acetonitrile [32] and 40% trifluoroacetic acid [31], 0.1%, in bidistilled water [15].

For radiochemical identity, the retention time of ^{68}Ga-DOTATOC is compared to that of the (nonradioactive) DOTATOC reference standard [17] and must be 13.0 min ±10%. For determination of radiochemical purity, the portion of ^{68}Ga-DOTATOC in relation to total radioactivity is assessed and has to be >90%. Representative analytical HPLC traces are displayed in Fig. 4.

3.3 Radiochemical Identity and Radiochemical Purity by ITLC Method

The instant thin layer chromatography (ITLC) method consisted of two different chromatographic systems. In both systems, the ITLC analyses were performed

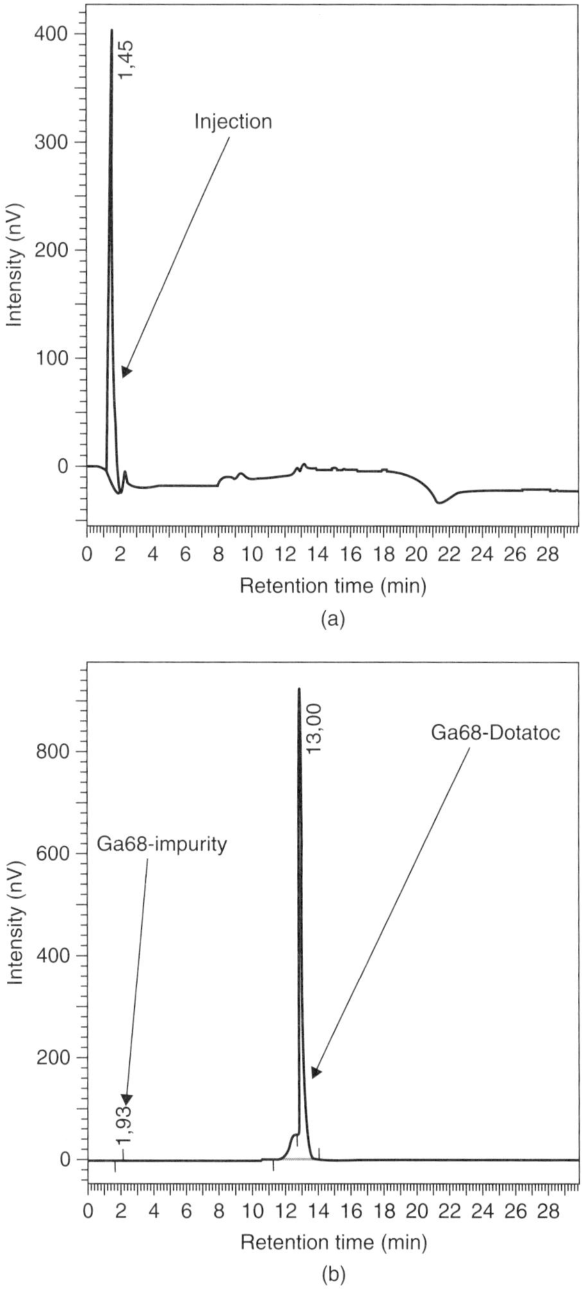

FIGURE 4 Analytical UV (a) and Radioactive (b) HPLC traces for ^{68}Ga-DOTATOC.

using silica gel impregnated glass fiber sheets ITLC SG [33] as the stationary phase. A suitable alternative might be ITLC SA plates [34], as discussed in detail elsewhere [35].

In ITLC system-1, a solution of 0.1 M sodium citrate buffer (pH 5.0) was used as the mobile phase. (The buffer was prepared by dissolving 29.41 g (0.1 mol) sodium citrate [36] in 1 L of bidistilled water [15] and then adjusting to pH 5.0 by addition of concentrated hydrochloric acid [37].) In ITLC system-1, R_f values were 0.1–0.3 for ^{68}Ga-DOTATOC and 0.7–1.0 for unbound ^{68}Ga species (e.g., [^{68}Ga]$GaCl_3$, [^{68}Ga]gallium acetate).

In ITLC system-2, an ammonium acetate/methanol solvent mixture was used as the mobile phase, which was prepared by (1 + 1) mixture of 1 M ammonium acetate [38] solution with methanol [39]. In ITLC system-2, R_f values were 0.7–1.0 for ^{68}Ga-DOTATOC and 0.1–0.3 for unbound ^{68}Ga species (e.g., [^{68}Ga]gallium colloid).

The radioactive spots of the developed ITLC strips were localized and quantified by a Raytest Gita Star Gamma Isotope Thin Layer Analyzer [40] equipped with a suitable collimator [41].

For passing the test for radiochemical purity, the portion of ^{68}Ga-DOTATOC in relation to total radioactivity must be >90%. Representative analytical ITLC traces are displayed in Figs. 5 and 6.

3.4 Specific Activity

The minimum specific activity (with respect to DOTATOC ligand) release limit in our institution is set to be >1990 MBq/μmol (or 1.4 MBq/μg) to surely avoid pharmacological effects of the nonradioactive ligands and is calculated by following Equation 1:

$$A_S = A[\text{MBq}]x\,1421.7[\mu\text{g}/\mu\text{mol}]/50[\mu\text{g}] \qquad (1)$$

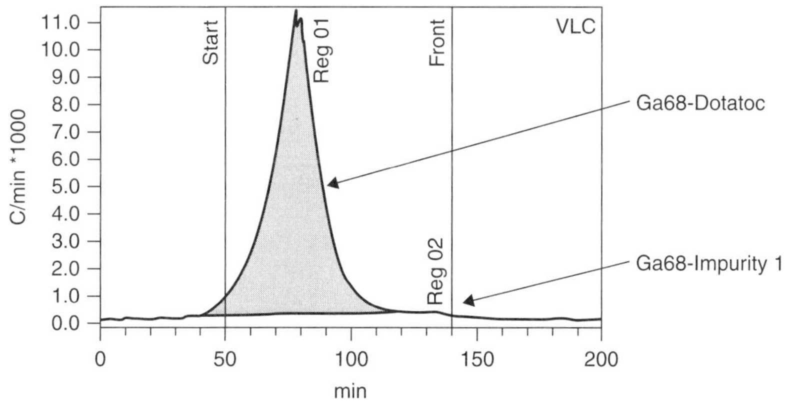

FIGURE 5 Analytical radioactive ITLC trace for ^{68}Ga-DOTATOC (ITLC-system 1; mobile phase: 0.1 M sodium citrate, pH 5).

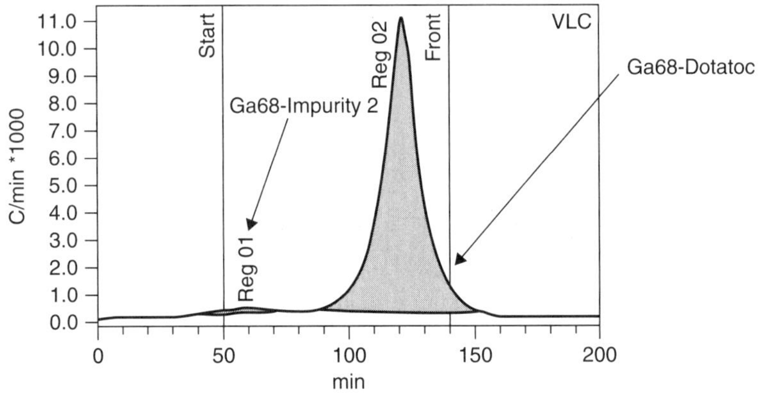

FIGURE 6 Analytical radioactive ITLC trace for ^{68}Ga-DOTATOC (ITLC-system 2; mobile phase: 1 M ammonium acetate/methanol (1 + 1 mixture)).

3.5 Dose pH

The pH of the ^{68}Ga-DOTATOC dose was analyzed with a WTW pH 537 Microprocessor pH meter [42]. The limits of the pH values for this parenteral preparation are set at 5.0–8.5.

3.6 Dose Osmolality

The osmolality of the ^{68}Ga-DOTATOC dose was analyzed by a Wescor osmometer [43]. A 10-μL aliquot of the dose was applied onto a paper filter inside the osmometer chamber. The limits of the osmolality for this parenteral preparation are set at 190–370 mOsm/kg.

3.7 Sterile Filter Integrity Test

The sterile filter [23] from the aseptic filling of the final ^{68}Ga-DOTATOC solution (with needle still attached) was connected to a nitrogen supply via a regulator. The needle was submerged in water, and the nitrogen pressure was gradually increased. If the pressure was raised above the filter acceptance pressure (50 psi) without seeing a stream of bubbles, the filter was considered intact.

3.8 Radionuclidic Identity/Purity

γ-Spectra were recorded within a Packard Cobra II Auto-Gamma counter [44] and peaks analyzed. Only at 511 and 1022 keV, peaks were allowed. Since all PET radioisotopes yield the same γ-trace, additionally, activities were measured at different time points using a Capintec CRC-15R Radioisotope Dose Calibrator [45], and the physical half-life was then calculated using Equation 2. The calculated half-life must be 67.9 ± 5 min.

$$t_{1/2} = -\ln 2 \times (t_2 - t_1)/\ln\left(\frac{A_2}{A_1}\right) \qquad (2)$$

where A_1 is the starting activity at t_1 and A_2 is the end activity at t_2.

3.9 ^{68}Ge Contamination

Since the decay of germanium-68 is not accompanied by any measurable form of radiation (because of 100% electron capture and direct decay into the ground state of gallium-68), the ^{68}Ge content has to be determined through its daughter nuclide, gallium-68, which grows steadily into the mother. Therefore, a 0.1-mL aliquot of the decayed ^{68}Ga-DOTATOC dose (after $\geq$20 half-lives of gallium-68, i.e., after $\sim$24 h) was analyzed in a Packard Cobra II Auto-Gamma counter [44], at full open window (15–2000 keV), and at 1-min measurement time. In this setup, the germanium-68 activity was determined according to a regression fit (counts per minute vs kilobecquerel) and was referred to the original activity of the ^{68}Ga-DOTATOC dose in the 0.1 mL aliquot at end of synthesis (EOS). The ^{68}Ge content must be $<$0.06%.

3.10 Endotoxin Analysis

Bacterial endotoxin content in an aliquot of ^{68}Ga-DOTATOC was analyzed (when the probes were physically decayed) by an external laboratory [46] where the analyses were performed according to the current Pharmacopoeia Europaea [47] and US Pharmacopeia [48] guidelines using the limulus amoebocyte lysate (LAL) test [49]: For this, an aliquot of the decayed ^{68}Ga-DOTATOC dose was added to the LAL and incubated at 37°C for 60 min, then the extent of clotting was compared against a negative control and a positive standard. The limit is $<$17.5 IU/mL. The determination of this parameter was always done after injection to the patient when radioactivity was decayed in accordance with the regulations laid down in the pharmacopoeias [26,27].

3.11 Sterility Testing

Sterility testing was performed (when the probes were physically decayed) according to current Pharmacopoeia Europaea [50] standard procedures by an external laboratory [46] whereby fluid thioglycolate media (FTM, sensitive for anaerobes and fungi) [50,51] and soybean casein digest media (SCDM, sensitive for aerobes) [50,51] were used and aliquots of the decayed ^{68}Ga-DOTATOC dose were added. The FTM samples were incubated at 30–35°C for 14 days, and the SCDM samples were incubated at 20–25°C and were visually compared against a negative control and a positive standard after the 14-day incubation period. The determination of this parameter was always done after injection to the patient when radioactivity was decayed in accordance with the regulations laid down in the pharmacopoeias [26,27].

Each of the three doses met all of the established QC criteria that are described in the text above.

CHEMICAL ABSTRACTS NOMENCLATURE (REGISTRY NUMBER)

Acetonitrile (75-05-8)
Ammonium acetate (631-61-8)

DOTATOC (= DOTA-[Tyr^3]-octreotide; = DOTA-D-Phe-Cys-Tyr-D-Trp-Lys-Thr-Cys-Thr(ol), cyclic disulfide Cys; =L-cysteinamide, *N*-[[4,7,10-tris(carboxymethyl)-1,4,7,10-tetraazacyclododec-1-yl]acetyl]-D-phenylalanyl-L-cysteinyl-L-tyronyl-D-tryptophyl-L-lysyl-L-threonyl-*N*-[(1*R*, 2*R*)-2-hydroxy-1-(hydroxymethyl)propyl]-, cyclic (2 → 7)-disulfide; = 2-[4-[[(1*R*)-1-[[(4*S*, 7*S*, 10*S*, 13*R*, 16*S*, 19*S*)-10-(4-aminobutyl)-4-[[(2*R*, 3*R*)-1,3-dihydroxybutan-2-yl]carbamoyl]-7-(1-hydroxyethyl)-16-[(4-hydroxyphenyl)methyl]-13-(1*H*-indol-3-ylmethyl)-6,9,12,15,18-pentaoxo-1,2-dithia-5,8,11,14,17-pentazacycloicos-19-yl]carbamoyl]-2-phenyl-ethyl]carbamoylmethyl]-7,10-bis(carboxymethyl)-1,4,7,10-tetrazacyclododec-1-yl]acetic acid) (204318-14-9)

Ethanol (64-17-5)

Hydrochloric acid (7647-01-0)

Methanol (67-56-1)

Sodium acetate trihydrate (6131-90-4)

Sodium chloride (7647-14-5)

Sodium citrate (68-04-2)

Trifluoroacetic acid (76-05-1)

REFERENCES AND NOTES

1. Hofmann M, Maecke H, Börner AR, Weckesser E, Schöffski P, Oei ML, Henze M, Heppeler A, Meyer GJ, Knapp WH. Eur J Nucl Med 2001;28:1751–1757.
2. Breeman WAP, de Jong M, Kwekkeboom DJ, Valkema R, Bakker WH, Kooij PMP, Visser TJ, Krenning EP. Eur J Nucl Med 2001;28:1421–1429.
3. Al-Nahhas A, Win Z, Szyszko T, Singh A, Nanni C, Fanti S, Rubello D. Anticancer Res 2007;27:4087–4094.
4. Green MW, Tucker WD. Int J Appl Radiat Isot 1961;12:62–63.
5. Nakayama M, Haratake M, Ono M, Koiso T, Harada K, Nakayama H, Yahara S, Ohmomo Y, Arano Y. Appl Radiat Isot 2003;58:9–14.
6. The tin(II)-oxide-based [^{68}Ge]Germanium/[^{68}Ga]Gallium generator is produced by iThemba Labs, RSA, and is available from I.D.B. Holland B.V., Baarle-Nassau, The Netherlands.
7. The titan(IV)-oxide-based [^{68}Ge]Germanium/[^{68}Ga]Gallium generator (Obninsk type, with the nominal activity of the fresh generator of 1850MBq [^{68}Ge]Germanium) was purchased from Eckert & Ziegler Eurotope GmbH, Berlin, Germany. Another equivalent titan(IV)-oxide-based [^{68}Ge]Germanium/[^{68}Ga]Gallium generator is produced from Eckert & Ziegler Isotope Products, Burbank (CA), USA, and is also available from Eckert & Ziegler Eurotope GmbH, Berlin, Germany.
8. Meyer G-J, Mäcke H, Schuhmacher J, Knapp WH, Hofmann M. Eur J Nucl Med Mol Imaging 2004;31:1097–1104.
9. Zhernosekov KP, Filosofov DV, Baum RP, Aschoff P, Bihl H, Razbash AA, Jahn M, Jennewein M, Rösch F. J Nucl Med 2007;48:1741–1748.
10. Breeman WAP, de Jong M, de Blois E, Bernard BF, Konijnenberg M, Krenning EP. Eur J Nucl Med Mol Imaging 2005;32:478–485.
11. Decristoforo C, Knopp R, von Guggenberg E, Rupprich M, Dreger T, Hess A, Virgolini I, Haubner R. Nucl Med Commun 2007;28(11):870–875.
12. Hydrochloric acid 0.1 M was purchased from Merck, Darmstadt, Germany, (part no. 1.09060.1000) and used as received.

13. Raytest SynChrom module, Straubenhardt, Germany.
14. Ethanol was purchased from Merck, Darmstadt, Germany, (part no. 1.00986.1000) and used as received.
15. Bidistilled water was purchased from Fresenius Kabi, Graz, Austria (part no.16.231) and used as received.
16. Sodium acetate was purchased from Fluka, Buchs, Switzerland (part no. 71188) and used as received.
17. DOTA-(Tyr^3)-octreotide acetate (DOTATOC) was purchased from Bachem, Bubendorf, Switzerland (part no. 4028058) and used as received.
18. Hydrochloric acid 0.05 M was prepared by diluting one part of 0.1 M hydrochloric acid [5] with one part of bidistilled water [8].
19. RP18 Light Sep-Pak cartridges were purchased from Waters, Milford, MA, USA (part no. WAT 023501) and conditioned with 10 mL ethanol [7] plus 10 mL air, and subsequently with 10 mL bidistilled water [8] plus 10 mL air.
20. Saline 0.9% was purchased from Noriderm Enterprises, Nicosia, Cyprus (part no. 1–28379), and used as received.
21. Sodium dihydrogenphosphate monohydrate was purchased from Merck, Darmstadt, Germany (part no. 1.06346.1000), and used as received.
22. Disodiumhydrogenphosphate dihydrate was purchased from Merck, Darmstadt, Germany (part no. 1.06576.1000), and used as received.
23. Millex GV Low Protein binding 0.22-μm filters were purchased from Millipore, Bedford, MA, USA (part no. SLGV 033RS), and used as received.
24. Sterile vials, 11 mL, were purchased from Mallinckrodt Medical, Petten, Holland (part no. DRN 4357), and used as received.
25. Millex GV low protein binding 0.22-μm filters were purchased from Millipore, Bedford, MA, USA (part no. SLGV 013SL), and used as received for the vented air needle.
26. Pharmacopoea Europea <6.0/0125>. Radioactive pharmaceuticals. Ph.Eur. 6; 2008.
27. U.S. Pharmacopeia <823>. Radiopharmaceuticals for Positron Emission Tomography Compounding. USP 32-NF 27; 2009.
28. Merck Hitachi, Darmstadt, Germany.
29. Packard, Meriden, CT, USA.
30. Macherey-Nagel, Düren, Germany (part no. 720120.46), and used as received.
31. Trifluoroacetic acid was purchased from Fluka, Buchs, Switzerland (part no. 91700), and used as received.
32. Acetonitrile in HPLC grade was purchased from Merck, Darmstadt, Germany (part no. 1.00029.1000), and used as received.
33. ITLC SG glass fiber sheets were purchased from Pall Life Sciences, Ann Arbor, MI, USA (part no. P/N 61886).
34. Varian, Darmstadt, Germany, Chromatography paper ITLC SA.
35. Wunderlich G, Herrling P, Zürn A, Anders P, Kotzerke J. Nuklearmedizin 2007;49:73–77.
36. Sodium citrate was purchased from Fluka, Buchs, Switzerland (part no. 71402), and used as received.
37. Concentrated aqueous hydrochloric acid solution (32%) was purchased from Merck, Darmstadt, Germany (part no. 100319), and used as received.
38. Ammonium acetate was purchased from Fluka, Buchs, Switzerland (part no. 1.01116.0500), and used as received.
39. Methanol in HPLC grade was purchased from Merck, Darmstadt, Germany (part no. 106008), and used as received.
40. The Gita Star Gamma Isotope Thin Layer Analyzer was purchased from Raytest, Straubenhardt, Germany.
41. So far at our institution a Gita Star 5 mm 60–150 keV tungsten PET collimator (part no. 02900005) was used, which was from Raytest, Straubenhardt, Germany. Even better suitable would be a Gita Star 20 mm >450 keV tungsten PET collimator (part no. 02900008), which is available from the same supplier.

42. The WTW pH 537 pH meter was purchased from Weilheim i. OB, Germany.
43. The Wescor osmometer 5520 was purchased from Sanova Medical Systems, Vienna, Austria.
44. The Cobra II Auto-Gamma counter was purchased from Packard, Meriden, CT, USA.
45. The activimeter was a Capintec CRC-15R Radioisotope Dose Calibrator, Ramsey, NJ, USA.
46. The external laboratory that performed both endotoxin analyses and sterility testings was the research laboratory of the Department for Infectious Diseases and Tropical Medicine of the Medical University of Vienna, Vienna, Austria.
47. Pharmacopoea Europea <6.0/2.06.14.00>. Determination of Bacterial Endotoxins. Ph.Eur. 6; 2008.
48. U.S. Pharmacopeia <85>. Bacterial Endotoxins Test. USP 32-NF 27; 2009.
49. The Pyrochrome® Limulus Amoebocyte Lysate (LAL) Test was purchased from Cape Cod, East Falmouth, MA, USA.
50. Pharmacopoea Europea <6.0/2.06.01.00>. Determination of Sterility. Ph.Eur. 6; 2008.
51. U.S. Pharmacopeia <71>. Sterility Tests. USP 32-NF 27; 2009.

APPENDIX 1

SUPPLIER INFORMATION

ABX Advanced Biochemical Compounds
www.abx.de

Acros
www.acros.com

Advanced Chromatography Technologies
www.ace-hplc.com

Agilent Corporation
www.agilent.com

Aguettant
www.aguettant.com

Alltech (a division of Grace Davison Discovery Sciences)
www.discoverysciences.com

Apoteket Production & Laboratories AB
www.apl.se

B. Braun
www.bbraun.com
www.bbraunusa.com

Beckman, Inc. USA
www.beckmancoulter.com

Radiochemical Syntheses, Volume 1: Radiopharmaceuticals for Positron Emission Tomography, First Edition. Edited by Peter J. H. Scott, Brian G. Hockley.

Becton, Dickinson and Company
www.bd.com

Biomerieux Laboratories
www.biomerieux.com

Bioscan, Inc.
www.bioscan.com

Branson Ultrasonics Corp., CT, USA.
www.bransonultrasonics.com

Capinetec
www.capintec.com

Carroll-Ramsey Associates
www.carroll-ramsey.com

Charles River Laboratories International, Inc.,
www.criver.com

Decon Laboratories Inc.
www.deconlabs.com

Dionex
www.dionex.com

Dade Moeller and Associates
www.moellerinc.com

EMSL Analytical Inc.
www.emsl.com

EMD Chemicals Inc., USA
www.emdchemicals.com

Fannin
www.fannin.ie

Fisher Scientific
www.fishersci.com

Fluka
Contact Sigma Aldrich (www.sigmaaldrich.com)

Fresenius Kabi, Graz, Austria
www.fresenius-kabi.at

GE Healthcare
www.gehealthcare.com/worldwide.html

Hameln Pharmaceuticals
www.hameln.co.uk

Hamilton Company, USA
www.hamiltoncompany.com

Hollister-Stier Laboratories LLC
www.hollister-stier.com

Hospira
www.hospira.com

IBA Molecular Imaging
www.iba-molecular.com

J M Loveridge Ltd
www.jmloveridge.com

Keystone Labs Inc.
www.keystonelabs.ca

Lavoisier
www.lavoisier.com

Macherey-Nagel
www.mn-net.com

Mallinckrodt
www.mallbaker.com

Merck
www.merck.de

Mettler Toledo
www.mt.com

Millipore
www.millipore.com

Packard Instrument Company
Contact: Perkin Elmer (www.perkinelmer.com)

Phenomonex
www.phenomenex.com

Ray Test Corporation
www.raytest.com

Restek
www.restek.com

Rotem Industries Ltd.
www.rotemi.co.il

Severn Biotech Ltd.
www.severnbiotech.com

Shimadzu
www.shimadzu.com

Sigma Aldrich
www.sigmaaldrich.com

Thomas Scientific
www.thomassci.com

Varian
www.varian.com

Veenstra Instruments
www.veenstranet.com

Waters
www.waters.com

INDEX

Radiochemical Syntheses, Volume 1: Radiopharmaceuticals for Positron Emission Tomography, First Edition. Edited by Peter J. H. Scott, Brian G. Hockley.